SCHAUM'S OUTLINE OF

THEORY AND PROBLEMS

of

COLLEGE PHYSICS

SEVENTH EDITION

•

by

FREDERICK J. BUECHE, Ph.D.

Professor of Physics
University of Dayton

SCHAUM'S OUTLINE SERIES

McGRAW-HILL BOOK COMPANY

New York   St. Louis   San Francisco   Auckland   Bogotá   Düsseldorf   Johannesburg
London   Madrid   Mexico   Montreal   New Delhi   Panama   Paris
São Paulo   Singapore   Sydney   Tokyo   Toronto

Schaum's Outline of Theory and Problems of
COLLEGE PHYSICS

0-07-008857-8

4 5 6 7 8 9 10 11 12 13 14 15 16 17 18 19 20   SH SH 8 7 6 5 4 3 2 1 0

Library of Congress Cataloging in Publication Data

Bueche, Frederick, 1923–
    Schaum's outline of college physics.

    (Schaum's outline series)
    Includes index.
    1. Physics. I. Title. II. Title: Outline of
college physics.
QC21.2.B842 1979   530   78-21551
ISBN 0-07-008857-8

Preface

Lord Kelvin said: "I often say that when you can measure what you are speaking about, and express it in numbers, you know something about it." Experienced teachers are aware of the wisdom in his remark, and they insist that physical understanding and problem-solving ability are firmly linked together. This book offers the student of general physics the opportunity to see clearly the principles of physics through their application to a large number of carefully selected problems.

Each chapter begins with a clear statement of the pertinent definitions, principles, and laws. This is followed by graduated sets of solved and supplementary problems. These are arranged so as to present a natural development of each topic, and they include a wide range of applications in both pure and applied physics. The solved problems illustrate and amplify the theory, provide the repetition of basic principles so vital to effective teaching, and bring into sharp focus those fine points which are often so worrisome to the student. The supplementary problems serve as a complete review of the material of each chapter. This book is no mere condensation of ordinary text material; it is a comprehensive approach to physics through problems.

Previous editions of this book have been favorably received and adopted by a multitude of colleges and technical schools. In this substantially revised seventh edition, major changes have been made to keep pace with the most recent concepts, methods, and terminology. Although the English gravitational and cgs systems are still included, so as to provide familiarity with them, the text now uses the SI as its fundamental units system. Many new problems have been included and many others have been revised in the interests of clarity and pedagogical value. The theory section of each chapter has been reexamined and revised where necessary to conform better with the needs of present-day students.

I am indebted to Daniel Schaum, the author of the original text, for the foundation upon which this revision is based. Professor J. Kepes was kind enough to recheck all problem solutions and I appreciate his help. A special thanks is due David Beckwith, of the Schaum's Outline Series, whose guidance and editorial expertise have been of considerable aid to me.

FREDERICK J. BUECHE

Contents

CONTENTS

CONTENTS

CONTENTS

CONTENTS

Chapter 1

Introduction to Vectors

A SCALAR QUANTITY has only magnitude. Typical scalar quantities are the number of students in a class, the quantity of sugar in a jar, the cost of a house, etc.

Scalars, being simple numbers, are added like any numbers. Two candies in one box plus seven in another give nine candies total.

A VECTOR QUANTITY has both magnitude and direction. For example, a *vector displacement* might be a change in position from one point to a second point 2 cm away and in the *x*-direction from the first point. As another example, a cord pulling northward on a post gives rise to a *vector force* on the post of 20 lb northward. Similarly, a car moving south at 40 km/h has a *vector velocity* of 40 km/h southward.

A vector quantity can be represented by an arrow drawn to scale. The length of the arrow is proportional to the magnitude of the vector quantity (2 cm, 20 lb, 40 km/h in the above examples). The direction of the arrow represents the direction of the vector quantity.

In printed material, vectors are represented by boldface type, such as **F**. When written by hand, the designations $\vec{F}$ and $\underset{\sim}{F}$ are often used.

THE RESULTANT of a number of similar vectors, force vectors for example, is that single vector which would have the same effect as all the original vectors taken together.

GRAPHICAL ADDITION OF VECTORS (POLYGON METHOD): This method for finding the resultant of several vectors consists in beginning at any convenient point and drawing (to scale) each vector arrow in turn. They may be taken in any order of succession. The tail end of each arrow is attached to the tip end of the preceding one.

The resultant is represented by an arrow with its tail end at the starting point and its tip end at the tip of the last vector added.

PARALLELOGRAM METHOD for adding two vectors: The resultant of two vectors acting at any angle may be represented by the diagonal of a parallelogram. The two vectors are drawn as the sides of the parallelogram and the resultant is its diagonal, as shown in Fig. 1-1. The direction of the resultant is away from the origin of the two vectors.

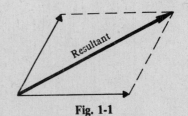

Fig. 1-1

SUBTRACTION OF VECTORS: To subtract a vector **B** from a vector **A**, reverse the direction of **B** and add it vectorially to vector **A**, i.e. $\mathbf{A} - \mathbf{B} = \mathbf{A} + (-\mathbf{B})$.

THE TRIGONOMETRIC FUNCTIONS are defined in relation to a right triangle. For the right triangle shown in Fig. 1-2, by definition

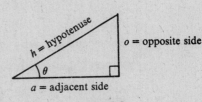

Fig. 1-2

$$\sin \theta = \frac{o}{h} \qquad \cos \theta = \frac{a}{h} \qquad \tan \theta = \frac{o}{a}$$

We often use these in the forms

$$o = h \sin \theta \qquad a = h \cos \theta \qquad o = a \tan \theta$$

A COMPONENT OF A VECTOR is its effective value in a given direction. For example, the x-component of a displacement is the displacement parallel to the x-axis caused by the given displacement. A vector may be considered as the resultant of its component vectors along the specified directions. It is customary, and useful, to resolve a vector into components along *mutually perpendicular* directions (*rectangular components*).

COMPONENT METHOD FOR ADDING VECTORS: Each vector is resolved into its x-, y-, and z-components, with negatively directed components taken as negative. The x-component of the resultant, R_x, is the algebraic sum of all the x-components. The y- and z-components of the resultant are found in a similar way. Knowing the components, the magnitude of the resultant is given by

$$R = \sqrt{R_x^2 + R_y^2 + R_z^2}$$

In two dimensions, the angle of the resultant with the x-axis can be found from the relation

$$\tan \theta = \frac{R_y}{R_x}$$

Solved Problems

1.1. Using the graphical method, find the resultant of the following two displacements: 2 m at 40° and 4 m at 127°, the angles being taken relative to the $+x$-axis.

Choose (x,y)-axes as shown in Fig. 1-3 and lay out the displacements to scale tip to tail from the origin. Notice that all angles are measured from the $+x$-axis. The resultant vector, **R**, points from starting point to end point as shown. We measure its length on the scale diagram to find its magnitude, 4.6 m. Using a protractor, we measure its angle θ to be 101°. The resultant displacement is therefore 4.6 m at 101°.

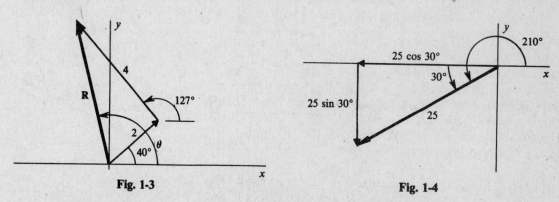

Fig. 1-3   Fig. 1-4

1.2. Find the x- and y-components of a 25 m displacement at an angle of 210°.

The vector displacement and its components are shown in Fig. 1-4. The components are

$$x\text{-component} = -25 \cos 30° = -21.7 \text{ m}$$
$$y\text{-component} = -25 \sin 30° = -12.5 \text{ m}$$

Notice in particular that each component points in the negative coordinate direction and must therefore be taken as negative.

In the above computation the components should properly have been written

$$-(25 \text{ m}) \cos 30°   \qquad -(25 \text{ m}) \sin 30°$$

We shall often omit units in situations such as this to save space.

1.3.   Solve Problem 1.1 by use of rectangular components.

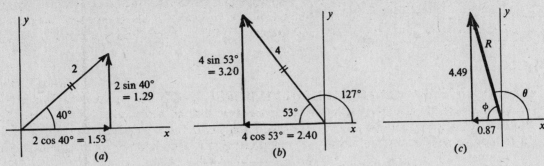

Fig. 1-5

We resolve each vector into rectangular components as shown in Fig. 1-5(a) and (b). (We place a cross hatch symbol on the original vector to show that it is replaced by its components.) The resultant has the components

$$R_x = 1.53 - 2.40 = -0.87 \text{ m} \qquad R_y = 1.29 + 3.20 = 4.49 \text{ m}$$

Notice that components pointing in the negative direction must be assigned a negative value.

The resultant is shown in Fig. 1-5(c); we see that

$$R = \sqrt{(0.87)^2 + (4.49)^2} = 4.57 \text{ m} \qquad \tan \phi = \frac{4.49}{0.87}$$

Hence, $\phi = 79°$, from which $\theta = 180° - \phi = 101°$.

1.4.   Add the following two force vectors by use of the parallelogram method: 30 pounds at 30° and 20 pounds at 140°. (A *pound of force* is chosen such that a 1 kg object weighs 2.21 lb on earth. One pound is equivalent to a force of 4.45 newtons, 4.45 N.)

The force vectors are shown in Fig. 1-6(a). We construct a parallelogram using them as sides, as shown in Fig. 1-6(b). The resultant, **R**, is then shown as the diagonal. By measurement, we find that **R** is 30 lb at 72°.

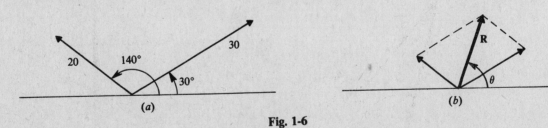

Fig. 1-6

1.5.   Four coplanar forces act on a body at point O as shown in Fig. 1-7(a). Find their resultant graphically. (In Fig. 1-7, the force unit N is newtons. A 1 kg object weighs 9.8 N on earth. A force of 1 N is equivalent to a force of 0.225 pounds.)

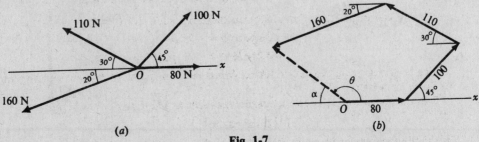

Fig. 1-7

Starting from O, the four vectors are plotted in turn as shown in Fig. 1-7(b). We place the tail end of one vector at the tip end of the preceding one. The arrow from O to the tip of the last vector represents the resultant of the vectors.

We measure R from the scale drawing in Fig. 1-7(b) and find it to be 119 N. Angle α is measured by protractor and is found to be 37°. Hence the resultant makes an angle $\theta = 180° - 37° = 143°$ with the positive x-axis. The resultant is 119 N at 143°.

1.6.   Solve Problem 1.5 by use of the rectangular component method.

The vectors and their components are:

Vector	x-Component	y-Component
80	80	0
100	$100 \cos 45° = \quad 71$	$100 \sin 45° = \quad 71$
110	$-110 \cos 30° = -95$	$110 \sin 30° = \quad 55$
160	$-160 \cos 20° = -150$	$-160 \sin 20° = -55$

Notice the sign of each component. To find the resultant, we have

$$R_x = 80 + 71 - 95 - 150 = -94 \text{ N}$$
$$R_y = 0 + 71 + 55 - 55 = 71 \text{ N}$$

The resultant is shown in Fig. 1-8; we see that

$$R = \sqrt{(94)^2 + (71)^2} = 118 \text{ N}$$

Further, $\tan \alpha = 71/94$, from which $\alpha = 37°$.
Therefore the resultant is 118 N at $180 - 37 = 143°$.

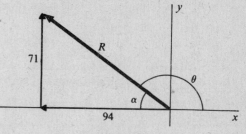

Fig. 1-8

1.7.   Perform graphically the following vector additions and subtractions, where **A**, **B**, and **C** are the vectors shown in Fig. 1-9:   (a) **A** + **B**;   (b) **A** + **B** + **C**;   (c) **A** − **B**;   (d) **A** + **B** − **C**.

See Fig. 1-9(a) through (d). In (c),   **A** − **B** = **A** + (−**B**); i.e. to subtract **B** from **A**, reverse the direction of **B** and add it vectorially to **A**. Similarly, in (d),   **A** + **B** − **C** = **A** + **B** + (−**C**),   where −**C** is equal in magnitude but opposite in direction to **C**.

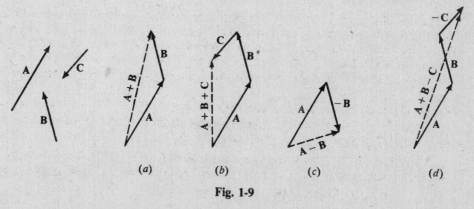

(a)   (b)   (c)   (d)

Fig. 1-9

1.8.   A force of 100 N makes an angle of θ with the x-axis and has a y-component of 30 N. Find both the x-component of the force and the angle θ.

The data are sketched roughly in Fig. 1-10. We wish to find F_x and θ. We know that

$$\sin \theta = \frac{o}{h} = \frac{30}{100} = 0.30$$

from which $\theta = 17.5°$. Then, since $a = h \cos \theta$, we have

$$F_x = 100 \cos 17.5° = 95.4 \text{ N}$$

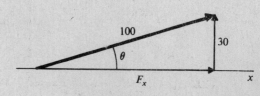

Fig. 1-10

1.9. A boat can travel at a speed of 8 km/h in still water on a lake. In the flowing water of a stream, it can move at 8 km/h relative to the water in the stream. If the stream speed is 3 km/h, how fast can the boat move past a tree on the shore in traveling (a) upstream? (b) downstream?

(a) If the water were standing still, the boat's speed past the tree would be 8 km/h. But the stream is carrying it in the opposite direction at 3 km/h. Therefore the boat's speed relative to the tree is 8 − 3 = 5 km/h.

(b) In this case, the stream is carrying the boat in the same direction the boat is trying to move. Hence its speed past the tree is 8 + 3 = 11 km/h.

1.10. A plane is traveling eastward at an airspeed of 500 km/h. But a 90 km/h wind is blowing southward. What are the direction and speed of the plane relative to the ground?

The plane's resultant velocity is the sum of two velocities, 500 km/h eastward and 90 km/h southward. These component velocities are shown in Fig. 1-11. The plane's resultant velocity is found by use of

$$R = \sqrt{(500)^2 + (90)^2} = 508 \text{ km/h}$$

The angle α is given by

$$\tan \alpha = \frac{90}{500} = 0.180$$

from which $\alpha = 10.2°$. The plane's velocity relative to the ground is 508 km/h at 10.2° south of east.

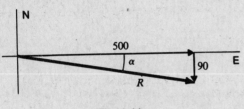

Fig. 1-11

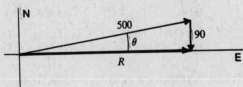

Fig. 1-12

1.11. With the same airspeed as in Problem 1.10, in what direction must the plane head in order to move due east relative to the earth?

The sum of the plane's velocity through the air and the velocity of the wind will be the resultant velocity of the plane relative to the earth. This is shown in the vector diagram of Fig. 1-12. Notice that, as required, the resultant velocity is eastward. It is seen that $\sin \theta = 90/500$, from which $\theta = 10.4°$. The plane should head 10.4° north of east if it is to move eastward on the earth.

If we wish to find the plane's eastward speed, the figure tells us that $R = 500 \cos \theta = 492$ km/h.

1.12. A child pulls a rope attached to a sled with a force of 60 N. The rope makes an angle of 40° to the ground. (a) Compute the effective value of the pull tending to move the sled along the ground. (b) Compute the force tending to lift the sled vertically.

As shown in Fig. 1-13, the components of the 60 N force are 39 N and 46 N. (a) The pull along the ground is the horizontal component, 46 N. (b) The lifting force is the vertical component, 39 N.

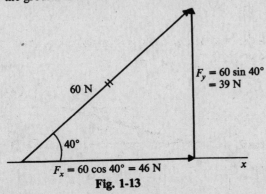

Fig. 1-13

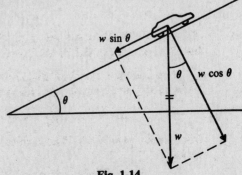

Fig. 1-14

1.13. A car whose weight is w is on a ramp which makes an angle θ to the horizontal. How large a perpendicular force must the ramp withstand if it is not to break under the car's weight?

As shown in Fig. 1-14, the car's weight is a force w that pulls straight down on the car. We take components of w along the incline and perpendicular to it. The ramp must balance the force component $w \cos \theta$ if the car is not to crash through the ramp.

1.14. The five coplanar forces shown in Fig. 1-15(a) act on an object. Find the resultant force due to them.

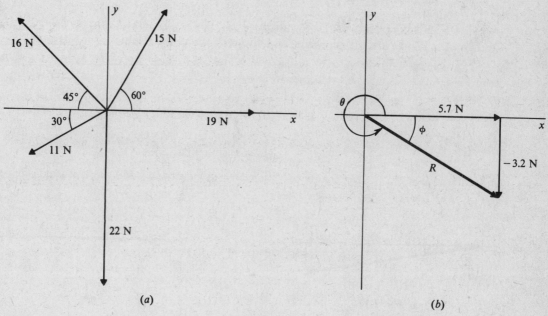

(a)   (b)

Fig. 1-15

(1) Find the x- and y-components of each force. These components are as follows:

Force	x-Component	y-Component
19 N	19.0	0
15 N	$15 \cos 60° = $ 7.5	$15 \sin 60° = $ 13.0
16 N	$-16 \cos 45° = -11.3$	$16 \sin 45° = $ 11.3
11 N	$-11 \cos 30° = -9.5$	$-11 \sin 30° = -5.5$
22 N	0	-22.0

Notice the signs to indicate $+$ and $-$ directions.

(2) The resultant **R** has components $R_x = \sum F_x$ and $R_y = \sum F_y$, where we read $\sum F_x$ as "the sum of all the x-force components." We then have

$$R_x = 19.0 + 7.5 - 11.3 - 9.5 + 0 = +5.7 \text{ N}$$
$$R_y = 0 + 13.0 + 11.3 - 5.5 - 22.0 = -3.2 \text{ N}$$

(3) Find the magnitude of the resultant from

$$R = \sqrt{R_x^2 + R_y^2} = 6.5 \text{ N}$$

(4) Sketch the resultant as shown in Fig. 1-15(b) and find its angle. We see that

$$\tan \phi = \frac{3.2}{5.7} = 0.56$$

from which $\phi = 29°$. Then we have that $\theta = 360° - 29° = 331°$. The resultant is 6.5 N at 331° (or $-29°$).

Supplementary Problems

1.15. A car goes 5.0 km east, 3.0 km south, 2.0 km west, and 1.0 km north.   (*a*) Determine how far north and how far east it has been displaced.   (*b*) Find the displacement vector both graphically and algebraically. *Ans.*   (*a*) 3 km east, −2 km north;   (*b*) 3.6 km at 34° south of east

1.16. Find the *x*- and *y*-components of a 400 N force at an angle of 125° to the *x*-axis. *Ans.*   −229 N, 328 N

1.17. Find the vector sum of the following four displacements on a map: 60 mm north; 30 mm west; 40 mm at 60° west of north; 50 mm at 30° west of south.   Solve graphically and also algebraically.   (mm stands for millimeter, 0.001 meter.)   *Ans.*   97 mm at 67.7° west of north

1.18. Two forces, 80 N and 100 N acting at an angle of 60° with each other, pull on an object.   (*a*) What single force would replace the two forces?   (*b*) What single force (called the *equilibrant*) would balance the two forces?   Solve algebraically. *Ans.*   (*a*) **R**: 156 N at 34° with the 80 N force;   (*b*) −**R**: 156 N at 214° with the 80 N force

1.19. Two forces act on a point object as follows: 100 N at 170° and 100 N at 50°.   Find their resultant. *Ans.*   100 N at 110°

1.20. Find graphically the resultant of each of the three coplanar force systems shown in Fig. 1-16. *Ans.*   (*a*) 35 lb at 34°;   (*b*) 59 lb at 236°;   (*c*) 172 lb at 315°

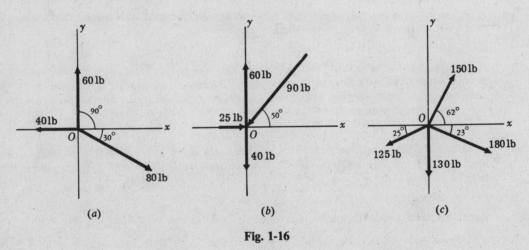

Fig. 1-16

1.21. Find algebraically the (*a*) resultant and (*b*) equilibrant (see Problem 1.18) of the following coplanar forces: 300 N at 0°, 400 N at 30°, and 400 N at 150°.   *Ans.*   (*a*) 500 N at 53°;   (*b*) 500 N at 233°

1.22. Compute algebraically the (*a*) resultant and (*b*) equilibrant of the following coplanar forces: 100 lb at 30°, 141.4 lb at 45°, and 100 lb at 240°.   Check your result graphically. *Ans.*   (*a*) 151 lb at 25°;   (*b*) 151 lb at 205°

1.23. Compute algebraically the resultant of the following displacements: 20 m at 30°, 40 m at 120°, 25 m at 180°, 42 m at 270°, and 12 m at 315°.   Check your answer by a graphical solution. *Ans.*   20 m at 197°

1.24.   Refer to Fig. 1-17.   In terms of vectors **A** and **B**, express the vectors (*a*) **P**;   (*b*) **R**;   (*c*) **S**;   (*d*) **Q**.
   Ans.   (*a*) **A** + **B**;   (*b*) **B**;   (*c*) −**A**;   (*d*) **A** − **B**

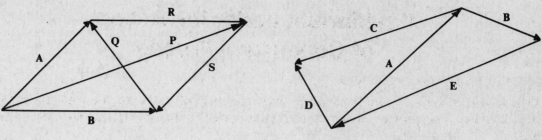

Fig. 1-17   Fig. 1-18

1.25.   Refer to Fig. 1-18.   In terms of vectors **A** and **B**, express the vectors (*a*) **E**;   (*b*) **D** − **C**;   (*c*) **E** + **D** − **C**.   *Ans.*   (*a*) −**A** − **B** or −(**A** + **B**);   (*b*) **A**;   (*c*) −**B**

1.26.   A reckless drunk is playing with a gun in an airplane that is going directly east at 500 km/h.   The drunk shoots the gun straight up at the ceiling of the plane.   The bullet leaves the gun at a speed of 1000 km/h.   According to someone standing on the earth, what angle does the bullet make to the vertical?   *Ans.*   26.6°

1.27.   A truck is moving north at a speed of 70 km/h.   The exhaust pipe above the truck cab sends out a trail of smoke that makes an angle of 20° east of south behind the truck.   If the wind is blowing directly toward the east, what is the wind speed at that location?   *Ans.*   25 km/h

1.28.   A ship is traveling due east at 10 km/h.   What must be the speed of a second ship heading 30° east of north if it is always due north from the first ship?   *Ans.*   20 km/h

1.29.   A boat, propelled so as to travel with a speed of 0.50 m/s in still water, moves directly across a river that is 60 m wide.   The river flows with a speed of 0.30 m/s.   (*a*) At what angle, relative to the straight-across direction, must the boat be pointed?   (*b*) How long does it take the boat to cross the river?   *Ans.*   (*a*) 37° upstream;   (*b*) 150 s

1.30.   A child is holding a wagon from rolling straight back down a driveway that is inclined at 20° to the horizontal.   If the wagon weighs 150 N, how hard must the child pull on the handle if the handle is parallel to the incline?   *Ans.*   51 N

1.31.   Repeat Problem 1.30 if the handle is at an angle of 30° above the incline.   *Ans.*   59 N

Chapter 2

Equilibrium under the Action
of Concurrent Forces

CONCURRENT FORCES are forces whose lines of action all pass through a common point. The forces acting on a point object are concurrent because they all pass through the same point, the point object.

AN OBJECT IS IN EQUILIBRIUM under concurrent forces if (*a*) it is at rest and remains at rest (called *static equilibrium*) or (*b*) it is in motion with constant vector velocity (called *translational equilibrium*).

THE FIRST CONDITION FOR EQUILIBRIUM requires that

$$\sum F_x = \sum F_y = \sum F_z = 0$$

that is, the resultant of all external forces acting on the object must be zero. This condition is sufficient for equilibrium when the external forces are concurrent. A second condition must also be satisfied if an object is to be in equilibrium under nonconcurrent forces; it is discussed in Chapter 3.

PROBLEM SOLUTION METHOD (CONCURRENT FORCES):
(1) Isolate the object for discussion.
(2) Show the forces acting on the isolated object in a diagram (the *free-body diagram*).
(3) Find the rectangular components of each force.
(4) Write the first condition for equilibrium in equation form.
(5) Solve for the required quantities.

THE WEIGHT OF AN OBJECT is the force with which gravity pulls downward upon it.

THE TENSION IN A STRING is the force with which the string pulls upon the object to which it is attached.

THE FRICTION FORCE (f) is a tangential force on a surface that opposes the sliding of the surface across an adjacent surface. The friction force is parallel to the surface and opposite in direction to its motion.

THE NORMAL FORCE (Y) on a surface resting (or sliding) on a second surface is the perpendicular component of the force exerted by the supporting surface on the surface being supported.

COEFFICIENT OF KINETIC FRICTION (μ_k) is defined for the case where one surface is sliding across another at constant speed. It is

$$\mu_k = \frac{\text{friction force}}{\text{normal force}} = \frac{f}{Y}$$

9

COEFFICIENT OF STATIC FRICTION (μ_s) is defined for the case where one surface is just on the verge of sliding across another surface. It is

$$\mu_s = \frac{\text{critical friction force}}{\text{normal force}} = \frac{f}{Y}$$

where the critical friction force is the friction force when the object is just on the verge of slipping.

Solved Problems

2.1. The object in Fig. 2-1(a) weighs 50 N and is supported by a cord. Find the tension in the cord.

We isolate the object shown for our discussion. Two forces act upon it, the upward pull of the cord and the downward pull of gravity. We shall represent the pull of the cord by T, the tension in the cord. The pull of gravity, the weight of the object, is $w = 50$ N. These two forces are shown in the free-body diagram, Fig. 2-1(b).

The forces are already in component form and so we can write the first condition for equilibrium at once.

$$\sum F_x = 0 \qquad \text{becomes} \qquad 0 = 0$$

$$\sum F_y = 0 \qquad \text{becomes} \qquad T - 50\ \text{N} = 0$$

from which $T = 50$ N.

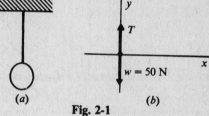

(a)   (b)

Fig. 2-1

2.2. As shown in Fig. 2-2(a), the tension in the horizontal cord is 30 N. Find the weight of the object.

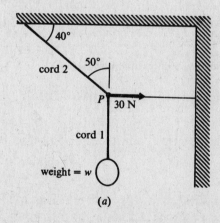

(a)

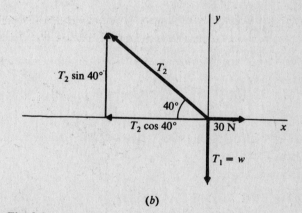

(b)

Fig. 2-2

As we saw in Problem 2.1, the tension in cord 1 is equal to the weight of the object hanging from it. Therefore $T_1 = w$, and we wish to find T_1 or w.

Notice that the unknown force, T_1, and the known force, 30 N, both pull on the knot at point P. It therefore makes sense to isolate the knot at P as our object. The free-body diagram showing the forces on the knot is drawn as Fig. 2-2(b). The force components are also found there.

We next write the first condition for equilibrium for the knot. From the free-body diagram,

$$\sum F_x = 0 \qquad \text{becomes} \qquad 30\ \text{N} - T_2 \cos 40° = 0$$

$$\sum F_y = 0 \qquad \text{becomes} \qquad T_2 \sin 40° - w = 0$$

Solving the first equation for T_2 gives $T_2 = 39.2$ N. Substituting this value in the second equation gives $w = 25.2$ N as the weight of the object.

2.3.   A rope extends between two poles.   A 90 N boy hangs from it as shown in Fig. 2-3(*a*).   Find the tensions in the two parts of the rope.

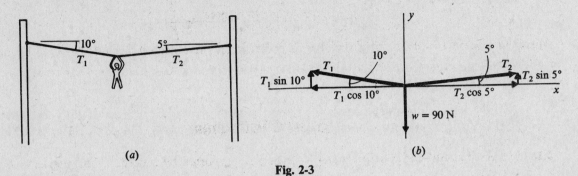

(*a*)   (*b*)

Fig. 2-3

We label the two tensions T_1 and T_2, and isolate the rope at the boy's hands as the object.   The free-body diagram for this object is shown in Fig. 2-3(*b*).

After resolving the forces into their components as shown, we can write the first condition for equilibrium.

$$\sum F_x = 0 \qquad \text{becomes} \qquad T_2 \cos 5° - T_1 \cos 10° = 0$$

$$\sum F_y = 0 \qquad \text{becomes} \qquad T_2 \sin 5° + T_1 \sin 10° - 90 \text{ N} = 0$$

Evaluating the sines and cosines, these equations become

$$0.996\, T_2 - 0.985\, T_1 = 0 \qquad \text{and} \qquad 0.087\, T_2 + 0.174\, T_1 - 90 = 0$$

Solving the first for T_2 gives $T_2 = 0.990\, T_1$.   Substituting this in the second equation gives

$$0.086\, T_1 + 0.174\, T_1 - 90 = 0$$

from which $T_1 = 346$ N.   Then, because $T_2 = 0.990\, T_1$,   we have $T_2 = 343$ N.

2.4.   A 200 N wagon is to be pulled up a 30° incline at constant speed.   How large a force parallel to the incline is needed if friction effects are negligible?

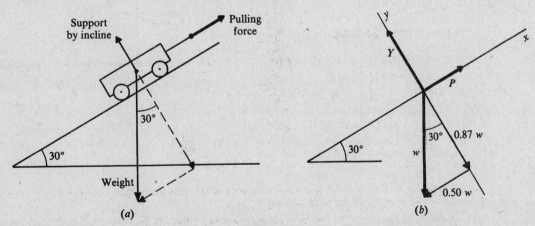

(*a*)   (*b*)

Fig. 2-4

The situation is shown in Fig. 2-4(*a*).   Because the wagon moves at constant speed along a straight line, its velocity vector is constant.   Therefore the wagon is in translational equilibrium, and the first condition for equilibrium applies to it.

We isolate the wagon as the object.   Three nonnegligible forces act on it: (1) the pull of gravity w (its weight), directed straight down; (2) the force P exerted on the wagon parallel to the incline to pull it up the incline; (3) the push Y of the incline that supports the wagon.   These three forces are shown in the free-body diagram, Fig. 2-4(*b*).

For situations involving inclines, it is convenient to take the x-axis parallel to the incline and the y-axis perpendicular to it. After taking components along these axes, we can write the first condition for equilibrium.

$$\sum F_x = 0 \qquad \text{becomes} \qquad P - 0.50\,w = 0$$

$$\sum F_y = 0 \qquad \text{becomes} \qquad Y - 0.87\,w = 0$$

Solving the first equation and recalling that $w = 200$ N, we find that $P = 0.50\,w = 100$ N. The required pulling force is 100 N.

2.5. A 50 N box is slid straight across the floor at constant speed by a force of 25 N, as shown in Fig. 2-5(a). How large a friction force impedes the motion of the box? How large is the normal force?

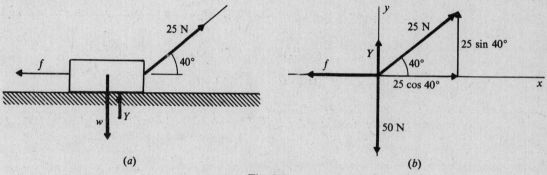

(a)   (b)

Fig. 2-5

Notice the forces acting on the box as shown in Fig. 2-5(a). The friction force is f and the normal force, the supporting force exerted by the floor, is Y. The free-body diagram and components are shown in Fig. 2-5(b). Because the box is moving with constant velocity, it is in equilibrium. The first condition for equilibrium tells us that

$$\sum F_x = 0 \qquad \text{or} \qquad 25\cos 40° - f = 0$$

We can solve for f at once to find that $f = 19$ N. The friction force is 19 N.
To find Y we use the fact that

$$\sum F_y = 0 \qquad \text{or} \qquad Y + 25\sin 40° - 50 = 0$$

Solving gives the normal force as $Y = 34$ N.

2.6. Find the tensions in the ropes shown in Fig. 2-6(a) if the supported object weighs 600 N.

Let us select as our object the knot at A because we know one force acting on it. The weight pulls down on it with a force of 600 N and so the free-body diagram for the knot is as shown in Fig. 2-6(b). Applying the first condition for equilibrium to that diagram, we have

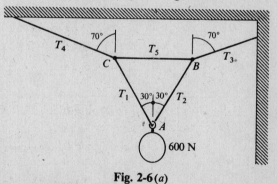

$$\sum F_x = 0 \quad \text{or} \quad T_2\cos 60° - T_1\cos 60° = 0$$

$$\sum F_y = 0 \quad \text{or} \quad T_1\sin 60° + T_2\sin 60° - 600 = 0$$

The first equation yields $T_1 = T_2$. Substitution of T_1 for T_2 in the second equation gives $T_1 = 346$ N, and this is also T_2.

Fig. 2-6(a)

Let us now isolate knot B as our object. Its free-body diagram is shown in Fig. 2-6(c). We have already found that $T_2 = 346$ N and so the equilibrium equations are

$$\sum F_x = 0 \qquad \text{or} \qquad T_3\cos 20° - T_5 - 346\sin 30° = 0$$

$$\sum F_y = 0 \qquad \text{or} \qquad T_3\sin 20° - 346\cos 30° = 0$$

The last equation yields $T_3 = 877$ N. Substituting this in the prior equation gives $T_5 = 651$ N.

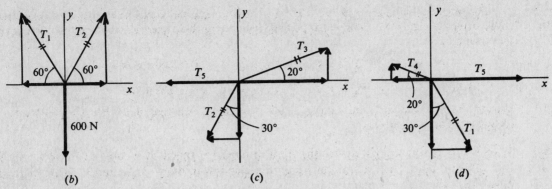

Fig. 2-6 (Cont.)

We can now proceed to the knot at C and the free-body diagram of Fig. 2-6(d). Recalling that $T_1 = 346$ N,

$$\sum F_x = 0 \qquad \text{becomes} \qquad T_5 + 346 \sin 30° - T_4 \cos 20° = 0$$

$$\sum F_y = 0 \qquad \text{becomes} \qquad T_4 \sin 20° - 346 \cos 30° = 0$$

The latter equation yields $T_4 = 877$ N.

You should have been able to guess from the symmetry of the physical situation that $T_1 = T_2$ and $T_4 = T_3$.

2.7. Each of the objects in Fig. 2-7 is in equilibrium. Find the normal force, Y, in each case.

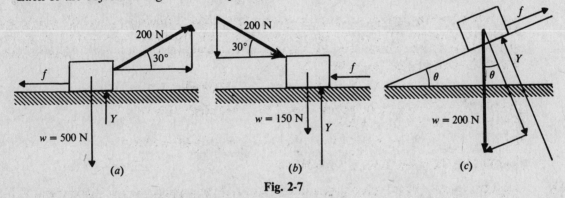

Fig. 2-7

We apply $\sum F_y = 0$ in each case.

(a)	$Y + 200 \sin 30° - 500 = 0$	from which	$Y = 400$ N
(b)	$Y - 200 \sin 30° - 150 = 0$	from which	$Y = 250$ N
(c)	$Y - 200 \cos \theta = 0$	from which	$Y = (200 \cos \theta)$ N

2.8. For the situations of Problem 2.7, find the coefficient of kinetic friction if the object is moving with constant speed.

We have already found Y for each case in Problem 2.7. To find f, the sliding friction force, we use $\sum F_x = 0$.

(a) $200 \cos 30° - f = 0$ and so $f = 173$ N

Then, $\mu_k = f/Y = 173/400 = 0.43$.

(b) $200 \cos 30° - f = 0$ and so $f = 173$ N

Then, $\mu_k = f/Y = 173/250 = 0.69$.

(c) $-200 \sin \theta + f = 0$ and so $f = (200 \sin \theta)$ N

Then, $\mu_k = f/Y = (200 \sin \theta)/(200 \cos \theta) = \tan \theta$.

2.9.   Suppose that in Fig. 2-7(c) the block is at rest.   The angle of the incline is slowly increased.   At an angle $\theta = 42°$,   the block begins to slide.   What is the coefficient of static friction between the block and the incline?   (The block and surface are not the same as in Problems 2.7 and 2.8.)

At the instant the block begins to slide, the friction force has its critical value.   Therefore, $\mu_s = f/Y$ at that instant.   Following the method of Problems 2.7 and 2.8, we have

$$Y = w \cos \theta \qquad \text{and} \qquad f = w \sin \theta$$

Therefore, when sliding just starts,

$$\mu_s = \frac{f}{Y} = \frac{w \sin \theta}{w \cos \theta} = \tan \theta$$

But θ was found by experiment to be 42°.   Therefore, $\mu_s = \tan 42° = 0.90$.

Supplementary Problems

2.10.   For the situation shown in Fig. 2-8, find the values of T_1 and T_2 if the weight is 600 N.   *Ans.*   503 N, 783 N

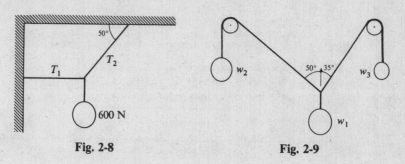

Fig. 2-8   Fig. 2-9

2.11.   The following coplanar forces pull on a ring: 200 N at 30°, 500 N at 80°, 300 N at 240°, and an unknown force.   Find the magnitude and direction of the unknown force if the ring is to be in equilibrium.   *Ans.*   350 N at 252°

2.12.   In Fig. 2-9, the pulleys are frictionless and the system hangs at equilibrium.   If w_3 is a 200 N weight, what are the values of w_1 and w_2?   *Ans.*   260 N, 150 N

2.13.   Suppose w_1 in Fig. 2-9 weighs 500 N.   Find the values of w_2 and w_3 if the system is to hang in equilibrium as shown.   *Ans.*   288 N, 384 N

2.14.   If in Fig. 2-10 the friction between the block and the incline is negligible, how heavy must weight w be if the 200 N block is to remain at rest?   *Ans.*   115 N

2.15.   The system in Fig. 2-10 remains at rest when the hanging weight w is 220 N.   What are the magnitude and direction of the friction force on the 200 N block?   *Ans.*   105 N down the incline

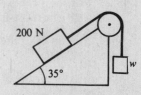

Fig. 2-10

2.16.   Find the normal force acting on the block in each of the equilibrium situations shown in Fig. 2-11.   *Ans.*   (a) 34 N;   (b) 46 N;   (c) 91 N

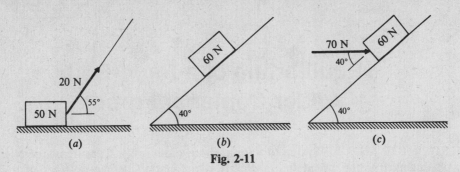

Fig. 2-11

2.17. The block shown in Fig. 2-11(a) slides with constant speed under the action of the force shown. (a) How large is the retarding friction force? (b) What is the coefficient of kinetic friction between the block and the floor?   *Ans.* (a) 11.5 N; (b) 0.34

2.18. The block shown in Fig. 2-11(b) slides at constant speed down the incline. (a) How large is the friction force that opposes its motion? (b) What is the coefficient of sliding (kinetic) friction between the block and plane?   *Ans.* (a) 38.6 N; (b) 0.84

2.19. The block in Fig. 2-11(c) just begins to slide up the incline when the pushing force shown is increased to 70 N. (a) What is the critical static friction force on it? (b) What is the value of the coefficient of static friction?   *Ans.* (a) 15 N; (b) 0.165

2.20. If $w = 40$ N in the equilibrium situation shown in Fig. 2-12, find T_1 and T_2.   *Ans.*   58 N, 31 N

2.21. Refer to the equilibrium situation shown in Fig. 2-12. The cords are strong enough to withstand a maximum tension of 80 N. What is the largest value of w that they can support as shown?   *Ans.*   55 N

2.22. The weight w in Fig. 2-13 is 80 N and is in equilibrium. Find T_1, T_2, T_3, and T_4.   *Ans.*   37 N, 88 N, 77 N, 139 N

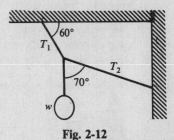

Fig. 2-12

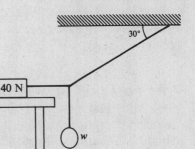

Fig. 2-13   Fig. 2-14   Fig. 2-15

2.23. The pulleys shown in Fig. 2-14 have negligible weight and friction. What is the value of w if it remains supported as shown by the 70 N weight?   *Ans.*   185 N

2.24. In Fig. 2-15, the system is in equilibrium. (a) What is the maximum value that w can have if the friction force on the 40 N block cannot exceed 12.0 N? (b) What is the coefficient of static friction between the block and tabletop?   *Ans.* (a) 6.9 N; (b) 0.30

2.25. The system of Fig. 2-15 is just on the verge of slipping. If $w = 8.0$ N,   what is the coefficient of static friction between the block and tabletop?   *Ans.*   0.346

Chapter 3

Equilibrium of a Rigid Body under Coplanar Forces

THE TORQUE (OR MOMENT) due to a force about an axis is a measure of the effectiveness of the force in producing rotation about that axis. It is defined to be the product of the force and the perpendicular distance from the axis of rotation to the line of action of the force. This perpendicular distance is called the *lever arm* (or *moment arm*):

$$\text{torque} = \tau = (\text{force}) \cdot (\text{lever arm})$$

The units of torque are N·m or lb·ft.

CONDITIONS FOR EQUILIBRIUM of a rigid object under the action of *coplanar forces* are

(1) The *force condition*: The vector sum of all forces acting on the body must be zero:

$$\sum F_x = 0 \qquad \sum F_y = 0$$

where the plane of the coplanar forces is taken to be the *xy*-plane.

(2) The *torque condition*: Take an axis perpendicular to the plane of the coplanar forces. Calling the torques that tend to cause clockwise rotation about the axis negative and counterclockwise torques positive, the sum of all the torques acting on the object must be zero:

$$\sum \tau = 0$$

THE CENTER OF GRAVITY of an object is the point at which the entire weight of the object may be considered concentrated, i.e. the line of action of the weight passes through the center of gravity. A single vertically upward force equal in magnitude to the weight of the object and applied through the center of gravity will keep the object in equilibrium.

THE POSITION OF THE AXIS IS ARBITRARY: If the sum of the torques is zero about one axis for a body that also obeys the force condition, it is zero about all other axes parallel to the first. We usually choose the axis in such a way that the line of an unknown force goes through the intersection point of the axis and the plane. The lever arm (and torque) of the unknown force is then zero and so this unknown force does not appear in the torque equation.

Solved Problems

3.1. In Fig. 3-1 are shown the four forces acting on a sheet of wood. (*a*) Find the lever arm for each about point *O* as axis (i.e. the axis is perpendicular to the plane at *O*). (*b*) Repeat for point *A* as axis.

(a) By definition, the lever arm is the length of the perpendicular dropped from the axis to the line of the force. For an axis at O, we have

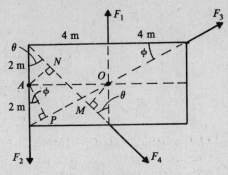

Fig. 3-1

Force	Lever Arm	Sense of Rotation
F_1	zero	
F_2	$\overline{OA} = 4$ m	ccw
F_3	zero	
F_4	$\overline{OM} = 2 \sin \theta = 1.41$ m	ccw

where "ccw" means counterclockwise. We find θ to be 45° by noting that $\tan \theta = 4/4 = 1.00$.

(b) For an axis at point A, we have

Force	Lever Arm	Sense of Rotation
F_1	$\overline{OA} = 4$ m	ccw
F_2	zero	
F_3	$\overline{AP} = 2 \cos \phi = 2 \cos 27° = 1.78$ m	ccw
F_4	$\overline{AN} = 2 \sin \theta = 1.41$ m	cw

where "cw" stands for clockwise. We found ϕ from the fact that $\tan \phi = 4/8 = 0.50$.

3.2. Find the torques due to the forces shown in Fig. 3-1 about (a) point O as axis and (b) point A as axis.

The lever arms were found in Problem 3.1. Because ccw torques are positive, we have

Force	(a) Torque about O	(b) Torque about A
F_1	0	$4F_1$
F_2	$4F_2$	0
F_3	0	$1.78 F_3$
F_4	$1.41 F_4$	$-1.41 F_4$

Only one torque is cw and that torque must be taken as negative.

3.3. A uniform beam weighs 200 N and holds a 450 N weight as shown in Fig. 3-2. Find the magnitudes of the forces exerted on the beam by the two supports at its ends.

Rather than draw a separate free-body diagram, the forces on the object being considered (the beam) are also shown in Fig. 3-2. Because the beam is uniform, its center of gravity is at its geometrical center. Thus the weight of the beam (200 N) is shown acting at the beam's center. The forces F_1 and F_2 are exerted on the beam by the supports.

We have two equations to write for this equilibrium situation: $\sum F_y = 0$ and $\sum \tau = 0$. There are no x-directed forces on the beam.

$$\sum F_y = 0 \quad \text{becomes} \quad F_1 + F_2 - 200 \text{ N} - 450 \text{ N} = 0$$

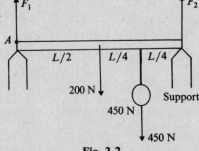

Fig. 3-2

Before writing the torque equation, an axis must be chosen. Choose it at A, because the unknown force F_1 will go through it and exert no torque. The torque equation is

$$-(200 \text{ N})(L/2) - (450 \text{ N})(3L/4) + (F_2)(L) = 0$$

Dividing through the equation by L and then solving for F_2 gives $F_2 = 438$ N.

To find F_1 we substitute the value of F_2 in the first equation and obtain $F_1 = 212$ N.

3.4. A uniform, 100 N pipe is used as a lever, as shown in Fig. 3-3. Where must the fulcrum (the support point) be placed if a 500 N weight at one end is to balance a 200 N weight at the other end? How much load must the support hold?

The forces in question are shown in Fig. 3-3. We assume that the support point is at a distance x from one end. Let us take the axis point to be at the position of the support. Then the torque equation, $\sum \tau = 0$, becomes

$$(200 \text{ N})(x) + (100 \text{ N})\left(x - \frac{L}{2}\right) - (500 \text{ N})(L - x) = 0$$

This simplifies to

$$(800 \text{ N})(x) = (550 \text{ N})(L)$$

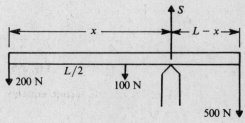

and so we find that $x = 0.69\,L$. The support should be placed 0.69 of the way from the lighter-loaded end.

To find the load S held by the support, we use $\sum F_y = 0$, which gives

$$S - 200 \text{ N} - 100 \text{ N} - 500 \text{ N} = 0$$

from which $S = 800$ N.

Fig. 3-3

3.5. Where must an 800 N weight be hung on a uniform, 100 N pole so that a girl at one end supports 1/3 as much as a woman at the other end?

The situation is shown in Fig. 3-4. We represent the force exerted by the girl by P and that by the woman by $3P$. Take the axis point at the left end. Then the torque equation becomes

$$-(800 \text{ N})(x) - (100 \text{ N})(L/2) + (P)(L) = 0$$

A second equation we can write is $\sum F_y = 0$, or

$$3P - 800 \text{ N} - 100 \text{ N} + P = 0$$

from which $P = 225$ N. Substitution of this value in the torque equation gives

$$(800 \text{ N})(x) = (225 \text{ N})(L) - (100 \text{ N})(L/2)$$

from which $x = 0.22\,L$. The load should be hung 0.22 of the way from the woman to the girl.

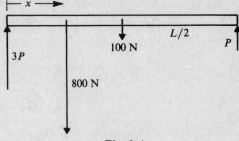

Fig. 3-4

3.6. A uniform, 200 N board of length L has two weights hanging from it, 300 N at $L/3$ from one end and 400 N at $3L/4$ from the same end. What single additional force acting on the board will cause the board to be in equilibrium?

The situation is shown in Fig. 3-5, where P is the force we wish to find. For equilibrium, $\sum F_y = 0$ and so

$$P = 400 \text{ N} + 200 \text{ N} + 300 \text{ N} = 900 \text{ N}$$

Because the board is to be in equilibrium, we are free to choose the axis anywhere. Choose it at point A. Then $\sum \tau = 0$ gives

$$(P)(x) - (400 \text{ N})(3L/4) - (200 \text{ N})(L/2) - (300 \text{ N})(L/3) = 0$$

Using $P = 900$ N, we find that $x = 0.56\,L$. The required force is 900 N upward at $0.56\,L$ from the left end.

3.7. The right-angle rule (or square) shown in Fig. 3-6 hangs at rest from a peg as shown. It is made of uniform metal sheet. One arm is L cm long while the other is $2L$ cm long. Find the angle θ at which it will hang.

If the rule is not too wide, we can approximate it as two thin rods of lengths L and $2L$ joined perpendicularly at A. Call the weight of each centimeter of arm γ. Then the forces acting on the rule are as indicated in Fig. 3-6, where P is the upward push of the peg.

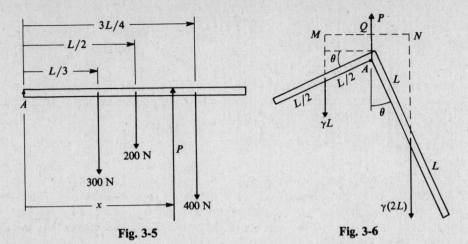

Fig. 3-5   Fig. 3-6

We now write the torque equation for the square, which is in equilibrium. Taking torques about point A, we have

$$(\gamma L)(\overline{MQ}) - (2\gamma L)(\overline{QN}) = 0$$

But $\overline{MQ} = \frac{1}{2}L\cos\theta$ and $\overline{QN} = L\sin\theta$. Substituting and rearranging gives

$$\frac{1}{2}\cos\theta = 2\sin\theta$$

and so $\tan\theta = 0.25$ and therefore $\theta = 14°$.

3.8. A *couple* consists of two equal and oppositely directed forces, not in the same straight line. It can produce only rotation. Show that the torque (or moment) due to a couple is independent of the axis position.

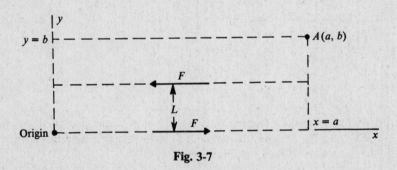

Fig. 3-7

As shown in Fig. 3-7, we take an axis at the arbitrary point A, with coordinates $x = a$, $y = b$. The torque about A is

$$\text{torque} = (F)(b) - (F)(b - L) = FL$$

Our answer for the torque (FL) due to the couple contains neither a nor b. Hence the axis could have been taken anywhere without altering the result.

3.9. Consider the situation shown in Fig. 3-8(a). The uniform, 600 N beam is hinged at P. Find the tension in the tie rope and the components of the force exerted by the hinge on the beam.

The forces acting on the beam are shown in Fig. 3-8(b). We represent the force exerted by the hinge by its components, H and V. We can work either with the tension T in the rope or with its components. In terms of T, the torque equation about P as axis is

$$(T)\left(\frac{3L}{4}\sin 40°\right) - (800\text{ N})(L) - (600\text{ N})\left(\frac{L}{2}\right) = 0$$

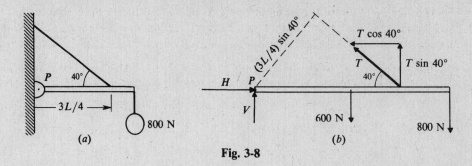

Fig. 3-8

We take the axis at P because then H and V do not appear in the torque equation.   Or, we could write the torque equation in terms of the components of T.   If we note that the component $T \cos 40°$ actually acts at the tie point and has zero lever arm, the torque equation in terms of the components of T is

$$(T \sin 40°)\left(\frac{3L}{4} \right) - (800 \text{ N})(L) - (600 \text{ N})\left(\frac{L}{2} \right) = 0$$

Notice that the two equations are identical.   Either one yields $T = 2280$ N.

   To find H and V we write

$$\sum F_x = 0 \qquad \text{or} \qquad -T \cos 40° + H = 0$$

$$\sum F_y = 0 \qquad \text{or} \qquad T \sin 40° + V - 600 - 800 = 0$$

Using the value found for T, these equations give $H = 1750$ N   and $V = 66$ N.

3.10.   A uniform, 400 N boom is supported as shown in Fig. 3-9(a).   Find the tension in the tie rope and the force exerted on the boom by the pin at P.

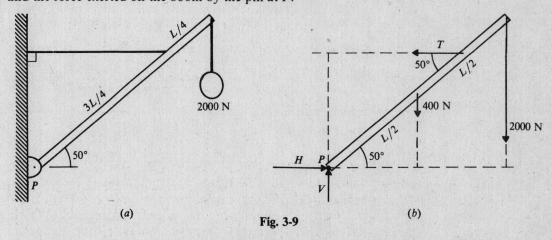

Fig. 3-9

   The forces acting on the boom are shown in Fig. 3-9(b).   Taking the pin as axis, the torque equation is

$$(T)\left(\frac{3L}{4} \sin 50° \right) - (400 \text{ N})\left(\frac{L}{2} \cos 50° \right) - (2000 \text{ N})(L \cos 50°) = 0$$

from which $T = 2460$ N.   We now write:

$$\sum F_x = 0 \qquad \text{or} \qquad H - T = 0$$

and so $H = 2460$ N.   Also

$$\sum F_y = 0 \qquad \text{or} \qquad V - 2000 \text{ N} - 400 \text{ N} = 0$$

and so $V = 2400$ N.   These are the components of the force at the pin.   The magnitude of this force is

$$\sqrt{(2400)^2 + (2460)^2} = 3440 \text{ N}$$

The tangent of the angle it makes with the horizontal is $\tan \theta = 2400/2460$,   and so $\theta = 44°$.

3.11.   As shown in Fig. 3-10, the hinges A and B hold a uniform, 400 N door in place. The upper hinge supports the entire weight of the door. Find the forces exerted on the door at the hinges. The width of the door is $h/2$, where h is the distance between the hinges.

The forces acting on the door are shown in Fig. 3-10. Only a horizontal force acts at B, because the upper hinge is assumed to support the door's weight. Let us take torques about point A as axis.

$$\sum \tau = 0 \qquad \text{becomes} \qquad (F_2)(h) - (400 \text{ N})(h/4) = 0$$

from which $F_2 = 100$ N. We also have

$$\sum F_x = 0 \qquad \text{or} \qquad F_2 - H = 0$$

$$\sum F_y = 0 \qquad \text{or} \qquad V - 400 \text{ N} = 0$$

We find from these that $H = 100$ N and $V = 400$ N.

To find the resultant force **R** on the hinge at A, we have

$$R = \sqrt{(400)^2 + (100)^2} = 412 \text{ N}$$

The tangent of the angle that **R** makes with the negative x-direction is V/H and so the angle is

$$\arctan 4 = 76°$$

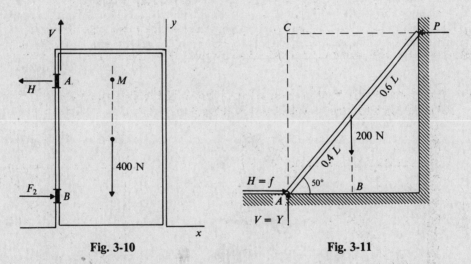

Fig. 3-10   Fig. 3-11

3.12.   A ladder leans against a smooth wall, as shown in Fig. 3-11. (By a "smooth" wall, we mean that the wall exerts on the ladder only a force that is perpendicular to the wall. There is no friction force.) The ladder weighs 200 N and its center of gravity is $0.4L$ from the base, where L is the ladder's length. (a) How large a friction force must exist at the base of the ladder if it is not to slip? (b) What is the necessary coefficient of static friction?

(a)   We wish to find the friction force H. Notice that no friction force exists at the top of the ladder. Taking torques about point A, the torque equation is

$$-(200 \text{ N})(\overline{AB}) + (P)(\overline{AC}) = 0$$

or $$-(200 \text{ N})(0.4 L \cos 50°) + (P)(L \cos 40°) = 0$$

Solving gives $P = 67$ N. We can also write

$$\sum F_x = 0 \qquad \text{or} \qquad H - P = 0$$

$$\sum F_y = 0 \qquad \text{or} \qquad V - 200 = 0$$

and so $H = 67$ N and $V = 200$ N.

(b) $$\mu_s = \frac{f}{Y} = \frac{H}{V} = \frac{67}{200} = 0.34$$

3.13. For the situation shown in Fig. 3-12(*a*), find T_1, T_2, and T_3. The boom is uniform and weighs 800 N.

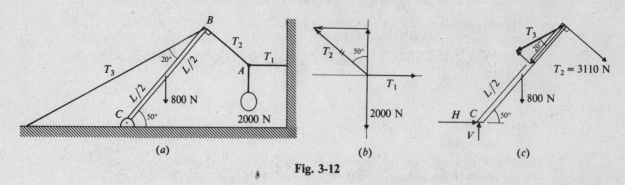

Fig. 3-12

Let us first apply the force condition to point A. The appropriate free-body diagram is shown in Fig. 3-12(*b*). We then have

$$T_2 \cos 50° - 2000 \text{ N} = 0 \qquad \text{and} \qquad T_1 - T_2 \sin 50° = 0$$

From the first of these we find $T_2 = 3110$ N; then the second equation gives $T_1 = 2380$ N.

Let us now isolate the boom and apply the equilibrium conditions to it. The appropriate free-body diagram is shown in Fig. 3-12(*c*). By resolving the vector $\mathbf{T}_3$ into the components shown we anticipate taking torques about point C. Then the component $T_3 \cos 20°$ exerts no torque about C. The torque equation is

$$(T_3 \sin 20°)(L) - (3110 \text{ N})(L) - (800 \text{ N})\left(\frac{L}{2} \cos 50° \right) = 0$$

Solving for T_3, we find it to be 9840 N. If it had been required, we could find H and V by using the x- and y-force equations.

Supplementary Problems

3.14. As shown in Fig. 3-13, the uniform, 1600 N beam is hinged at one end and held by a tie rope at the other. Determine the tension T in the rope and the force components at the hinge.
Ans. $T = 670$ N, $H = 670$ N, $V = 1600$ N

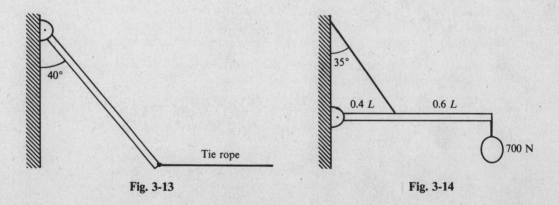

Fig. 3-13   Fig. 3-14

3.15. The uniform beam shown in Fig. 3-14 weighs 500 N and supports a 700 N load. Find the tension in the tie rope and the force of the hinge on the beam.   *Ans.* 2900 N, 2040 N at 35° below the horizontal

3.16. The foot of a ladder rests against a wall and its top is held by a tie rope, as shown in Fig. 3-15. The ladder weighs 100 N and its center of gravity is 0.4 of its length from the foot. A 150 N child hangs from a rung that is 0.2 of the length from the top. Determine the tension in the tie rope and the components of the force on the foot of the ladder.   *Ans.* $T = 120$ N, $H = 120$ N, $V = 250$ N

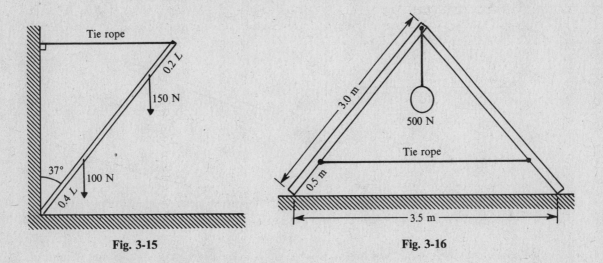

Fig. 3-15   Fig. 3-16

3.17. A truss is made by hinging two uniform, 150 N rafters as shown in Fig. 3-16. They rest on an essentially frictionless floor and are held together by a tie rope. A 500 N load is held at their apex. Find the tension in the tie rope.   *Ans.*   280 N

3.18. The hinges of a uniform door weighing 200 N are 2.50 m apart. One hinge is at a distance d from the top of the door, while the other is a distance d from the bottom. The door is 1.00 m wide. The weight of the door is supported by the lower hinge. Determine the forces exerted by the hinges on the door. *Ans.* The horizontal force at the upper hinge is 40 N. The force at the lower hinge is 204 N at 79° above the horizontal.

3.19. The uniform bar shown in Fig. 3-17 weighs 40 N and is subjected to the forces shown. Find the magnitude, location, and direction of the force needed to keep the bar in equilibrium.
Ans.   106 N; 0.675 L from right end; at 49°

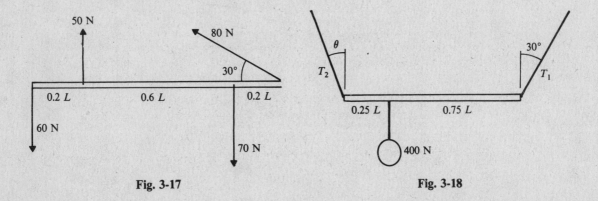

Fig. 3-17   Fig. 3-18

3.20. The uniform, 120 N board shown in Fig. 3-18 is supported by two ropes as shown. A 400 N weight is suspended 1/4 of the way from the left end. Find T_1, T_2, and the angle θ made by the left rope.
Ans.   185 N, 372 N, 14.4°

3.21.   In Fig. 3-19, the uniform beam weighs 500 N.   If the tie rope can support 1800 N, what is the maximum value the load *w* can have?   *Ans.*   930 N

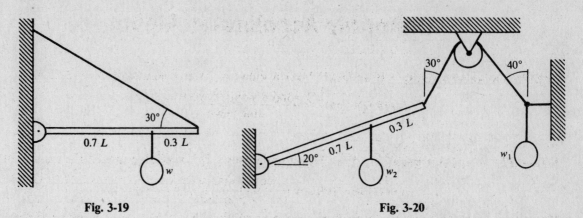

Fig. 3-19   Fig. 3-20

3.22.   The beam in Fig. 3-20 has negligible weight.   If the system hangs in equilibrium when $w_1 = 500$ N, what is the value of w_2?   *Ans.*   638 N

3.23.   Repeat Problem 3.22, but now find w_1 if w_2 is 500 N.   The beam weighs 300 N and is uniform. *Ans.*   560 N

3.24.   An object is subjected to the forces shown in Fig. 3-21.   What single force applied at a point on the *x*-axis will balance these forces?   (First find its components and then find the force.)   Where on the *x*-axis should the force be applied? *Ans.* $F_x = 232$ N, $F_y = -338$ N; $F = 410$ N at $-55.5°$;   at $x = 2.14$ m

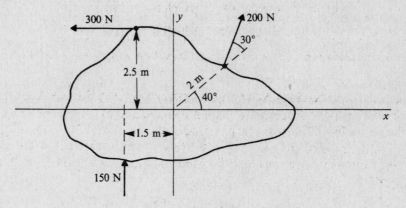

Fig. 3-21

Chapter 4

Uniformly Accelerated Motion

SPEED is a scalar quantity. If an object takes a time t to cover a certain distance d, then

$$average\ speed = \frac{total\ distance\ covered}{time\ taken} = \frac{d}{t}$$

VELOCITY is a vector quantity. If an object undergoes a vector displacement **s** in a time t, then

$$\bar{\mathbf{v}} = average\ velocity = \frac{vector\ displacement}{time\ taken} = \frac{\mathbf{s}}{t}$$

The direction of the velocity vector is the same as that of the displacement vector. The units of velocity (and speed) consist of a distance unit divided by a time unit. Typical are m/s, km/h, cm/d (centimeter per day).

ACCELERATION measures the time rate of change of velocity. We have

$$\bar{\mathbf{a}} = average\ acceleration = \frac{change\ in\ velocity\ vector}{time\ taken} = \frac{\mathbf{v}_f - \mathbf{v}_0}{t}$$

where $\mathbf{v}_0$ is the original velocity, $\mathbf{v}_f$ is the final velocity, and t is the time taken for the change. The units of acceleration are a velocity unit divided by a time unit. Typical examples are

$$(m/s) \div s \quad\quad \text{which is} \quad\quad m/s^2$$
$$(cm/s) \div s \quad\quad \text{which is} \quad\quad (cm/s)^2$$
$$(km/h) \div s \quad\quad \text{which is} \quad\quad km/h \cdot s$$

Notice that acceleration is a vector. It has the direction of $\mathbf{v}_f - \mathbf{v}_0$, the change in velocity.

UNIFORMLY ACCELERATED MOTION ALONG A STRAIGHT LINE is an exceptionally important case. In this case, the *acceleration vector is constant* and along the line of the displacement vector, so that the vector nature of displacement, **v**, and **a** can be handled by plus and minus signs. Calling the displacement x (positive if in the positive direction and negative if in the negative direction), the motion can be described by the *five motion equations* for uniformly accelerated motion:

$$x = \bar{v}t$$

$$\bar{v} = \frac{v_f + v_0}{2}$$

$$a = \frac{v_f - v_0}{t}$$

$$v_f^2 = v_0^2 + 2ax$$

$$x = v_0t + \frac{at^2}{2}$$

Often x is replaced by y or s, and sometimes v_f is written simply as v.

INSTANTANEOUS VELOCITY is the average velocity evaluated for a time interval that approaches zero. For all sufficiently small time intervals, the displacements will be essentially along the same straight line—say, the x-axis.   Thus, if an object undergoes a displacement Δx in a time Δt,

$$v = instantaneous\ velocity = \lim_{\Delta t \to 0} \frac{\Delta x}{\Delta t}$$

where the notation means that the ratio $\Delta x / \Delta t$ is to be evaluated for a time interval Δt that approaches zero.

GRAPHICAL INTERPRETATIONS for motion on a straight line (the x-axis) are as follows.

The *instantaneous velocity* of an object at a certain time is the slope of the x versus t graph at that time.

The *instantaneous acceleration* of an object at a certain time is the slope of the v versus t graph at that time.

For constant-velocity motion, the x versus t graph is a straight line.   For constant-acceleration motion, the v versus t graph is a straight line.

ACCELERATION DUE TO GRAVITY (g):   The acceleration of a body moving only under the force of gravity is g, the gravitational acceleration, which is directed vertically downward.   On earth, $g = 9.8$ m/s$^2 = 32.2$ ft/s^2;   the value varies slightly from place to place.   On the moon, the free-fall acceleration is 1.6 m/s^2.

DIRECTION IS IMPORTANT and a positive direction must be chosen when analyzing motion along a line.   Either direction may be chosen as the positive direction.   If a displacement, velocity, or acceleration is in the opposite direction, it must be taken negative.

PROJECTILE PROBLEMS can be solved easily if air friction can be ignored.   One simply considers the motion to consist of two independent parts: horizontal motion with $a = 0$ and $v_f = v_0 = \bar{v}$ (i.e. constant velocity), and vertical motion with $a = g = 9.8$ m/s^2 downward.

Solved Problems

4.1.   Change a speed of 0.200 cm/s to km/year.

$$0.200\ \frac{cm}{s} = \left(0.200\ \frac{cm}{s}\right)\left(10^{-5}\ \frac{km}{cm}\right)\left(3600\ \frac{s}{h}\right)\left(24\ \frac{h}{d}\right)\left(365\ \frac{d}{year}\right) = 63\ \frac{km}{year}$$

4.2.   A runner makes one lap around a 200 m track in a time of 25 s.   What were the runner's (a) average speed and (b) average velocity?

(a)   From the definition,

$$average\ speed = \frac{distance\ traveled}{time\ taken} = \frac{200\ m}{25\ s} = 8.0\ m/s$$

(b)   Because the endpoint for the run was at the starting point, the displacement vector from start to endpoint has zero length.   Therefore,

$$\bar{v} = \frac{displacement}{time} = \frac{0\ m}{25\ s} = 0\ m/s$$

4.3.   An object starts from rest with constant acceleration 8 m/s^2 along a straight line.   Find (a) the speed at the end of 5 s, (b) the average speed for the 5 s interval, (c) the distance traveled in the 5 s.

We are interested in the motion for the first 5 s.   For this, we know that $v_0 = 0$, $t = 5$ s, $a = 8$ m/s^2.   Because the motion is uniformly accelerated, the five motion equations apply.

(a) $$v_f = v_0 + at = 0 + (8 \text{ m/s}^2)(5 \text{ s}) = 40 \text{ m/s}$$

(b) $$\bar{v} = \frac{v_0 + v_f}{2} = \frac{0 + 40}{2} \text{ m/s} = 20 \text{ m/s}$$

(c) $$s = v_0 t + \tfrac{1}{2}at^2 = 0 + \tfrac{1}{2}(8 \text{ m/s}^2)(5 \text{ s})^2 = 100 \text{ m}$$

or $$s = \bar{v}t = (20 \text{ m/s})(5 \text{ s}) = 100 \text{ m}$$

4.4. A truck's speed increases uniformly from 15 km/h to 60 km/h in 20 s. Determine (a) the average speed, (b) the acceleration, (c) the distance traveled, all in the units of meters and seconds.

For the 20 s trip under discussion,

$$v_0 = \left(15 \ \frac{\text{km}}{\text{h}}\right)\left(1000 \ \frac{\text{m}}{\text{km}}\right)\left(\frac{1}{3600} \ \frac{\text{h}}{\text{s}}\right) = 4.17 \text{ m/s}$$

$$v_f = 60 \text{ km/h} = 16.7 \text{ m/s}$$

$$t = 20 \text{ s}$$

(a) $$\bar{v} = \tfrac{1}{2}(v_0 + v_f) = \tfrac{1}{2}(4.17 + 16.7) \text{ m/s} = 10.4 \text{ m/s}$$

(b) $$a = \frac{v_f - v_0}{t} = \frac{(16.7 - 4.2) \text{ m/s}}{20 \text{ s}} = 0.63 \text{ m/s}^2$$

(c) $$x = \bar{v}t = (10.4 \text{ m/s})(20 \text{ s}) = 208 \text{ m}$$

4.5. A ball is dropped from rest at a height of 50 m above the ground. (a) What is its speed just before it hits the ground? (b) How long does it take to reach the ground?

If we ignore air friction, the ball is uniformly accelerated until it reaches the ground. Its acceleration is downward and is 9.8 m/s². Taking *down* as positive, we have for the trip:

$$y = 50 \text{ m} \qquad a = 9.8 \text{ m/s}^2 \qquad v_0 = 0$$

(a) $$v_f^2 = v_0^2 + 2ay = 0 + 2(9.8 \text{ m/s}^2)(50 \text{ m}) = 980 \text{ m}^2/\text{s}^2$$

and so $v_f = 31$ m/s.

(b) From $a = (v_f - v_0)/t$,

$$t = \frac{v_f - v_0}{a} = \frac{(31 - 0) \text{ m/s}}{9.8 \text{ m/s}^2} = 3.2 \text{ s}$$

4.6. A skier starts from rest and slides 9 m down a slope in 3 s. In what time after starting will the skier acquire a velocity of 24 m/s? Assume constant acceleration.

We must first find the skier's acceleration from the data concerning the 3 s trip. For it, we have $t = 3$ s, $v_0 = 0$, and $s = 9$ m. Then $s = v_0 t + \tfrac{1}{2}at^2$ gives

$$a = \frac{2s}{t^2} = \frac{18 \text{ m}}{(3 \text{ s})^2} = 2 \text{ m/s}^2$$

We can now use this value of a for the longer trip, from the starting point to the place where $v = 24$ m/s. For this trip, $v_0 = 0$, $v_f = 24$ m/s, $a = 2$ m/s². Then, from $v_f = v_0 + at$,

$$t = \frac{v_f - v_0}{a} = \frac{24 \text{ m/s}}{2 \text{ m/s}^2} = 12 \text{ s}$$

4.7. A bus moving at a speed of 20 m/s begins to slow at a rate of 3 m/s each second. Find how far it goes before stopping.

For the trip under consideration, $v_0 = 20$ m/s, $v_f = 0$ m/s, $a = -3$ m/s². Notice that the bus is not speeding up in the positive motion direction. Instead, it is slowing in that direction and so its acceleration is negative (a deceleration). Use

$$v_f^2 = v_0^2 + 2ax$$

to find

$$x = \frac{-(20 \text{ m/s})^2}{2(-3 \text{ m/s}^2)} = 67 \text{ m}$$

4.8. A car moving at 30 m/s slows uniformly to a speed of 10 m/s in a time of 5 s. Determine (a) the acceleration of the car and (b) the distance it moves in the third second.

(a) For the 5 s interval, we have $t = 5$ s, $v_0 = 30$ m/s, $v_f = 10$ m/s. Using $v_f = v_0 + at$ gives

$$a = \frac{(10 - 30) \text{ m/s}}{5 \text{ s}} = -4 \text{ m/s}^2$$

(b) $$x = (\text{distance covered in 3 s}) - (\text{distance covered in 2 s})$$
$$= \left(v_0 t_3 + \tfrac{1}{2} a t_3^2\right) - \left(v_0 t_2 + \tfrac{1}{2} a t_2^2\right)$$
$$= v_0(t_3 - t_2) + \tfrac{1}{2} a \left(t_3^2 - t_2^2\right)$$

Using $v_0 = 30$ m/s, $a = -4$ m/s^2, $t_2 = 2$ s, $t_3 = 3$ s, gives

$$x = (30 \text{ m/s})(1 \text{ s}) - (2 \text{ m/s}^2)(5 \text{ s}^2) = 20 \text{ m}$$

4.9. The velocity of a train is reduced uniformly from 15 m/s to 7 m/s while traveling a distance of 90 m. (a) Compute the acceleration. (b) How much farther will the train travel before coming to rest, provided the acceleration remains constant?

(a) We have $v_0 = 15$ m/s, $v_f = 7$ m/s, $x = 90$ m. Then $v_f^2 = v_0^2 + 2ax$ gives

$$a = -0.98 \text{ m/s}^2$$

(b) We now have the new conditions $v_0 = 7$ m/s, $v_f = 0$, $a = -0.98$ m/s^2. Then

$$v_f^2 = v_0^2 + 2ax$$

gives

$$x = \frac{0 - (7 \text{ m/s})^2}{-1.96 \text{ m/s}^2} = 25 \text{ m}$$

4.10. A stone is thrown straight upward and it rises to a height of 20 m. With what speed was it thrown?

Let us take *up* as positive. The stone's velocity is zero at the top of its path. Then $v_f = 0$, $y = 20$ m, $a = -9.8$ m/s^2. The minus sign arises because the acceleration due to gravity is always downward and we have taken *up* to be positive. Use $v_f^2 = v_0^2 + 2ay$ to find

$$v_0 = \sqrt{-2(-9.8 \text{ m/s}^2)(20 \text{ m})} = 19.8 \text{ m/s}$$

4.11. A stone is thrown straight upward with a speed of 20 m/s. It is caught on its way down at a point 5.0 m above where it was thrown. (a) How fast was it going when it was caught? (b) How long did the trip take?

The situation is shown in Fig. 4-1. Let us take *up* as positive. Then, for the trip that lasts from the instant after throwing to the instant before catching, $v_0 = 20$ m/s, $y = +5$ m (since it is an upward displacement), $a = -9.8$ m/s^2.

(a) Use $v_f^2 = v_0^2 + 2ay$ to find

$$v_f^2 = (20 \text{ m/s})^2 + 2(-9.8 \text{ m/s}^2)(5 \text{ m}) = 302 \text{ m}^2/\text{s}^2$$

$$v_f = \pm\sqrt{302 \text{ m}^2/\text{s}^2} = -17.4 \text{ m/s}$$

We take the negative sign because the stone is going downward, in the negative direction.

(b) Use $a = (v_f - v_0)/t$ to find

$$t = \frac{(-17.4 - 20) \text{ m/s}}{-9.8 \text{ m/s}^2} = 3.8 \text{ s}$$

Notice the need for using the negative sign on v_f.

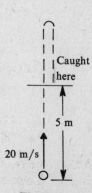

Fig. 4-1

4.12.   A ball thrown vertically upward returns to its starting point in 4 s.   Find its initial speed.

Let us take *up* as positive.   For the trip from beginning to end, $y = 0$, $a = -9.8$ m/s^2, $t = 4$ s. Notice that the start and the endpoint for the trip are the same, so the displacement is zero.   Use $y = v_0 t + \frac{1}{2} at^2$ to find

$$0 = v_0(4 \text{ s}) + \tfrac{1}{2}(-9.8 \text{ m/s}^2)(4 \text{ s})^2$$

from which $v_0 = 19.6$ m/s.

4.13.   An antiaircraft shell is fired vertically upward with an initial velocity of 500 m/s.   Neglecting friction, compute (*a*) the maximum height it can reach, (*b*) the time taken to reach that height, (*c*) the instantaneous velocity at the end of 60 s.   (*d*) When will its height be 10 km?

Take *up* as positive.   At the highest point, the velocity of the shell will be zero.

(*a*) $v_f^2 = v_0^2 + 2ay$ or $0 = (500 \text{ m/s})^2 + (-9.8 \text{ m/s}^2)y$ or $y = 12.8$ km

(*b*) $v_f = v_0 + at$ or $0 = 500 \text{ m/s} + (-9.8 \text{ m/s}^2)t$ or $t = 51$ s

(*c*) $v_f = v_0 + at$ or $v_f = 500 \text{ m/s} + (-9.8)(60 \text{ s}) = -88$ m/s

Because v_f is negative, and because we are taking *up* as positive, the velocity is directed downward. The shell is on its way down at $t = 60$ s.

(*d*) $y = v_0 t + \frac{1}{2} at^2$ or $10\,000 \text{ m} = (500 \text{ m/s})t + \frac{1}{2}(-9.8 \text{ m/s}^2)t^2$

or $4.9\, t^2 - 500t + 10\,000 = 0$

The quadratic formula,

$$x = \frac{-b \pm \sqrt{b^2 - 4ac}}{2a}$$

gives $t = 27$ s   and   75 s.   At $t = 27$ s,   the shell is at 10 km and ascending; at $t = 75$ s, it is at the same height but descending.

4.14.   A ballast bag is dropped from a balloon that is 300 m above the ground and rising at 13 m/s. For the bag, find (*a*) the maximum height reached, (*b*) its position and velocity 5 s after being released, (*c*) the time before it hits the ground.

The initial velocity of the bag when released is the same as that of the balloon, 13 m/s upward. Let us choose *up* as positive and take $y = 0$ at the point of release.

(*a*)   At the highest point, $v_f = 0$.   From $v_f^2 = v_0^2 + 2ay$,

$$0 = (13 \text{ m/s})^2 + 2(-9.8 \text{ m/s}^2)y   \text{or}   y = 8.6 \text{ m}$$

The maximum height is $300 + 8.6 = 308.6$ m.

(*b*)   Take the end point to be its position at $t = 5$ s.   Then, from $y = v_0 t + \frac{1}{2} at^2$,

$$y = (13 \text{ m/s})(5 \text{ s}) + \tfrac{1}{2}(-9.8 \text{ m/s}^2)(5 \text{ s})^2 = -58 \text{ m}$$

So its height is $300 - 58 = 242$ m.   Also, from $v_f = v_0 + at$,

$$v_f = 13 \text{ m/s} + (-9.8 \text{ m/s}^2)(5 \text{ s}) = -36 \text{ m/s}$$

It is on its way down with a velocity of 36 m/s downward.

(*c*)   Just before it hits the ground, the bag's displacement is -300 m.

$$y = v_0 t + \tfrac{1}{2} at^2   \text{becomes}   -300 \text{ m} = (13 \text{ m/s})t + \tfrac{1}{2}(-9.8 \text{ m/s}^2)t^2$$

or $4.9\, t^2 - 13t - 300 = 0$.   The quadratic formula gives $t = 9.3$ s   and -6.6 s.   Only the positive time has physical meaning, so the required answer is 9.3 s.

4.15.   As shown in Fig. 4-2, a projectile is fired with a horizontal velocity of 330 m/s from the top of a cliff 80 m high.   (*a*) How long will it take to strike the level ground at the base of the cliff? (*b*) How far from the foot of the cliff will it strike?   (*c*) With what velocity will it strike?

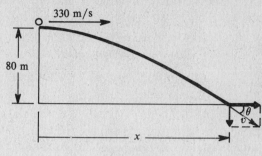

Fig. 4-2

(a) The horizontal and vertical motions are independent of each other. Consider first the vertical motion. Taking *down* as positive, we have

$$y = v_{0y}t + \tfrac{1}{2}a_y t^2$$

or

$$80 \text{ m} = 0 + \tfrac{1}{2}(9.8 \text{ m/s}^2)t^2$$

from which $t = 4.04$ s. Notice that the initial velocity had zero vertical component and so $v_0 = 0$ in the vertical motion.

(b) Now consider the horizontal motion. For it, $a = 0$ and so $\bar{v}_x = v_{0x} = v_{fx} = 330$ m/s. Then, using the value of t found in (a),

$$x = \bar{v}_x t = (330 \text{ m/s})(4.04 \text{ s}) = 1330 \text{ m}$$

(c) The final velocity has a horizontal component of 330 m/s. But its vertical velocity at $t = 4.04$ s is given by $v_{fy} = v_{0y} + a_y t$ as

$$v_{fy} = 0 + (9.8 \text{ m/s}^2)(4.04 \text{ s}) = 40 \text{ m/s}$$

The resultant of these two components is labeled v in Fig. 4-2; we have

$$v = \sqrt{(40 \text{ m/s})^2 + (330 \text{ m/s})^2} = 332 \text{ m/s}$$

Angle θ shown is given by $\tan \theta = 40/1330$ to be 6.9°.

4.16. A stunt flier is moving at 15 m/s parallel to the flat ground 100 m below, as shown in Fig. 4-3. How large must the distance x from plane to target be if a sack of flour released from the plane is to strike the target?

Following the same procedure as in Problem 4.15, we use $y = v_{0y}t + \tfrac{1}{2}a_y t^2$ to get

$$100 \text{ m} = 0 + \tfrac{1}{2}(9.8 \text{ m/s}^2)t^2 \qquad \text{or} \qquad t = 4.5 \text{ s}$$

Now use $x = \bar{v}_x t = (15 \text{ m/s})(4.5 \text{ s}) = 68$ m.

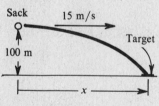

Fig. 4-3

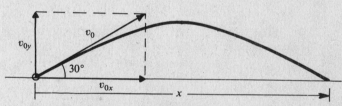

Fig. 4-4

4.17. A baseball is thrown with an initial velocity of 100 m/s at an angle of 30° above the horizontal, as shown in Fig. 4-4. How far from the throwing point will the baseball attain its original level?

We divide the problem into horizontal and vertical parts, for which

$$v_{0x} = v_0 \cos 30° = 86.6 \text{ m/s} \qquad v_{0y} = v_0 \sin 30° = 50 \text{ m/s}$$

where *up* is being taken positive.

In the vertical problem, $y = 0$ since the ball returns to its original height. Then

$$y = v_{0y}t + \tfrac{1}{2}a_y t^2 \qquad \text{or} \qquad 0 = (50 \text{ m/s}) + \tfrac{1}{2}(-9.8 \text{ m/s}^2)t$$

and $t = 10.2$ s.

In the horizontal problem, $v_{0x} = v_{fx} = \bar{v}_x = 87$ m/s. Therefore

$$x = \bar{v}_x t = (87 \text{ m/s})(10.2 \text{ s}) = 890 \text{ m}$$

4.18. As shown in Fig. 4-5, a ball is thrown from the top of one building toward a tall building 50 ft away. The initial velocity of the ball is 20 ft/s at 40° above the horizontal. How far above or below its original level will the ball strike the opposite wall?

We have

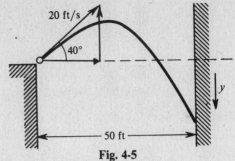

$$v_{0x} = (20 \text{ ft/s}) \cos 40° = 15.3 \text{ ft/s}$$

$$v_{0y} = (20 \text{ ft/s}) \sin 40° = 12.9 \text{ ft/s}$$

Consider first the horizontal motion. For it,

$$v_{0x} = v_{fx} = \bar{v}_x = 15.3 \text{ ft/s}$$

Then $x = \bar{v}_x t$ gives

$$50 \text{ ft} = (15.3 \text{ ft/s})t \qquad \text{or} \qquad t = 3.27 \text{ s}$$

In the vertical motion, taking *down* as positive,

$$y = v_{0y}t + \tfrac{1}{2}a_y t^2 = (-12.9 \text{ ft/s})(3.27 \text{ s}) + \tfrac{1}{2}(32.2 \text{ ft/s}^2)(3.27 \text{ s})^2 = 130 \text{ ft}$$

Fig. 4-5

Since y is positive, and since *down* is positive, the ball will hit at 130 ft below the original level.

4.19. (*a*) Find the range x of a gun which fires a shell with muzzle velocity v at an angle of elevation θ. (*b*) Find the angle of elevation θ of a gun which fires a shell with a muzzle velocity of 1200 ft/s at a target on the same level but 15 000 ft distant. See Fig. 4-6.

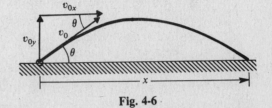

Fig. 4-6

(*a*) Let t be the time it takes the shell to hit the target. Then, $x = v_{0x}t$ or $t = x/v_{0x}$.
Consider the vertical motion alone, and take *up* as positive. When the shell strikes the target,

$$\text{vertical displacement} = 0 = v_{0y}t + \tfrac{1}{2}(-g)t^2$$

where $g = 32.2 \text{ ft/s}^2$. Solving this equation gives $t = 2v_{0y}/g$. But $t = x/v_{0x}$, so

$$\frac{x}{v_{0x}} = \frac{2v_{0y}}{g} \qquad \text{or} \qquad x = \frac{2v_{0x}v_{0y}}{g} = \frac{2(v_0 \cos \theta)(v_0 \sin \theta)}{g}$$

The formula $2 \sin \theta \cos \theta = \sin 2\theta$ can be used to simplify this. After substitution,

$$x = \frac{v_0^2 \sin 2\theta}{g}$$

The maximum range corresponds to $\theta = 45°$, since $\sin 2\theta$ has a maximum value, 1, when $2\theta = 90°$ or $\theta = 45°$.

(*b*) From the range equation found in (*a*), we have

$$\sin 2\theta = \frac{gx}{v_0^2} = \frac{32(15\,000)}{(1200)^2} = 0.333$$

Therefore, $2\theta = \arcsin 0.333 = 19.5°$ and so $\theta = 9.7°$.

4.20. The graph of an object's motion along a line is shown in Fig. 4-7. Find the instantaneous velocity of the object at points A and B. What is the object's average velocity? Its acceleration?

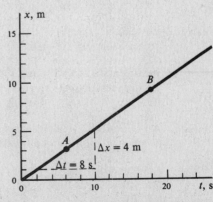

Because the velocity is given by the slope, $\Delta x / \Delta t$, of the tangent line, we take a tangent to the curve at point A. The tangent line is the curve itself in this case. For the triangle shown at A, we have

$$\frac{\Delta x}{\Delta t} = \frac{4 \text{ m}}{8 \text{ s}} = 0.50 \text{ m/s}$$

This is also the velocity at point B and at every other point on the straight-line graph. It follows that $a = 0$ and $\bar{v}_x = v_x = 0.50 \text{ m/s}$.

Fig. 4-7

4.21. Refer to Fig. 4-8. Find the instantaneous velocity at point F for the object whose motion the curve represents.

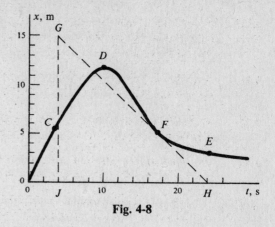

The tangent at F is the dashed line GH. Taking triangle GHJ, we have

$$\Delta t = 24 - 4 = 20 \text{ s} \qquad \Delta x = 0 - 15 = -15 \text{ m}$$

Hence

$$\text{slope at } F = v_F = \frac{\Delta x}{\Delta t} = \frac{-15 \text{ m}}{20 \text{ s}} = -0.75 \text{ m/s}$$

The negative sign tells us the object is moving in the $-x$-direction.

Fig. 4-8

Supplementary Problems

4.22. A car's speedometer reads 22 687 km at the start of a trip and 22 791 km at the end. The trip took 4 hours. What was the car's average speed in km/h? In m/s? *Ans.* 26 km/h, 7.2 m/s

4.23. An auto travels at a rate of 25 km/h for 4 minutes, then at 50 km/h for 8 minutes, and finally at 20 km/h for 2 minutes. Find (*a*) the total distance covered in km and (*b*) the average speed for the complete trip in m/s. *Ans.* (*a*) 9 km; (*b*) 10.7 m/s

4.24. A runner travels 1.5 laps around a circular track in a time of 50 s. The diameter of the track is 40 m and its circumference is 126 m. Find (*a*) the average speed of the runner and (*b*) the magnitude of the runner's average velocity. *Ans.* (*a*) 3.78 m/s; (*b*) 0.80 m/s

4.25. A body with initial velocity 8 m/s moves along a straight line with constant acceleration and travels 640 m in 40 s. For the 40 s interval, find (*a*) the average velocity, (*b*) the final velocity, and (*c*) the acceleration. *Ans.* (*a*) 16 m/s; (*b*) 24 m/s; (*c*) 0.40 m/s^2

4.26. A truck starts from rest and moves with a constant acceleration of 5 m/s^2. Find its speed and the distance traveled after 4 s has elapsed. *Ans.* 20 m/s, 40 m

4.27. A box slides down an incline with uniform acceleration. It starts from rest and attains a speed of 2.7 m/s in 3 s. Find (*a*) the acceleration and (*b*) the distance moved in the first 6 s.
Ans. (*a*) 0.90 m/s^2; (*b*) 16.2 m

4.28. A car is accelerating uniformly as it passes two check points that are 30 m apart. The time taken between check points is 4.0 s and the car's speed at the first check point is 5.0 m/s. Find the car's acceleration and its speed at the second check point. *Ans.* 1.25 m/s^2, 10 m/s

4.29. An auto's velocity increases uniformly from 6.0 m/s to 20 m/s while covering 70 m. Find the acceleration and the time taken. *Ans.* 2.6 m/s^2, 5.4 s

4.30. A plane starts from rest and accelerates along the ground before takeoff. It moves 600 m in 12 s. Find (*a*) the acceleration, (*b*) speed at the end of 12 s, (*c*) distance moved during the twelfth second.
Ans. (*a*) 8.3 m/s^2; (*b*) 100 m/s; (*c*) 96 m

4.31. A train running at 30 m/s is slowed uniformly to a stop in 44 s. Find the acceleration and the stopping distance. *Ans.* -0.68 m/s^2, 660 m

4.32. An object moving at 13 m/s slows uniformly at the rate of 2 m/s each second for a time of 6 s. Determine (*a*) its final speed, (*b*) its average speed during the 6 s, and (*c*) the distance moved in the 6 s.
Ans. (*a*) 1 m/s; (*b*) 7 m/s; (*c*) 42 m

4.33. A body falls freely from rest. Find (*a*) its acceleration, (*b*) the distance it falls in 3 s, (*c*) its speed after falling 70 m, (*d*) the time required to reach a speed of 25 m/s, (*e*) the time taken to fall 300 m. *Ans.* (*a*) 9.8 m/s²; (*b*) 44 m; (*c*) 37 m/s; (*d*) 2.55 s; (*e*) 7.8 s

4.34. A marble dropped from a bridge strikes the water in 5 s. Calculate (*a*) the speed with which it strikes and (*b*) the height of the bridge. *Ans.* (*a*) 49 m/s; (*b*) 123 m

4.35. A stone is thrown straight downward with initial speed 8 m/s from a height of 25 m. Find (*a*) the time it takes to reach the ground and (*b*) the speed with which it strikes. *Ans.* (*a*) 1.59 s; (*b*) 23.5 m/s

4.36. The hammer of a pile driver strikes the pile with a speed of 25 ft/s. From what height above the top of the pile did it fall? Neglect friction forces. *Ans.* 9.7 ft

4.37. A baseball is thrown straight upward with a speed of 30 m/s. (*a*) How long will it rise? (*b*) How high will it rise? (*c*) How long after it leaves the hand will it return to the starting point? (*d*) When will its speed be 16 m/s? *Ans.* (*a*) 3.06 s; (*b*) 46 m; (*c*) 6.1 s; (*d*) 1.43 s and 4.7 s

4.38. A bottle dropped from a balloon reaches the ground in 20 s. Determine the height of the balloon if (*a*) it was at rest in the air and (*b*) it was ascending with a speed of 50 m/s when the bottle was dropped. *Ans.* (*a*) 1960 m; (*b*) 960 m

4.39. A stone is shot straight upward with a speed of 80 ft/s from a tower 224 ft high. Find the speed with which it strikes the ground. *Ans.* 144 ft/s

4.40. A nut comes loose from a bolt on the bottom of an elevator as the elevator is moving up the shaft at 3 m/s. The nut strikes the bottom of the shaft in 2 s. (*a*) How far from the bottom of the shaft was the elevator when the nut fell off? (*b*) How far above the bottom was the nut 0.25 s after it fell off? *Ans.* (*a*) 13.6 m; (*b*) 14.0 m

4.41. A marble with speed 20 cm/s rolls off the edge of a table 80 cm high. (*a*) How long does it take to drop to the floor? (*b*) How far, horizontally, from the table edge does the marble strike the floor? *Ans.* (*a*) 0.404 s; (*b*) 8.1 cm

4.42. A body projected upward from the level ground at an angle of 50° with the horizontal has an initial speed of 40 m/s. (*a*) How long will it be before it hits the ground? (*b*) How far from the starting point will it strike? (*c*) At what angle with the horizontal will it strike? *Ans.* (*a*) 6.3 s; (*b*) 161 m; (*c*) 50°

4.43. A body is projected downward at an angle of 30° with the horizontal from the top of a building 170 m high. Its initial speed is 40 m/s. (*a*) How long will it take before striking the ground? (*b*) How far from the foot of the building will it strike? (*c*) At what angle with the horizontal will it strike? *Ans.* (*a*) 4.2 s; (*b*) 145 m; (*c*) 60°

4.44. A hose lying on the ground shoots a stream of water upward at an angle of 40° to the horizontal. The speed of the water is 20 m/s as it leaves the hose. How high up will it strike a wall which is 8 m away? *Ans.* 5.4 m

4.45. A World Series batter hits a home run ball with a velocity of 132 ft/s at an angle of 26° above the horizontal. A fielder who has a reach of 7 ft above the ground is backed up against the bleacher wall, which is 386 ft from home plate. The ball was 3 ft above the ground when hit. How high above the fielder's glove does the ball pass? *Ans.* 15 ft

4.46. Prove that a gun will shoot three times as high when its angle of elevation is 60° as when it is 30°, but will carry the same horizontal distance.

4.47. A ball is thrown upward at an angle of 30° to the horizontal and lands on the top edge of a building that is 20 m away. The top edge is 5 m above the throwing point. How fast was the ball thrown? *Ans.* 20 m/s

4.48. Refer back to Fig. 4-8 for the motion of an object along the *x*-axis. (*a*) What is the instantaneous velocity of the object at point *D*? (*b*) At point *C*? (*c*) At point *E*? *Ans.* (*a*) 0; (*b*) 1.33 m/s; (*c*) −0.13 m/s

Chapter 5

Newton's Laws

A FORCE is a push or pull exerted on a body. It is a vector quantity, having magnitude and direction.

AN UNBALANCED FORCE on an object causes the object to accelerate in the direction of the force. The acceleration is proportional to the force and inversely proportional to the mass of the object.

THE MASS of an object is a measure of the inertia of the object. *Inertia* is the tendency of an object at rest to remain at rest and of an object in motion to continue moving with unchanged vector velocity.

THE STANDARD KILOGRAM is an object whose mass is defined to be one kilogram. The masses of other objects are found by comparison with this mass. A *gram mass* is equivalent to 0.001 kg. The British unit of mass is the *slug*; one slug is 14.594 kg.

THE NEWTON is the SI (or mks) unit of force. One newton (1 N) is that unbalanced force which will give a 1 kg mass an acceleration of 1 m/s^2. The *dyne force unit* is 10^{-5} N. The *pound force unit* is 4.45 N. (The pound mass unit will not be used in this book.)

NEWTON'S FIRST LAW: If the resultant external force acting on an object is zero, then the vector velocity of the object will not change. An object at rest will remain at rest; an object in motion will continue in motion with constant velocity. A body accelerates only if an unbalanced force acts on it. This is often called the *inertia law*.

NEWTON'S SECOND LAW: If **F** is the resultant external force acting on an object of mass m, then the acceleration **a** of the object is related to **F** by **F** = m**a** *provided proper units are used*. In the SI system, F is in newtons, m is in kg, and a is in m/s^2. In the British system, F is in pounds, m is in slugs, and a is in ft/s^2. Mixed sets of units should never be used in **F** = m**a**.

The vector equation **F** = m**a** can be written in terms of components. One has

$$\sum F_x = ma_x \qquad \sum F_y = ma_y \qquad \sum F_z = ma_z$$

where these forces are the components of the external forces acting on the object.

NEWTON'S THIRD LAW: For every force exerted on one body, there is an equal, but oppositely directed, force acting on some other body. This is often called the *law of action and reaction*. Notice that the action and reaction forces act on different objects.

LAW OF UNIVERSAL GRAVITATION: Two masses, m and m', attract each other with forces of equal magnitude. For point masses (or spherically symmetric bodies), the attractive force is given by

$$\text{force} = G\,\frac{mm'}{r^2}$$

where r is the distance between mass centers, and where $G = 6.67 \times 10^{-11}$ N·m^2/kg^2 when F is in newtons, m and m' are in kilograms, and r is in meters.

THE WEIGHT of an object is the force of gravitation which pulls on the object. On the earth, it is the gravitational force exerted on the object by the earth. Its units are newtons (in the SI) and pounds (in the British system).

RELATION BETWEEN MASS AND WEIGHT: Mass is a measure of the inertia of an object, while weight is the force of gravity on the object. They are not the same in either concept or units. However, they are related. If an object of mass m falls freely toward the earth under the pull of gravity (a force w, the object's weight), then the acceleration of the object is the free-fall acceleration. Therefore, for a freely falling body, $F = ma$ becomes $w = mg$. *The following relation exists between the weight and mass of an object*: $w = mg$.

Because $g = 9.8$ m/s^2 = 32.2 ft/s^2 on earth, a 1 kg object weighs 9.8 N and a 1 slug object weighs 32.2 lb on earth.

Those people who use the British system of units often replace m in all formulas by w/g. In this way, the mass unit, the slug, is avoided.

General Considerations in Solving Problems

IN ANY SYSTEM OF UNITS a single equation must have all terms alike in physical significance, and numerical equality must be supplemented by physical equality. It is obvious that we cannot equate or add area to volume, just as we cannot add cows to trees. For example, in the equation

$$s = v_0 t + \tfrac{1}{2} a t^2$$

for motion with constant acceleration a and initial velocity v_0, all the terms represent distances:

$$v_0 t = \frac{\text{distance}}{\text{time}} \times \text{time} = \text{distance} \qquad a t^2 = \frac{\text{distance}}{\text{time}^2} \times \text{time}^2 = \text{distance}$$

This helps us to check equations, because all terms must have the same dimensions.

It is also obvious that the same system of units must be used consistently throughout the solution of a problem. Thus, in the equation $s = v_0 t + \tfrac{1}{2} a t^2$, if s is in meters, then v_0 must be in m/s, t in seconds, and a in m/s^2; if s is in ft, then v_0 must be in ft/s, t in s, and a in ft/s^2.

DIMENSIONAL FORMULAS: The dimensions of all mechanical quantities may be expressed in terms of the fundamental dimensions, which are length L, mass M, and time T. Thus a velocity has the dimensions L/T, or $[LT^{-1}]$ as it is usually written, since a velocity is made up of a length divided by an amount of time; an acceleration has the dimensions L/T^2 or $[LT^{-2}]$; a force, being a mass multiplied by an acceleration, has the dimensions $[MLT^{-2}]$; a volume has the dimensions $[L^3]$. This method of distinguishing physical quantities is helpful in checking equations, since every equation must have the same dimensions on both sides. For example, in the equation

$$s = v_0 t + \tfrac{1}{2} a t^2$$

the dimensions are

$$[L] \rightarrow [LT^{-1}][T] + [LT^{-2}][T^2] \qquad \text{or} \qquad [L] \rightarrow [L] + [L] \rightarrow [L]$$

MATHEMATICAL OPERATIONS WITH UNITS: In every mathematical operation, the units terms (e.g. lb, cm, ft^3, mi/h, m/s^2) must be carried along with the numbers and must undergo the same mathematical operations as the numbers.

Quantities cannot be added or subtracted directly unless they have the same units (as well as the same dimensions). For example, if we are to add algebraically 5 m (length) and 8 cm (length), we must first convert m to cm or cm to m. However, quantities of any sort can be combined in multiplication or division, in which the units as well as the numbers obey the algebraic laws of squaring, cancellation, etc. Thus:

$$(1) \quad 6 \text{ m}^2 + 2 \text{ m}^2 = 8 \text{ m}^2 \qquad\qquad (\text{m}^2 + \text{m}^2 \rightarrow \text{m}^2)$$

$$(2) \quad 5 \text{ cm} \times 2 \text{ cm}^2 = 10 \text{ cm}^3 \qquad (\text{cm} \times \text{cm}^2 = \text{cm}^3)$$

$$(3) \quad 2 \text{ m}^3 \times 1500 \, \frac{\text{kg}}{\text{m}^3} = 3000 \text{ kg} \quad \left(\text{m}^3 \times \frac{\text{kg}}{\text{m}^3} = \text{kg} \right)$$

$$(4) \quad 2 \text{ s} \times 3 \, \frac{\text{ft}}{\text{s}^2} = 6 \, \frac{\text{ft}}{\text{s}} \qquad\qquad \left(\text{s} \times \frac{\text{ft}}{\text{s}^2} = \frac{\text{ft}}{\text{s}} \right)$$

$$(5) \quad \frac{15 \text{ g}}{3 \text{ g/cm}^3} = 5 \text{ cm}^3 \qquad\qquad \left(\frac{\text{g}}{\text{g/cm}^3} = \text{g} \times \frac{\text{cm}^3}{\text{g}} = \text{cm}^3 \right)$$

Solved Problems

5.1. Find the weight of a body whose mass is (*a*) 3 kg, (*b*) 200 g, (*c*) 0.70 slug.

The general relation between mass m and weight w is $w = mg$. In this relation, the following units combinations are to be used:

In the SI (or mks)

$$(m \text{ in kg}), \ (g = 9.8 \text{ m/s}^2), \ (w \text{ in newtons})$$

In the British system

$$(m \text{ in slugs}), \ (g = 32.2 \text{ ft/s}^2), \ (w \text{ in pounds})$$

The values given for g apply only on earth; the acceleration due to gravity varies from place to place in the universe.

(*a*)
$$w = (3 \text{ kg})(9.8 \text{ m/s}^2) = 29.4 \text{ kg} \cdot \text{m/s}^2 = 29.4 \text{ N}$$

(*b*)
$$w = (0.20 \text{ kg})(9.8 \text{ m/s}^2) = 1.96 \text{ N}$$

(*c*)
$$w = (0.70 \text{ slug})(32.2 \text{ ft/s}^2) = 22.5 \text{ slug} \cdot \text{ft/s}^2 = 22.5 \text{ lb}$$

5.2. Find the mass of an object which on earth weighs (*a*) 25 N, (*b*) 5000 dynes, (*c*) 80 lb.

As in Problem 5.1, use $w = mg$.

(*a*)
$$m = \frac{25 \text{ N}}{9.8 \text{ m/s}^2} = 2.55 \text{ N} \cdot \text{s}^2/\text{m} = 2.55 \text{ kg}$$

(*b*)
$$m = \frac{5000 \times 10^{-5} \text{ N}}{9.8 \text{ m/s}^2} = 5.1 \times 10^{-3} \text{ kg} = 5.1 \text{ g}$$

(*c*)
$$m = \frac{80 \text{ lb}}{32.2 \text{ ft/s}^2} = 2.48 \text{ lb} \cdot \text{s}^2/\text{ft} = 2.48 \text{ slugs}$$

5.3. The only force acting on a 5 kg object has components $F_x = 20 \text{ N}$ and $F_y = 30 \text{ N}$. Find the acceleration of the object.

We make use of $\sum F_x = ma_x$ and $\sum F_y = ma_y$ to obtain

$$a_x = \frac{\sum F_x}{m} = \frac{20 \text{ N}}{5 \text{ kg}} = 4 \text{ m/s}^2$$

$$a_y = \frac{\sum F_y}{m} = \frac{30 \text{ N}}{5 \text{ kg}} = 6 \text{ m/s}^2$$

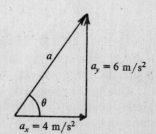

These components of the acceleration are shown in Fig. 5-1. From the figure, we see that

$$a = \sqrt{(4)^2 + (6)^2} \ \text{m/s}^2 = 7.2 \text{ m/s}^2$$

and $\theta = \arctan(6/4) = 56°$.

Fig. 5-1

5.4.   A 600 N object is to be given an acceleration of 0.70 m/s². How large an unbalanced force must act upon it?

Assuming the weight to be measured on the earth, we use $w = mg$ to find

$$m = \frac{w}{g} = \frac{600 \text{ N}}{9.8 \text{ m/s}^2} = 61 \text{ kg}$$

Now that we know the mass of the object (61 kg) and the desired acceleration (0.70 m/s²), we have

$$F = ma = (61 \text{ kg})(0.70 \text{ m/s}^2) = 43 \text{ N}$$

5.5.   How large a force is required to give a 2000 lb car an acceleration of 8 ft/s² along level ground? Neglect friction forces trying to slow the car.

We are given the weight of the car, 2000 lb. Following customary procedure when using the British system, we replace m by w/g. Then, the accelerating force is

$$F = ma = \left(\frac{w}{g}\right)a = \left(\frac{2000 \text{ lb}}{32.2 \text{ ft/s}^2}\right)(8 \text{ ft/s}^2) = 497 \text{ lb}$$

5.6.   A constant force acts on a 5 kg object and reduces its velocity from 7 m/s to 3 m/s in a time of 3 s. Find the force.

We must first find the acceleration of the object, which is constant because the force is constant. We have, from Chapter 4,

$$a = \frac{v_f - v_0}{t} = \frac{-4 \text{ m/s}}{3 \text{ s}} = -1.33 \text{ m/s}^2$$

Now we can use $F = ma$ with $m = 5$ kg.

$$F = (5 \text{ kg})(-1.33 \text{ m/s}^2) = -6.7 \text{ N}$$

The minus sign indicates that the force is a retarding force, directed opposite to the motion.

5.7.   A 600 kg car is moving on a level road at 30 m/s. (a) How large a retarding force (assumed constant) is required to stop it in a distance of 70 m? (b) What is the minimum coefficient of friction between tires and roadway if this is to be possible?

(a) We must first find the car's acceleration from a motion equation. It is known that $v_0 = 30$ m/s, $v_f = 0$, and $x = 70$ m. Use $v_f^2 = v_0^2 + 2ax$ to find

$$a = \frac{v_f^2 - v_0^2}{2x} = \frac{0 - 900 \text{ m}^2/\text{s}^2}{140 \text{ m}} = -6.43 \text{ m/s}^2$$

Now we can write $F = ma = (600 \text{ kg})(-6.43 \text{ m/s}^2) = -3860 \text{ N}$

(b) The force found in (a) is supplied as the friction force between the tires and roadway. Therefore, the magnitude of the friction force on the tires is $f = 3860$ N. The coefficient of friction is given by $\mu = f/Y$, where Y is the normal force. In the present case, the roadway pushes up on the car with a force equal to the car's weight. Therefore,

$$Y = w = mg = (600 \text{ kg})(9.8 \text{ m/s}^2) = 5900 \text{ N}$$

We then have $\mu = \dfrac{f}{Y} = \dfrac{3860}{5900} = 0.66$

The coefficient of friction must be at least 0.66 if the car is to stop within 70 m.

5.8.   An 8000 kg engine pulls a 40 000 kg train along a level track and gives it an acceleration $a_1 = 1.20$ m/s². What acceleration (a_2) would the engine give a 16 000 kg train?

For a given engine force, the acceleration is inversely proportional to the total mass. Thus

$$a_2 = \frac{m_1}{m_2} a_1 = \frac{8000 \text{ kg} + 40\,000 \text{ kg}}{8000 \text{ kg} + 16\,000 \text{ kg}} (1.20 \text{ m/s}^2) = 2.40 \text{ m/s}^2$$

5.9.   As shown in Fig. 5-2, a 6 kg object hangs by a rope from a scale. What is the tension in the rope (or reading on the scale) if the acceleration of the object is (*a*) zero, (*b*) 3 m/s² upward, (*c*) 5 m/s² downward, (*d*) 9.8 m/s² downward?

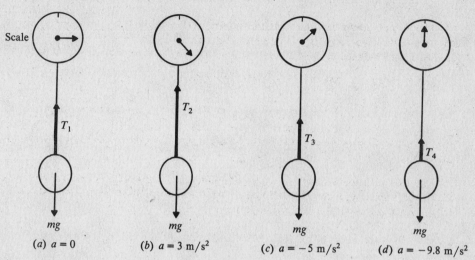

Fig. 5-2

The various situations are shown in Fig. 5-2. In each case, the weight of the object pulls the object downward and the rope pulls it upward. These are the only external forces acting on the object. The downward force in each case is

$$mg = (6 \text{ kg})(9.8 \text{ m/s}^2) = 58.8 \text{ N}$$

Let us apply $F = ma$ to each case in turn, where F is the resultant external force acting on the object. We take *up* as positive.

(*a*) $T_1 - 58.8 \text{ N} = (6 \text{ kg})(0 \text{ m/s}^2)$ or $T_1 = 58.8 \text{ N}$

(*b*) $T_2 - 58.8 \text{ N} = (6 \text{ kg})(3 \text{ m/s}^2)$ or $T_2 = 77 \text{ N}$

(*c*) $T_3 - 58.8 \text{ N} = (6 \text{ kg})(-5 \text{ m/s}^2)$ or $T_3 = 29 \text{ N}$

(*d*) $T_4 - 58.8 \text{ N} = (6 \text{ kg})(-9.8 \text{ m/s}^2)$ or $T_4 = 0 \text{ N}$

Notice that, in (*d*), the acceleration is that for free fall. Of course, nothing can be pulling upward on the object in this case and so $T = 0$. The object appears to be weightless. Also notice that T must be greater than mg for an upward acceleration and less than mg if the acceleration is to be downward.

5.10.   Suppose that the object in Fig. 5-2 weighs 50 lb. Find the acceleration of the object if the tension in the rope is (*a*) 50 lb, (*b*) 75 lb, (*c*) 40 lb, (*d*) zero.

In contrast to Problem 5.9, $w = mg = 50$ lb, the accelerations are unknown, and the tensions are known. Applying $F = ma$ to the object and using $m = w/g$ gives:

(*a*) $50 \text{ lb} - 50 \text{ lb} = \dfrac{50 \text{ lb}}{32 \text{ ft/s}^2} a$ or $a = 0 \text{ ft/s}^2$

(*b*) $75 \text{ lb} - 50 \text{ lb} = \dfrac{50 \text{ lb}}{32 \text{ ft/s}^2} a$ or $a = 16 \text{ ft/s}^2$ (upward)

(*c*) $40 \text{ lb} - 50 \text{ lb} = \dfrac{50 \text{ lb}}{32 \text{ ft/s}^2} a$ or $a = -6.4 \text{ ft/s}^2$ (downward)

(*d*) $0 - 50 \text{ lb} = \dfrac{50 \text{ lb}}{32 \text{ ft/s}^2} a$ or $a = -32 \text{ ft/s}^2$ (the free-fall acceleration)

5.11.   A tow rope will break if the tension in it exceeds
1500 N. It is used to tow a 700 kg car along level
ground. What is the largest acceleration the rope can
give to the car?

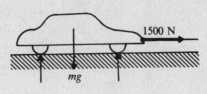

The forces acting on the car are shown in Fig. 5-3.
Only the x-directed force is of importance, because the y-
directed forces balance each other.

Fig. 5-3

$$\sum F_x = ma_x \qquad \text{becomes} \qquad 1500 \text{ N} = (700 \text{ kg})a$$

from which $a = 2.14 \text{ m/s}^2$.

5.12.   Compute the least acceleration with which a 45 kg woman can slide down a rope if the rope
can withstand a tension of only 300 N.

The weight of the woman is $mg = (45 \text{ kg})(9.8 \text{ m/s}^2) = 441 \text{ N}$. Because the rope can only support
300 N, the unbalanced downward force, F, on the woman must be at least   441 N − 300 N = 141 N.
Her minimum downward acceleration is then

$$a = \frac{F}{m} = \frac{141 \text{ N}}{45 \text{ kg}} = 3.1 \text{ m/s}^2$$

5.13.   A cord passing over an easily turned pulley has a 7 kg mass hanging
from one end and a 9 kg mass hanging from the other, as shown in
Fig. 5-4. (This arrangement is called *Atwood's machine*.) Find the
acceleration of the masses and the tension in the cord.

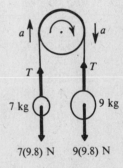

Because the pulley is easily turned, the tension in the cord will be the
same on each side. Notice that the forces acting on each of the two masses
are drawn in Fig. 5-4. Recall that the weight of an object is mg.

It is convenient in situations involving objects connected by cords to
take the motion direction as the positive direction. In the present case, we
take *up* positive for the 7 kg mass and *down* positive for the 9 kg mass. (If
we do this, the acceleration will be positive for each mass. Because the cord
doesn't stretch, the accelerations are numerically equal.) Writing $\sum F_y$
$= ma_y$ for each mass in turn, we have

Fig. 5-4

$$T - (7)(9.8) \text{ N} = (7 \text{ kg})a \qquad \text{and} \qquad (9)(9.8) \text{ N} - T = (9 \text{ kg})a$$

If we add these two equations, the unknown T drops out, giving

$$(9 - 7)(9.8) \text{ N} = (16 \text{ kg})a$$

from which $a = 1.23 \text{ m/s}^2$. We can now substitute 1.23 m/s² for a in either equation and obtain
$T = 77 \text{ N}$.

Another Method

Consider the two masses as a single system. Then the unbalanced force on the system is

$$F = (9 - 7)(9.8) \text{ N}$$

Using this in $F = ma$,

$$(9 - 7)(9.8) \text{ N} = [(7 + 9) \text{ kg}]a$$

where the total mass being moved by this unbalanced force is 16 kg. Solving for a, we find it to be
1.23 m/s².

In order to find T, we would need to isolate one mass and write $F = ma$ for it, as in the previous
method. Hence this latter method is most valuable if the tension is not required.

5.14.   A 70 kg box is slid along the floor by a 400 N force as shown in Fig. 5-5. The coefficient of
friction between the box and floor is 0.50 when the box is sliding. Find the acceleration of
the box.

Since the y-directed forces must balance,

$$Y = mg = (70 \text{ kg})(9.8 \text{ m/s}^2) = 686 \text{ N}$$

But the friction force f is given by

$$f = \mu Y = (0.50)(686 \text{ N}) = 343 \text{ N}$$

Now write $\sum F_x = ma_x$ for the box, taking the motion direction as positive:

$$400 \text{ N} - 343 \text{ N} = (70 \text{ kg})a \quad \text{or} \quad a = 0.81 \text{ m/s}^2$$

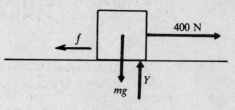

Fig. 5-5

5.15. Suppose, as shown in Fig. 5-6, that a 70 kg box is pulled by a 400 N force at an angle of 30° to the horizontal. The coefficient of sliding friction is 0.50. Find the acceleration of the box.

Because the box does not move up or down, we have $\sum F_y = ma_y = 0$. From Fig. 5-6, we see that this equation is

$$Y + 200 \text{ N} - mg = 0$$

But $mg = (70 \text{ kg})(9.8 \text{ m/s}^2) = 686 \text{ N}$. It follows that $Y = 486 \text{ N}$.

We can find the friction force acting on the box by writing

$$f = \mu Y = (0.50)(486 \text{ N}) = 243 \text{ N}$$

Now let us write $\sum F_x = ma_x$ for the box. It is

$$(346 - 243) \text{ N} = (70 \text{ kg})a_x$$

from which $a_x = 1.47 \text{ m/s}^2$.

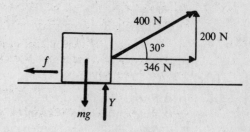

Fig. 5-6

5.16. Refer to Fig. 5-7. A 100 lb block rests on the floor and the coefficient of sliding friction between block and floor is 0.25. A horizontal force of 40 lb acts on the block for 3 s and accelerates it. Compute the velocity of the block at the end of 3 s.

The forces acting on the block are shown in Fig. 5-7. Since the block does not move up or down,

$$\sum F_y = ma_y = 0 \quad \text{or} \quad -100 \text{ lb} + Y = 0$$

from which $Y = 100 \text{ lb}$. Therefore, $f = \mu Y = 25 \text{ lb}$.

Writing $\sum F_x = ma_x$, we have

$$40 \text{ lb} - 25 \text{ lb} = (100/32)a_x$$

where m is replaced by w/g. From this, $a_x = 4.8 \text{ ft/s}^2$.

To find the speed at the end of 3 s, we know that $v_0 = 0$ and so

$$v_f = v_0 + at = 0 + (4.8 \text{ ft/s}^2)(3 \text{ s}) = 14.4 \text{ ft/s}$$

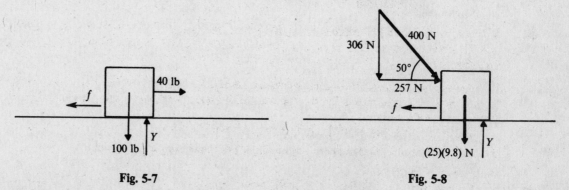

Fig. 5-7   **Fig. 5-8**

5.17. As shown in Fig. 5-8, a force of 400 N pushes on a 25 kg box. Starting from rest, the box achieves a velocity of 2.0 m/s in a time of 4 s. Find the coefficient of sliding friction between box and floor.

We will need to find f by use of $F = ma$. But first we must find a from a motion problem. We know that $v_0 = 0$, $v_f = 2$ m/s, $t = 4$ s. Using $v_f = v_0 + at$ gives

$$a = \frac{v_f - v_0}{t} = \frac{2 \text{ m/s}}{4 \text{ s}} = 0.50 \text{ m/s}^2$$

Now we can write $\sum F_x = ma_x$, where $a_x = a = 0.50$ m/s². From Fig. 5-8, this equation is

$$257 \text{ N} - f = (25 \text{ kg})(0.50 \text{ m/s}^2) \qquad \text{or} \qquad f = 245 \text{ N}$$

We now wish to use $\mu = f/Y$. To find Y, we write $\sum F_y = ma_y = 0$, since no vertical motion occurs. From Fig. 5-8,

$$Y - 306 \text{ N} - (25)(9.8) \text{ N} = 0 \qquad \text{or} \qquad Y = 551 \text{ N}$$

Then

$$\mu = \frac{f}{Y} = \frac{245}{551} = 0.44$$

5.18. A 20 kg box sits on an incline as shown in Fig. 5-9. The coefficient of sliding friction between box and incline is 0.30. Find the acceleration of the box down the incline.

In solving inclined plane problems, we take (x, y)-axes as shown in the figure, parallel and perpendicular to the incline. We shall find the acceleration by writing $\sum F_x = ma_x$. But first we must find the friction force f.

$$F_y = ma_y = 0 \qquad \text{gives} \qquad Y - 0.87 \, mg = 0$$

from which $Y = (0.87)(20 \text{ kg})(9.8 \text{ m/s}^2) = 171$ N. Now we can find f from

$$f = \mu Y = (0.30)(171 \text{ N}) = 51 \text{ N}$$

Writing $\sum F_x = ma_x$, we have

$$f - 0.5 \, mg = ma_x \qquad \text{or} \qquad 51 \text{ N} - (0.5)(20)(9.8) \text{ N} = (20 \text{ kg})a_x$$

from which $a_x = -2.35$ m/s². The box accelerates down the incline at 2.35 m/s².

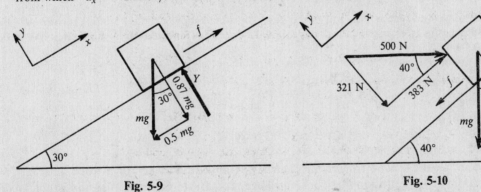

Fig. 5-9   Fig. 5-10

5.19. When a force of 500 N pushes on a 25 kg box as shown in Fig. 5-10, the acceleration of the box up the incline is 0.75 m/s². Find the coefficient of sliding friction between box and incline.

The acting forces and their components are shown in Fig. 5-10. Notice how the x- and y-axes are taken. Since the box moves up the incline, the friction force (which tries to retard the motion) is directed down the incline.

Let us first find f by writing $\sum F_x = ma_x$. From Fig. 5-10, it is

$$383 \text{ N} - f - (0.64)(25)(9.8) \text{ N} = (25 \text{ kg})(0.75 \text{ m/s}^2)$$

from which $f = 207$ N.

We also need Y. Writing $\sum F_y = ma_y = 0$, we have

$$Y - 321 \text{ N} - (0.77)(25)(9.8) \text{ N} = 0 \qquad \text{or} \qquad Y = 510 \text{ N}$$

Then

$$\mu = \frac{f}{Y} = \frac{207}{510} = 0.41$$

5.20. In Fig. 5-11, the coefficient of sliding friction between block A and the table is 0.20. Also, $m_A = 25$ kg, $m_B = 15$ kg. How far will block B drop in the first 3 s after the system is released?

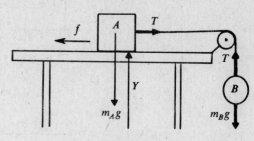

Fig. 5-11

Since, for block A, there is no motion vertically, the normal force is

$$Y = m_A g = (25 \text{ kg})(9.8 \text{ m/s}^2) = 245 \text{ N}$$

Then

$$f = \mu Y = (0.20)(245 \text{ N}) = 49 \text{ N}$$

We must first find the acceleration of the system and then we can describe its motion. Let us apply $F = ma$ to each block in turn. Taking the motion direction as positive, we have

$$T - f = m_A a \qquad \text{or} \qquad T - 49 \text{ N} = (25 \text{ kg})a$$

and

$$m_B g - T = m_B a \qquad \text{or} \qquad -T + (15)(9.8) \text{ N} = (15 \text{ kg})a$$

We can eliminate T by adding the two equations. Then, solving for a, we find $a = 2.45 \text{ m/s}^2$. Now we can work a motion problem with $a = 2.45 \text{ m/s}^2$, $v_0 = 0$, $t = 3$ s.

$$y = v_0 t + \tfrac{1}{2}at^2 \qquad \text{gives} \qquad y = 0 + \tfrac{1}{2}(2.45 \text{ m/s}^2)(3 \text{ s})^2 = 11.0 \text{ m}$$

as the distance B falls in the first 3 s.

5.21. How large a horizontal force in addition to T must pull on block A in Fig. 5-11 to give it an acceleration of 0.75 m/s^2 *toward the left*? Assume, as in Problem 5.20, that $\mu = 0.20$, $m_A = 25$ kg, and $m_B = 15$ kg.

If we were to redraw Fig. 5-11 in this case, we should show a force P pulling toward the left on A. In addition, the retarding friction force f should be reversed in direction in the figure. As in Problem 5.20, $f = 49$ N.

We write $F = ma$ for each block in turn, taking the motion direction to be positive. We have

$$P - T - 49 \text{ N} = (25 \text{ kg})(0.75 \text{ m/s}^2) \qquad \text{and} \qquad T - (15)(9.8) \text{ N} = (15 \text{ kg})(0.75 \text{ m/s}^2)$$

Solve the last equation for T and substitute in the previous equation. We can then solve for the single unknown, P, and find it to be 226 N.

5.22. The coefficient of static friction between a box and the flat bed of a truck is 0.60. What is the maximum acceleration the truck can have along level ground if the box is not to slide?

The box experiences only one x-directed force, the friction force. When the box is on the verge of slipping, $f = \mu_s w$, where w is the weight of the box.

As the truck accelerates, the friction force must cause the box to have the same acceleration as the truck; otherwise, the box will slip. When the box is just on the verge of slipping, $\sum F_x = ma_x$ applied to the box becomes

$$f = m_{\text{box}} a \qquad \text{or} \qquad \mu_s w = m_{\text{box}} a$$

But $m_{\text{box}} = w/g$ and so

$$a = \frac{\mu_s w}{w/g} = \mu_s g = (0.60)(9.8 \text{ m/s}^2) = 5.9 \text{ m/s}^2$$

5.23. In Fig. 5-12, the two boxes have identical masses, 40 kg. Both experience a sliding friction force with $\mu = 0.15$. Find the acceleration of the boxes and the tension in the tie cord.

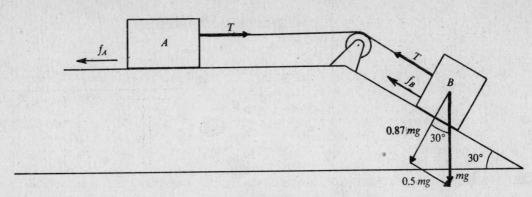

Fig. 5-12

Using $f = \mu Y$, we find that the friction forces on the two boxes are

$$f_A = (0.15)(mg) \qquad f_B = (0.15)(0.87\,mg)$$

But $m = 40$ kg and so $f_A = 59$ N and $f_B = 51$ N.

Let us apply $\sum F_x = ma_x$ to each block in turn, taking the direction of motion as positive.

$$T - 59\text{ N} = (40\text{ kg})a \qquad \text{and} \qquad 0.5\,mg - T - 51\text{ N} = (40\text{ kg})a$$

Solving these two equations for a and T gives $a = 1.08$ m/s^2 and $T = 102$ N.

5.24. In Fig. 5-13, the weights of the objects are 200 N and 300 N. The pulleys are essentially frictionless and massless. Pulley P_1 has a stationary axle but pulley P_2 is free to move up and down. Find the tensions T_1 and T_2, and the acceleration of each body.

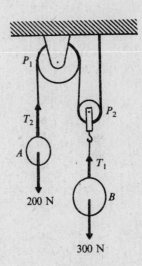

Mass B will rise and mass A will fall. You can see this by noticing that the forces acting on pulley P_2 are $2T_2$ up and T_1 down. Therefore $T_1 = 2T_2$ (the inertialess object transmits the tension). Twice as large a force is pulling upward on B as on A.

Let $a =$ downward acceleration of A. Then $\frac{1}{2}a =$ upward acceleration of B. (Why?)

Write $\sum F_y = ma_y$ for each mass in turn, taking the direction of motion as positive in each case. We have

$$T_1 - 300\text{ N} = m_B\left(\tfrac{1}{2}a\right) \qquad \text{and} \qquad 200\text{ N} - T_2 = m_A a$$

But $m = w/g$ and so $m_A = (200/9.8)$ kg and $m_B = (300/9.8)$ kg. Further, $T_1 = 2T_2$. Substitution of these values in the two equations allows us to compute T_2 and then T_1 and a. The results are

$$T_1 = 327\text{ N} \qquad T_2 = 164\text{ N} \qquad a = 1.78\text{ m/s}^2$$

Fig. 5-13

5.25. Compute the mass of the earth, assuming it to be a sphere of radius 6370 km.

Let M be the mass of the earth, and m the mass of a certain object on the earth's surface. The weight of the object is equal to mg. It is also equal to the gravitational force $G(Mm)/r^2$, where r is the earth's radius. Hence,

$$mg = G\frac{Mm}{r^2}$$

from which

$$M = \frac{gr^2}{G} = \frac{(9.8\text{ m/s}^2)(6.37 \times 10^6\text{ m})^2}{6.67 \times 10^{-11}\text{ N}\cdot\text{m}^2/\text{kg}^2} = 6.0 \times 10^{24}\text{ kg}$$

Supplementary Problems

5.26. A force acts on a 2 kg mass and gives it an acceleration of 3 m/s^2. What acceleration is produced by the same force when acting on a mass of (a) 1 kg? (b) 4 kg? (c) How large is the force? *Ans.* (a) 6 m/s^2; (b) 1.5 m/s^2; (c) 6 N

5.27. (Fill in the blanks.) The mass of a 300 g object is ... (a). Its weight on earth is ... (b). An object that weighs 20 N on earth has a mass on the moon equal to ... (c). The mass of an object that weighs 5 lb on earth is ... (d). *Ans.* (a) 0.300 kg; (b) 2.94 N; (c) 2.04 kg; (d) 0.155 slug

5.28. A resultant external force of 7.0 lb acts on an object that weighs 40 lb on earth. What is the object's acceleration (a) on earth? (b) on the moon? *Ans.* (a) 5.6 ft/s^2; (b) 5.6 ft/s^2

5.29. A horizontal cable pulls a 200 kg cart along a horizontal track. The tension in the cable is 500 N. Starting from rest, (a) how long will it take the cart to reach a speed of 8 m/s? (b) How far will it have gone? *Ans.* (a) 3.2 s; (b) 12.8 m

5.30. A 900 kg car is going 20 m/s along a level road. How large a constant retarding force is required to stop it in a distance of 30 m? (*Hint:* First find its deceleration.) *Ans.* 6000 N

5.31. A 12.0 g bullet is accelerated from rest to a speed of 700 m/s as it travels 20 cm in a gun barrel. Assuming the acceleration to be constant, how large was the accelerating force? (*Be careful of units.*) *Ans.* 14 700 N

5.32. A 20 kg crate hangs at the end of a long rope. Find its acceleration when the tension in the rope is (a) 250 N, (b) 150 N, (c) zero, (d) 196 N. *Ans.* (a) 2.7 m/s^2 up; (b) 2.3 m/s^2 down; (c) 9.8 m/s^2 down; (d) zero

5.33. A 5 kg mass hangs at the end of a cord. Find the tension in the cord if the acceleration of the mass is (a) 1.5 m/s^2 up, (b) 1.5 m/s^2 down, (c) 9.8 m/s^2 down. *Ans.* (a) 56.5 N; (b) 41.5 N; (c) zero

5.34. A 700 N man stands on a scale on the floor of an elevator. The scale records the force it exerts on whatever is on it. What is the scale reading if the elevator has an acceleration of (a) 1.8 m/s^2 up? (b) 1.8 m/s^2 down? (c) 9.8 m/s^2 down? *Ans.* (a) 829 N; (b) 571 N; (c) zero

5.35. Using the scale described in Problem 5.34, a 65 kg astronaut weighs himself on the moon, where $g = 1.60$ m/s^2. What does the scale read? *Ans.* 104 N

5.36. A cord passing over a frictionless, massless pulley has a 4 kg object tied to one end and a 12 kg object tied to the other. Compute the acceleration and the tension in the cord. *Ans.* 4.9 m/s^2, 59 N

5.37. An elevator starts from rest with a constant upward acceleration. It moves 2.0 m in the first 0.60 s. A passenger in the elevator is holding a 3 kg package by a vertical string. What is the tension in the string during the accelerating process? *Ans.* 63 N

5.38. Just as her parachute opens, a 150 lb parachutist is falling at a speed of 160 ft/s. After 0.80 s has passed, the chute is fully open and her speed has dropped to 35 ft/s. Find the average retarding force exerted upon the chutist during this time if the deceleration is uniform. *Ans.* 730 + 150 = 880 lb

5.39. A 20 kg wagon is pulled along the level ground by a rope inclined at 30° above the horizontal. A friction force of 30 N opposes the motion. How large is the pulling force if the wagon is moving with (a) constant speed, (b) an acceleration of 0.40 m/s^2? *Ans.* (a) 34.6 N; (b) 43.9 N

5.40. A 12 kg box is released from the top of an incline that is 5.0 m long and makes an angle of 40° to the horizontal. A 60 N friction force impedes the motion of the box. (a) What will be the acceleration of the box and (b) how long will it take to reach the bottom of the incline? *Ans.* (a) 1.30 m/s^2; (b) 2.8 s

5.41. For the situation outlined in Problem 5.40, what is the coefficient of friction between box and incline?
Ans. 0.67

5.42. An inclined plane makes an angle of 30° with the horizontal. Find the constant force, applied parallel to the plane, required to cause a 15 kg box to slide (*a*) up the plane with acceleration 1.2 m/s², (*b*) down the incline with acceleration 1.2 m/s². Neglect friction forces.   *Ans.* (*a*) 91.5 N; (*b*) 55.5 N

5.43. A horizontal force P is exerted on a 20 kg box in order to slide it up a 30° incline. The friction force retarding the motion is 80 N. How large must P be if the acceleration of the moving box is to be (*a*) zero and (*b*) 0.75 m/s²?   *Ans.* (*a*) 206 N; (*b*) 223 N

5.44. An inclined plane making an angle of 25° with the horizontal has a pulley at its top. A 30 kg block on the plane is connected to a freely hanging 20 kg block by means of a cord passing over the pulley. Compute the distance the 20 kg block will fall in 2 s starting from rest. Neglect friction.
Ans. 2.87 m

5.45. Repeat Problem 5.44 if the coefficient of friction between block and plane is 0.20.   *Ans.* 0.74 m

5.46. A horizontal force of 200 N is required to cause a 15 kg block to slide up a 20° incline with an acceleration of 25 cm/s². Find (*a*) the friction force on the block and (*b*) the coefficient of friction.
Ans. (*a*) 134 N; (*b*) 0.65

5.47. (*a*) What is the smallest force parallel to a 37° incline needed to keep a 100 N weight from sliding down the incline if the coefficients of static and kinetic friction are both 0.30? (*b*) What parallel force is required to keep the weight moving up the incline at constant speed? (*c*) If the parallel pushing force is 94 N, what will be the acceleration of the object? (*d*) If the object in (*c*) starts from rest, how far will it move in 10 s?   *Ans.* (*a*) 36 N; (*b*) 84 N; (*c*) 0.98 m/s² up the plane; (*d*) 49 m

5.48. A 5 kg block rests on a 30° incline. The coefficient of static friction between the block and incline is 0.20. How large a horizontal force must push on the block if the block is to be on the verge of sliding (*a*) up the incline, (*b*) down the incline?   *Ans.* (*a*) 43 N; (*b*) 16.6 N

5.49. Three blocks with masses 6 kg, 9 kg, and 10 kg are connected as shown in Fig. 5-14. The coefficient of friction between the table and the 10 kg block is 0.2. Find (*a*) the acceleration of the system and (*b*) the tensions in the cord on the left and in the cord on the right.   *Ans.* (*a*) 0.39 m/s²; (*b*) 61 N, 85 N

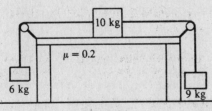

Fig. 5-14

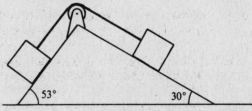

Fig. 5-15

5.50. The two blocks shown in Fig. 5-15 have equal masses. The coefficients of static and dynamic friction are equal, 0.30 for both blocks. (*a*) Show that the system, when released, remains at rest. (*b*) If the system is given an initial speed of 0.90 m/s to the left, how far will it move before coming to rest if the inclines are quite long?   *Ans.* (*b*) 0.296 m

5.51. The earth's radius is about 6370 km. An object that has a mass of 20 kg is taken to a height of 160 km above the earth's surface. (*a*) What is the object's mass at this height? (*b*) How much does the object weigh (i.e. how large a gravitational force does it experience) at this height?
Ans. (*a*) 20 kg; (*b*) 186.5 N

5.52. The radius of the earth is about 6370 km, while that of Mars is about 3440 km. If an object weighs 200 N on earth, what would it weigh, and what would be the acceleration due to gravity, on Mars? Mars has a mass 0.11 that of earth.   *Ans.* 75 N, 3.7 m/s²

Chapter 6

Work, Energy, Power

THE WORK done by a force **F** is defined in the following way. As shown in Fig. 6-1, a force **F** acts on a body. The body undergoes a vector displacement **s**. The component of **F** in the direction of **s** is $F \cos \theta$. The work W done by the force **F** is defined to be the component of **F** in the direction of the displacement, multiplied by the displacement:

$$W = (F \cos \theta)s = Fs \cos \theta$$

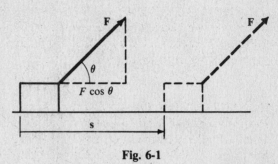

Fig. 6-1

Notice that θ is the angle between the force and displacement vectors. Work is a scalar quantity.

If **F** and **s** are in the same direction, $\cos \theta = \cos 0° = 1$ and $W = Fs$. But if **F** and **s** are in opposite directions, then $\cos \theta = \cos 180° = -1$ and $W = -Fs$, the work is negative. Forces such as friction often slow the motion of an object and are opposite in direction to the displacement. In such cases, they do negative work.

UNITS OF WORK = (a force unit) × (a length unit). One *newton-meter*, called a *joule* (J), is the work done by a 1 N force when it displaces an object 1 m in the direction of the force.

One *dyne-centimeter*, called an *erg*, is the work done by a 1 dyne force (10^{-5} N) when it displaces an object 1 cm in the direction of the force. 1 erg = 10^{-7} J.

One *foot-pound* is the work done by a force of 1 lb when it displaces an object 1 ft in the direction of the force. 1 ft·lb = 1.36 J.

THE ENERGY of a body is its ability to do work. Since the energy of a body is measured in terms of the work it can do, it has the same units as work. Energy, like work, is a scalar quantity.

THE CENTER OF MASS of an object is the single point that moves in the same way as a point mass would move when subjected to the same external forces that act on the object. That is,

$$\mathbf{F} = m\mathbf{a}_{cm}$$

where **F** is the resultant external force on the object, m is the mass of the object, and $\mathbf{a}_{cm}$ is the acceleration of the center of mass.

If the object is considered to be composed of tiny masses m_1, m_2, m_3, and so on, at coordinates (x_1, y_1, z_1), (x_2, y_2, z_2), etc., then the coordinates of the center of mass are given by

$$x_{cm} = \frac{\sum x_i m_i}{\sum m_i} \qquad y_{cm} = \frac{\sum y_i m_i}{\sum m_i} \qquad z_{cm} = \frac{\sum z_i m_i}{\sum m_i}$$

where the sums are to extend over all masses composing the object. In a uniform gravitational field, the center of mass and the center of gravity (Chapter 3) coincide.

THE KINETIC ENERGY (KE) of an object is its ability to do work because of its motion. If an object of mass m has a velocity v, its translational KE is

$$KE = \tfrac{1}{2}mv^2$$

The units of KE are J if m is in kg and v is in m/s. In the British system, KE is in ft·lb if m is in slugs and v is in ft/s.

46

THE GRAVITATIONAL POTENTIAL ENERGY (GPE) of an object is its ability to do work because of its position in a gravitational field.   In falling through a vertical distance h, a mass m can do work in the amount mgh.   We define the GPE of an object relative to an arbitrary zero level, often the earth's surface.   If the object is at a height h above the zero (or reference) level, its

$$GPE = mgh$$

where g is the acceleration due to gravity.   Notice that mg is the weight of the object.   The units of GPE are J when m is in kg, g is in m/s^2, and h is in m.   Or, in the British system, GPE is in ft · lb if the weight mg is in lb and h is in ft.

CONSERVATION OF ENERGY: Energy can neither be created nor destroyed, but only transformed from one kind to another.   (This implies that mass can be regarded as one form of energy. Ordinarily the conversion of mass into energy, and vice versa, predicted by the theory of relativity, can be ignored.   This subject is treated in Chapter 43.)

WORK-ENERGY CONVERSION: When a force does work on an object, it must increase the energy of the object by a like amount (or decrease it if the work is negative).   When an object loses energy of any form, it must experience a like increase in energy of some other form, or it must do a like amount of work.

POWER is the time rate of doing work:

$$\text{average power} = \frac{\text{work done by a force}}{\text{time taken to do this work}}$$

$$= (\text{force}) \times (\text{velocity})$$

where "velocity" means the component of the object's velocity in the direction of the force applied to it.   Equivalently, we could take the product of the object's velocity and the component of the applied force in the direction of the velocity.

In the SI, the unit of power is the *watt* (W) and this is 1 J/s.

In the British system, the unit of power is the ft · lb/s.

Another unit of power often used (but not in our basic equation) is the *horsepower*:

$$1 \text{ hp} = 746 \text{ W} = 550 \text{ ft} \cdot \text{lb/s}$$

THE KILOWATT HOUR is a unit of work.   If a force is doing work at a rate of 1 kilowatt (which is 1000 J/s), then in 1 hour it will do 1 kW · h of work.

$$1 \text{ kW} \cdot \text{h} = 3.6 \times 10^6 \text{ J}$$

Solved Problems

6.1.   In a situation similar to that shown in Fig. 6-1, an object is pulled along the ground by a 75 N force directed 28° above the horizontal.   How much work does the force do in pulling the object 8 m?

The work done is equal to the product of the displacement, 8 m, and the component of the force that is parallel to the displacement, (75 N) cos 28°.

$$\text{work} = [(75 \text{ N}) \cos 28°](8 \text{ m}) = 530 \text{ J}$$

6.2.   A block moves up a 30° incline under the action of certain forces, three of which are shown in Fig. 6-2. F_1 is horizontal and of magnitude 40 N. F_2 is normal to the plane and of magnitude 20 N. F_3 is parallel to the plane and of magnitude 30 N. Determine the work done by each force as the block (and point of application of each force) moves 80 cm up the incline.

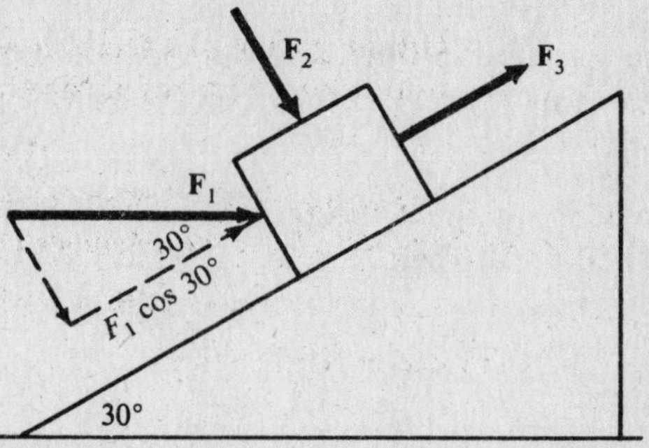

Fig. 6-2

The component of F_1 along the direction of the displacement is

$$F_1 \cos 30° = (40 \text{ N})(0.866) = 34.6 \text{ N}$$

Hence the work done by F_1 is $(34.6 \text{ N})(0.80 \text{ m}) = 28$ J. (Notice that the distance must be expressed in meters.)

Because it has no component in the direction of the displacement, F_2 does no work.

The component of F_3 in the direction of the displacement is 30 N. Hence the work done by F_3 is $(30 \text{ N})(0.80 \text{ m}) = 24$ J.

6.3.   How much work is done against gravity in lifting a 3 kg object through a distance of 40 cm?

To lift a 3 kg object at constant speed, an upward force equal in magnitude to its weight, $mg = (3)(9.8)$ N, must be exerted on the object. The work done by this force is what we refer to as the work done against gravity. By conservation of energy, the same amount of work would be done by a variable force.

In lifting the object a distance h, the work done against gravity will be $(mg)(h)$:

$$\text{work against gravity} = mgh = [(3)(9.8) \text{ N}](0.40 \text{ m}) = 11.8 \text{ J}$$

6.4.   How much work is done against gravity in lifting a 20 lb object through a distance of 4.0 ft?

As in Problem 6.3,

$$\text{work against gravity} = (\text{weight})(h) = (20 \text{ lb})(4 \text{ ft}) = 80 \text{ ft} \cdot \text{lb}$$

6.5.   A ladder 3.0 m long and weighing 200 N has its center of gravity 120 cm from the bottom. At its top end is a 50 N weight. Compute the work required to raise the ladder from a horizontal position on the ground to a vertical position.

The work done (against gravity) consists of two parts, the work to raise the center of gravity 1.20 m and the work to raise the weight at the end through 3 m. Therefore

$$\text{work done} = (200 \text{ N})(1.20 \text{ m}) + (50 \text{ N})(3 \text{ m}) = 390 \text{ J}$$

6.6.   Compute the work done against gravity by a pump that discharges 600 liters of fuel oil into a tank 20 m above the intake. One cubic centimeter of fuel oil has a mass of 0.82 grams. One liter is 1000 cm^3.

The mass lifted is

$$(600 \text{ liters})\left(1000 \ \frac{\text{cm}^3}{\text{liter}}\right)\left(0.82 \ \frac{\text{g}}{\text{cm}^3}\right) = 492\,000 \text{ g} = 492 \text{ kg}$$

The lifting work is then

$$\text{work} = (mg)(h) = [(492)(9.8) \text{ N}](20 \text{ m}) = 96\,400 \text{ J}$$

6.7.   A 2 kg mass falls 400 cm. (*a*) How much work was done on it by the gravitational force? (*b*) How much GPE did it lose?

   Gravity pulls with a force mg on the object and the displacement is 4 m in the direction of the force. The work done by gravity is therefore

$$(mg)(4 \text{ m}) = [(2)(9.8) \text{ N}](4 \text{ m}) = 78 \text{ J}$$

   The change in GPE of the object is $mgh_f - mgh_0$, where h_0 and h_f are the initial and final heights of the object above the reference level. We then have

$$\text{change in GPE} = mgh_f - mgh_0 = mg(h_f - h_0) = [(2)(9.8) \text{ N}](-4 \text{ m}) = -78 \text{ J}$$

The loss in GPE is 78 J.

6.8.   A force of 1.50 N acts on a 0.20 kg cart so as to accelerate it along an air track. The track and force are horizontal and in line. How fast is the cart going after acceleration from rest through 30 cm if friction is negligible?

   The work done by the force causes an equal increase in KE of the cart. Therefore,

$$\text{work done} = (\text{KE})_{\text{end}} - (\text{KE})_{\text{start}} \qquad \text{or} \qquad Fs \cos 0° = \tfrac{1}{2}mv_f^2 - 0$$

Substituting gives

$$(1.50 \text{ N})(0.30 \text{ m}) = \tfrac{1}{2}(0.20 \text{ kg})v_f^2$$

from which $v_f = 2.1$ m/s.

6.9.   A 0.50 kg block is sliding across a tabletop with an initial velocity of 20 cm/s. It slides to rest in a distance of 70 cm. Find the average friction force that retarded its motion.

   The initial KE of the block is lost doing work against the friction force. (The friction force does work on the block. The equal and opposite force given by Newton's third law does an equal amount of work "against the friction force," i.e. on the tabletop.) We can therefore write

$$\text{loss in KE} = \text{work done against friction}$$

$$\tfrac{1}{2}mv_0^2 - 0 = fs$$

Substituting gives

$$\tfrac{1}{2}(0.50 \text{ kg})(0.20 \text{ m/s})^2 = f(0.70 \text{ m}) \qquad \text{or} \qquad f = 0.0143 \text{ N}$$

6.10.   A 3000 lb car going 40 ft/s is brought to rest in a distance of 8 ft as it hits a wall. Find the average stopping force exerted on the car by the wall.

   The KE of the car is lost doing work against the wall. We can write

$$\text{loss in KE} = \text{work done against stopping force}$$

$$\tfrac{1}{2}mv_0^2 - 0 = Px$$

where P is the wall's stopping force, $x = 8$ ft, $v_0 = 40$ ft/s, and

$$m = \frac{w}{g} = \frac{3000 \text{ lb}}{32.2 \text{ ft/s}^2}$$

Substitution gives $P = 9400$ lb.

6.11.   A projectile is shot upward from the earth with a speed of 20 m/s. How high is it when its speed is 8.0 m/s? Ignore air friction.

$$\text{KE loss of projectile} = \text{GPE gain of projectile}$$

$$\tfrac{1}{2}mv_0^2 - \tfrac{1}{2}mv_f^2 = mgh_f - mgh_0 = mg(h_f - h_0) = mgh$$

where h is the required height. Notice that m cancels from the equation. Substituting $v_0 = 20$ m/s, $v_f = 8.0$ m/s, $g = 9.8$ m/s^2 and solving for h gives $h = 17.1$ m.

6.12.   A force of 70 N is applied continuously to a 5 kg object as shown in Fig. 6-3.   (a) If friction can be neglected, what will be the object's speed after it has moved 6 m from rest?   (b) Repeat if the coefficient of friction between block and floor is 0.40.

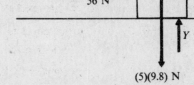

(a)   As we see from Fig. 6-3, if $f = 0$, then the component of the force in the direction of motion is 56 N.   Then

work done by force = KE gain of object

$$[(70 \text{ N}) \cos 37°](s) = \tfrac{1}{2} m v_f^2 - \tfrac{1}{2} m v_0^2$$

$$(56 \text{ N})(6 \text{ m}) = \tfrac{1}{2}(5 \text{ kg}) v_f^2 - 0$$

from which $v_f = 11.6$ m/s.

Fig. 6-3

(b)   When f is not zero, the work done by the forces acting on the object is

$$(70 \cos 37° - f)(s)$$

because the net force in the direction of motion is $70 \cos 37° - f$.   We must find Y to obtain f. Since $a_y = 0$,

$$\sum F_y = ma_y \qquad \text{becomes} \qquad -42 \text{ N} - (5)(9.8) \text{ N} + Y = 0$$

and so $Y = 91$ N,   whence

$$f = \mu Y = (0.40)(91) = 36 \text{ N}$$

We then have

work done = KE gain

$$(70 \cos 37° - 36)(6) = \tfrac{1}{2}(5) v_f^2 - 0$$

from which $v_f = 6.9$ m/s.

6.13.   As shown in Fig. 6-4, a bead slides on a wire.   If friction forces are negligible and the bead has a speed of 200 cm/s at A, (a) what will be its speed at point B?   (b) at point C?

We can write for the bead:

gain in KE = loss in GPE

$$\tfrac{1}{2} m v_f^2 - \tfrac{1}{2} m v_0^2 = mgh_0 - mgh_f$$

(a)   Here, $v_0 = 2.0$ m/s, $h_0 = 0.8$ m,   and $h_f = 0$.   Using these values, and noticing that m cancels out, gives $v_f = 4.44$ m/s.

(b)   Here, $v_0 = 2.0$ m/s, $h_0 = 0.8$ m, $h_f = 0.5$ m.   Using these values gives $v_f = 3.14$ m/s.

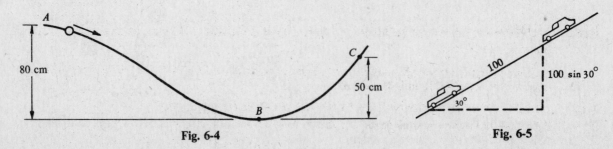

Fig. 6-4   Fig. 6-5

6.14.   A 1200 kg car is coasting down a 30° hill as shown in Fig. 6-5.   At a time when the car's speed is 12 m/s, the driver applies the brakes.   What constant force F (parallel to the road) must result if the car is to stop after traveling 100 m?

loss in KE + loss in GPE = work done against stopping force

$$\left(\tfrac{1}{2}mv_0^2 - \tfrac{1}{2}mv_f^2\right) + (mg)(100 \sin 30°) = Fs$$

We have $m = 1200$ kg, $v_0 = 12$ m/s, $v_f = 0$, $s = 100$ m.   Solving gives $F = 6700$ N.

6.15.   Refer back to Fig. 6-4.   Suppose the bead has a mass of 15 g, a speed of 2.0 m/s at A, and stops as it reaches point C.   The length of the wire from A to C is 250 cm.   How large an average friction force opposed the motion of the bead?

When the bead goes from A to C, we have:

loss in KE + loss in PE = work done against friction

$$\left(\tfrac{1}{2}mv_0^2 - 0\right) + mg(0.30\text{ m}) = f(2.5\text{ m})$$

Using $m = 0.015$ kg   and $v_0 = 2.0$ m/s,   we find upon substitution that $f = 0.0296$ N.

6.16.   A 60 000 kg train is pulled up a 1% grade by a steady drawbar pull of 3000 N.   The friction force opposing the motion of the train is 4000 N.   The train's initial speed is 12 m/s.   What distance s will the train go before its speed is reduced to 9 m/s?

We can write:

loss in KE + work by drawbar pull = gain in GPE + work against friction

$$\left(\tfrac{1}{2}mv_0^2 - \tfrac{1}{2}mv_f^2\right) + (3000\text{ N})s = mg(h_f - h_0) + (4000\text{ N})s$$

By definition, a 1% grade is one that rises 0.01 m for each meter along its length.   Therefore, $h_f - h_0 = 0.01\,s$.   Using $m = 60\,000$ kg, $v_0 = 12$ m/s, $v_f = 9$ m/s,   we solve for s and find it to be 277 m.

6.17.   An advertisement claims that a certain 1200 kg car can accelerate from rest to a speed of 25 m/s in a time of 8.0 s.   What average power must the motor produce in order to cause this acceleration?   Ignore friction losses.

The work done in accelerating the car is given by

$$\text{work done} = \text{increase in KE} = \tfrac{1}{2}mv_f^2$$

The time taken for this work is 8 s.   Therefore

$$\text{power} = \frac{\text{work}}{\text{time}} = \frac{\tfrac{1}{2}(1200\text{ kg})(25\text{ m/s})^2}{8\text{ s}} = 46\,900\text{ W}$$

Converting from watts to horsepower, we have

$$\text{power} = (46\,900\text{ W})\left(\frac{1\text{ hp}}{746\text{ W}}\right) = 63\text{ hp}$$

6.18.   A 0.25 hp motor is used to lift a load at a rate of 5.0 cm/s.   How great a load can it lift at this constant speed?

The power output of the motor will be assumed to be   0.25 hp = 186.5 W.   In 1 s, the load mg is lifted a distance of 0.050 m.   Therefore,

work done in 1 s = (weight)(height change in 1 s) = $(mg)(0.050$ m)

By definition,   power = work/time,   so that

$$186.5\text{ W} = \frac{(mg)(0.050\text{ m})}{1\text{ s}}$$

Using $g = 9.8$ m/s^2,   we find that $m = 381$ kg.   The motor can lift a load of about 380 kg at this speed.

6.19. A motor having an efficiency of 90% operates a crane having an efficiency of 40%. With what constant speed does the crane lift an 880 lb bale if the power supplied to the motor is 5 kW?

$$\text{overall efficiency} = (40\%)(90\%) = 36\%$$

$$\text{useful power} = (0.36)(5 \text{ kW}) = 1.8 \text{ kW}$$

We could work directly from the definition of power, as in Problem 6.18. Instead, let us use the fact that

$$\text{power} = (\text{force})(\text{speed})$$

$$(1.8 \times 10^3 \text{ W})\left(\frac{1 \text{ hp}}{746 \text{ W}}\right)\left(\frac{550 \text{ ft} \cdot \text{lb/s}}{1 \text{ hp}}\right) = (880 \text{ lb})(\text{speed})$$

where the relations   1 hp = 746 W = 550 ft·lb/s   have been applied. Solving gives the speed to be 1.51 ft/s.

6.20. In unloading grain from the hold of a ship, an elevator lifts the grain through a distance of 12 m. Grain is discharged at the top of the elevator at a rate of 2.0 kg each second and the discharge speed of each grain particle is 3.0 m/s. Find the minimum-horsepower motor that can elevate grain in this way.

The work done by the motor each second is

$$\text{power} = mgh + \tfrac{1}{2}mv^2$$

where m is the mass of grain discharged (and lifted) each second. We are given that $m = 2.0$ kg, $v = 3.0$ m/s,   and $h = 12$ m.   Substitution gives

$$\text{power} = 244 \text{ W} = 0.33 \text{ hp}$$

The motor must have an output of at least 0.33 hp.

Supplementary Problems

6.21. A force of 3 N acts through a distance of 12 m in the direction of the force. Find the work done. *Ans.* 36 J

6.22. A 4 kg object is lifted 1.5 m. (*a*) How much work is done against gravity? (*b*) Repeat if the object is lowered instead of lifted.   *Ans.* (*a*) 58.8 J; (*b*) −58.8 J

6.23. A uniform rectangular marble slab is 3.4 m long and 2.0 m wide. It has a mass of 180 kg. If it is originally lying on the flat ground, how much work is needed to stand it on end?   *Ans.* 3000 J

6.24. A 400 lb load of bricks is to be lifted to the top of a scaffold 28 ft high. How much work must be done against gravity to lift it?   *Ans.* 11 200 ft·lb

6.25. Compute the useful work done by an engine as it lifts 40 liters of tar 20 m. One cm³ of tar has a mass of 1.07 grams.   *Ans.* 8390 J

6.26. How large a force is required to accelerate a 1300 kg car from rest to a speed of 20 m/s in a distance of 80 m?   *Ans.* 3250 N

6.27. A 1200 kg car going 30 m/s applies its brakes and skids to rest. If the friction force between the sliding tires and the pavement is 6000 N, how far does the car skid before coming to rest?   *Ans.* 90 m

6.28. A 200 kg cart is pushed slowly up an incline. How much work does the pushing force do in moving the object up along the incline to a platform 1.5 m above the starting point if friction is negligible? *Ans.* 2940 J

6.29. Repeat Problem 6.28 if the distance along the incline to the platform is 7 m and a friction force of 150 N opposes the motion. *Ans.* 3990 J

6.30. A 100 000 lb freight car is pulled 2500 ft up along a 1.2% grade at constant speed. (*a*) Find the work done against gravity by the drawbar pull. (*b*) If the friction force retarding the motion is 440 lb, find the total work done. *Ans.* (*a*) 3.0×10^6 ft·lb; (*b*) 4.1×10^6 ft·lb

6.31. A 60 kg woman walks up a flight of stairs that connects two floors 3.0 m apart. (*a*) How much lifting work is done by the woman? (*b*) By how much does the woman's GPE change? *Ans.* (*a*) 1760 J; (*b*) 1760 J

6.32. A pump lifts water from a lake to a large tank 20 m above the lake. How much work against gravity does the pump do as it transfers 5 m³ of water to the tank? One cubic meter of water has a mass of 1000 kg. *Ans.* 9.8×10^5 J

6.33. Just before striking the ground, a 2.0 kg mass has 400 J of KE. If friction can be ignored, from what height was it dropped? *Ans.* 20.4 m

6.34. A 0.5 kg ball falls past a window that is 1.50 m in vertical length. (*a*) How much did the KE of the ball increase as it fell past the window? (*b*) If its speed was 3.0 m/s at the top of the window, what was it at the bottom? *Ans.* (*a*) 7.35 J; (*b*) 6.2 m/s

6.35. At sea level a nitrogen molecule in the air has an average translational KE of 6.2×10^{-21} J. Its mass is 4.7×10^{-26} kg. (*a*) If the molecule could shoot straight up without striking other air molecules, how high would it rise? (*b*) What is the molecule's initial speed? *Ans.* (*a*) 13.5 km; (*b*) 510 m/s

6.36. The coefficient of sliding friction between a 900 kg car and pavement is 0.80. If the car is moving at 25 m/s along level pavement when it begins to skid to a stop, how far will it go before stopping? *Ans.* 40 m

6.37. Consider the simple pendulum shown in Fig. 6-6. (*a*) If it is released from point A, what will be the speed of the ball as it passes through point C? (*b*) What is the ball's speed at point B? *Ans.* (*a*) 3.83 m/s; (*b*) 3.43 m/s

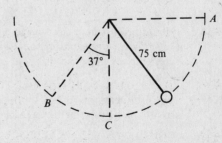

6.38. A 1200 kg car coasts from rest down a driveway that is inclined 20° to the horizontal and is 15 m long. How fast is the car going at the end of the driveway if (*a*) friction is negligible and (*b*) a friction force of 3000 N opposes the motion? *Ans.* (*a*) 10.0 m/s; (*b*) 5.06 m/s

Fig. 6-6

6.39. A driver of a 1200 kg car notices that the car slows from 20 m/s to 15 m/s as it coasts a distance of 130 m along level ground. How large a force opposes the motion? *Ans.* 810 N

6.40. A 10 lb weight slides from rest down a rough inclined plane, 100 ft long and inclined 30° to the horizontal, and gains a speed of 52 ft/s. Find the work done against friction. *Ans.* 80 ft·lb

6.41. A 2000 kg elevator rises from rest in the basement to the fourth floor, a distance of 25 m. As it passes the fourth floor, its speed is 3.0 m/s. There is a constant frictional force of 500 N. Calculate the work done by the lifting mechanism. *Ans.* 5.115×10^5 J

6.42. Figure 6-7 shows a bead sliding on a wire. How large must height h_1 be if the bead, starting at rest at A, is to have a speed of 200 cm/s at point B? Ignore friction. *Ans.* 20.4 cm

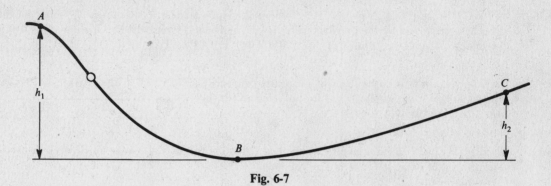

Fig. 6-7

6.43. In Fig. 6-7, $h_1 = 50$ cm, $h_2 = 30$ cm, and the length along the wire from A to C is 400 cm. A 3.0 gram bead released at A coasts to point C and stops. How large an average friction force opposed its motion? *Ans.* 1.47×10^{-3} N

6.44. Calculate the average horsepower required to raise a 150 kg drum to a height of 20 m in a time of 1 minute. *Ans.* 0.66 hp

6.45. Compute the power output of a machine that lifts a 500 kg crate through a height of 20 m in a time of 60 s. *Ans.* 1.63 kW

6.46. An engine expends 40 hp in propelling a car along a level track at 15 m/s. How large is the total retarding force acting on the car? *Ans.* 1990 N

6.47. A 1000 kg auto travels up a 3% grade at 20 m/s. Find the horsepower required, neglecting friction. *Ans.* 7.9 hp

6.48. Water flows from a reservoir to a turbine 330 ft below. The efficiency of the turbine is 80% and it takes 100 ft^3 of water each minute. Neglecting friction in the pipe and the small KE of the exiting water, compute the horsepower output of the turbine. Water weighs 62.4 lb/ft^3. *Ans.* 50 hp

6.49. Find the mass (in kg) of the block that a 40 hp engine can pull along a level road at 15 m/s if the friction coefficient between block and road is 0.15. *Ans.* 1350 kg

6.50. What power must a man expend on a 100 kg log that he is dragging down a hillside at a speed of 0.50 m/s? The hillside makes an angle of 20° with the horizontal and the coefficient of friction is 0.9. *Ans.* 247 W

Chapter 7

Simple Machines

A MACHINE is any device by which the magnitude, direction, or method of application of a force is changed in order to achieve some advantage. Examples of simple machines are the lever, inclined plane, pulley, crank and axle, and jackscrew.

PRINCIPLE OF WORK that applies to a continuously operating machine:

$$\text{work input} = \text{useful work output} + \text{friction work}$$

In machines that operate for only a short time, some of the input work may be used to store energy within the machine. An internal spring might be stretched or a movable pulley might be raised, for example.

MECHANICAL ADVANTAGE: The *actual mechanical advantage* (AMA) of a machine is

$$\text{AMA} = \text{force ratio} = \frac{\text{force exerted by machine on load}}{\text{force used to operate machine}}$$

The *ideal mechanical advantage* (IMA) of a machine is

$$\text{IMA} = \text{distance ratio} = \frac{\text{distance moved by input force}}{\text{distance moved by load}}$$

Because friction is always present, the AMA is always less than the IMA.

EFFICIENCY of a machine $= \dfrac{\text{work output}}{\text{work input}} = \dfrac{\text{power output}}{\text{power input}}$

The efficiency is also equal to the ratio AMA/IMA.

Solved Problems

7.1. A hoisting machine lifts a 3000 kg load a height of 8 m in a time of 20 s. The power supplied to the engine is 18 hp. Compute (a) the work output, (b) the power output and the power input, (c) the efficiency of the engine and hoist system.

(a) work output = (lifting force) × (height) = [(3000 × 9.8) N](8 m) = 235 000 J = 235 kJ

(b) $\text{power output} = \dfrac{\text{work output}}{\text{time taken}} = \dfrac{235 \text{ kJ}}{20 \text{ s}} = 11.8 \text{ kW}$

$\text{power input} = (18 \text{ hp})\left(\dfrac{0.746 \text{ kW}}{1 \text{ hp}} \right) = 13.4 \text{ kW}$

(c) $\text{efficiency} = \dfrac{\text{power output}}{\text{power input}} = \dfrac{11.8 \text{ kW}}{13.4 \text{ kW}} = 0.88 = 88\%$

or $\text{efficiency} = \dfrac{\text{work output}}{\text{work input}} = \dfrac{235 \text{ kJ}}{(13.4 \text{ kJ/s})(20 \text{ s})} = 0.88 = 88\%$

7.2. What power in kW is supplied to a 12 hp motor, having an efficiency of 90%, when it is delivering its full rated output?

From the definition of efficiency,

$$\text{power input} = \frac{\text{power output}}{\text{efficiency}} = \frac{(12 \text{ hp})(0.746 \text{ kW/hp})}{0.90} = 9.95 \text{ kW}$$

7.3. For the three levers shown in Fig. 7-1, determine the vertical forces F_1, F_2, F_3 required to support the load $w = 90$ N. Neglect the weights of the levers. Also find the IMA, AMA and efficiency for each system.

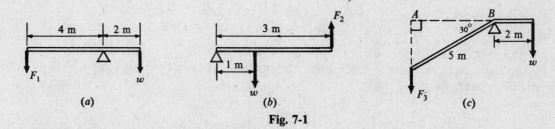

Fig. 7-1

In each case, take torques about the fulcrum point as axis. [Notice in Fig. 7-1(c) that the lever arm is $(5 \text{ m}) \cos 30°$.] If we assume that the lifting is occurring slowly at constant speed, then the systems are in equilibrium. The clockwise torques balance the counterclockwise torques.

clockwise torques = counterclockwise torques

(a) $(90 \text{ N})(2 \text{ m}) = (F_1)(4 \text{ m})$ $\qquad F_1 = 45 \text{ N}$

(b) $(90 \text{ N})(1 \text{ m}) = (F_2)(3 \text{ m})$ $\qquad F_2 = 30 \text{ N}$

(c) $(90 \text{ N})(2 \text{ m}) = (F_3)(4.33 \text{ m})$ $\qquad F_3 = 41.6 \text{ N}$

The IMA, AMA, and efficiency of the three levers are

	Lever (a)	Lever (b)	Lever (c)
IMA	$\dfrac{4 \text{ m}}{2 \text{ m}} = 2$	$\dfrac{3 \text{ m}}{1 \text{ m}} = 3$	$\dfrac{4.33 \text{ m}}{2 \text{ m}} = 2.16$
AMA	$\dfrac{90 \text{ N}}{45 \text{ N}} = 2$	$\dfrac{90 \text{ N}}{30 \text{ N}} = 3$	$\dfrac{90 \text{ N}}{41.6 \text{ N}} = 2.16$
Eff.	1.00	1.00	1.00

The efficiencies are 1 because we have neglected friction at the fulcrums.

7.4. Determine the force F required to lift a 100 lb load, w, with each of the pulley systems shown in Fig. 7-2. Neglect friction and the weights of the pulleys.

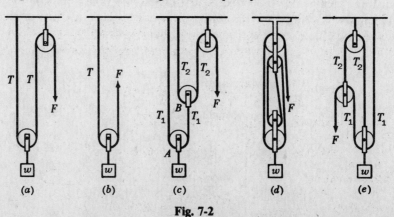

Fig. 7-2

(a) Load w is supported by two ropes; each rope exerts an upward pull of $T = \frac{1}{2}w$. But, because the rope is continuous and the pulleys are frictionless, $T = F$. Then

$$F = T = \tfrac{1}{2}w = \tfrac{1}{2}(100 \text{ lb}) = 50 \text{ lb}$$

(b) Here, too, the load is supported by the tensions in two ropes, T and F, where $T = F$. Then

$$T + F = w \qquad \text{or} \qquad F = \tfrac{1}{2}w = 50 \text{ lb}$$

(c) Let T_1 and T_2 be the tensions around pulleys A and B respectively. Pulley A is in equilibrium and so

$$T_1 + T_1 - w = 0 \qquad \text{or} \qquad T_1 = \tfrac{1}{2}w$$

Consider next pulley B. It, too, is in equilibrium and so

$$T_2 + T_2 - T_1 = 0 \qquad \text{or} \qquad T_2 = \tfrac{1}{2}T_1 = \tfrac{1}{4}w$$

But $F = T_2$ and so $F = \tfrac{1}{4}w = 25$ lb.

(d) Four ropes, each with the same tension T, support the load w. Therefore

$$4T_1 = w \qquad \text{and so} \qquad F = T_1 = \tfrac{1}{4}w = 25 \text{ lb}$$

(e) We see at once that $F = T_1$. Because the pulley on the left is in equilibrium, we have

$$T_2 - T_1 - F = 0$$

But $T_1 = F$ and so $T_2 = 2F$. The pulley on the right is also in equilibrium and so

$$T_1 + T_2 + T_1 - w = 0$$

Recalling that $T_1 = F$ and that $T_2 = 2F$ gives $4F = w$, so $F = 25$ lb.

7.5. Using the wheel and axle shown in Fig. 7-3, a 400 N load can be raised by a force of 50 N applied to the rim of the wheel. The radii of the wheel and axle are 85 cm and 6 cm respectively. Determine the IMA, AMA, and efficiency of the machine.

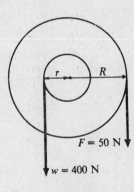

We know that in one turn of the wheel-axle system, a length of cord equal to the circumference of the wheel or axle will be wound or unwound. Therefore

$$\text{IMA} = \frac{\text{distance moved by } F}{\text{distance moved by } w} = \frac{2\pi R}{2\pi r} = \frac{85 \text{ cm}}{6 \text{ cm}} = 14.2$$

$$\text{AMA} = \text{force ratio} = \frac{400 \text{ N}}{50 \text{ N}} = 8$$

$$\text{efficiency} = \frac{\text{AMA}}{\text{IMA}} = \frac{8}{14.2} = 0.56 = 56\%$$

Fig. 7-3

7.6. The inclined plane shown in Fig. 7-4 is 15 m long and rises 3 m. (a) What force F parallel to the plane is required to slide a 20 kg box up the plane if friction is neglected? (b) What is the IMA of the plane? (c) Find the AMA and efficiency if a 64 N force is actually required.

(a) There are several ways to approach this. Let us consider energy. Since there is no friction, the work done by the pushing force, $(F)(15 \text{ m})$, must equal the lifting work done, $(20 \text{ kg})(9.8 \text{ m/s}^2)(3 \text{ m})$. Equating these two expressions and solving for F gives $F = 39$ N.

(b) $$\text{IMA} = \frac{\text{distance moved by } F}{\text{distance } w \text{ is lifted}} = \frac{15 \text{ m}}{3 \text{ m}} = 5.0$$

(c) $$\text{AMA} = \text{force ratio} = \frac{w}{F} = \frac{196 \text{ N}}{64 \text{ N}} = 3.1$$

$$\text{efficiency} = \frac{\text{AMA}}{\text{IMA}} = \frac{3.1}{5.0} = 0.62 = 62\%$$

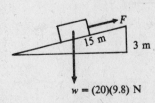

Or, as a check,

$$\text{efficiency} = \frac{\text{work output}}{\text{work input}} = \frac{(w)(3 \text{ m})}{(F)(15 \text{ m})} = 0.62 = 62\%$$

Fig. 7-4

7.7. As shown in Fig. 7-5, a jackscrew has a lever arm of 40 cm and a pitch of 5 mm. If the efficiency is 30%, what force F is required to lift a load w of 270 kg?

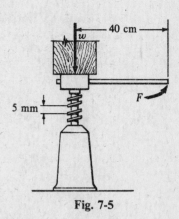

When the jack handle is moved around one complete circle, the input force moves a distance

$$2\pi r = 2\pi(0.40 \text{ m})$$

while the load is lifted a distance of 0.005 m. The IMA is therefore

$$\text{IMA} = \text{distance ratio} = \frac{2\pi(0.40 \text{ m})}{0.005 \text{ m}} = 500$$

Since efficiency = AMA/IMA, we have

$$\text{AMA} = (\text{efficiency})(\text{IMA}) = (0.30)(500) = 150$$

But AMA = (load lifted)/(input force) and so

$$F = \frac{\text{load lifted}}{\text{AMA}} = \frac{(270)(9.8) \text{ N}}{150} = 17.6 \text{ N}$$

Fig. 7-5

7.8. A differential pulley (chain hoist) is shown in Fig. 7-6. Two toothed pulleys of radii $r = 10$ cm and $R = 11$ cm are fastened together and turn on the same axle. A continuous chain passes over the smaller (10 cm) pulley, then around the movable pulley at the bottom, and finally around the 11 cm pulley. The operator exerts a downward force F on the chain to lift the load w. (a) Determine the IMA. (b) What is the efficiency of the machine if an applied force of 50 lb is required to lift a load of 700 lb?

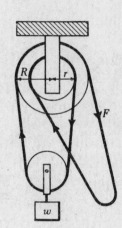

(a) Suppose that the force F moves down a distance sufficient to cause the upper rigid system of pulleys to turn one revolution. Then the smaller upper pulley unwinds a length of chain equal to its circumference, $2\pi r$, while the larger upper pulley winds a length $2\pi R$. As a result, the chain supporting the lower pulley is shortened by a length $2\pi R - 2\pi r$. The load w is lifted half this distance,

$$\tfrac{1}{2}(2\pi R - 2\pi r) = \pi(R - r)$$

Fig. 7-6

when the input force moves a distance $2\pi R$. Therefore

$$\text{IMA} = \frac{\text{distance moved by } F}{\text{distance moved by } w} = \frac{2\pi R}{\pi(R - r)} = \frac{2R}{R - r} = \frac{22 \text{ cm}}{1 \text{ cm}} = 22$$

(b) From the data,

$$\text{AMA} = \frac{\text{load lifted}}{\text{input force}} = \frac{700 \text{ N}}{50 \text{ N}} = 14$$

Then

$$\text{efficiency} = \frac{\text{AMA}}{\text{IMA}} = \frac{14}{22} = 0.64 = 64\%$$

Supplementary Problems

7.9. A motor furnishes 120 hp to a device that lifts a 5000 kg load to a height of 13.0 m in a time of 20 s. Find the efficiency of the machine. *Ans.* 36%

7.10. Refer back to Fig. 7-2(*d*). If a force of 200 N is required to lift a 50 kg load, find the IMA, AMA, and efficiency for the system.   *Ans.*   4, 2.45, 61%

7.11. In Fig. 7-7, the 300 N load is balanced by a force *F*. Assuming the efficiency of either system to be 100%, how large is *F* in each case? Assume all ropes to be vertical.
Ans. (*a*) 100 N;   (*b*) 75 N

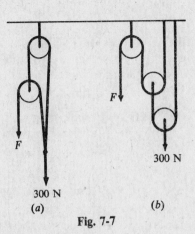

7.12. With a certain machine, the applied force moves 3.3 m to raise a load 8.0 cm. Find (*a*) IMA; (*b*) AMA if the efficiency is 60%. (*c*) What load can be lifted by an applied force of 50 N if the efficiency is 100%? (*d*) if the efficiency is 60%?
Ans. (*a*) 41;   (*b*) 25;   (*c*) 2060 N;   (*d*) 1240 N

7.13. With a wheel and axle, a force of 30 lb applied to the rim of the wheel can lift a load of 240 lb. The diameters of the wheel and axle are 36 inches and 4 inches respectively. Determine the AMA, IMA, and efficiency of the machine.   *Ans.* 8, 9, 89%

Fig. 7-7

7.14. A certain hydraulic jack in a gas station lifts a 900 kg car a distance of 0.25 cm when a force of 150 N pushes a piston through a distance of 20 cm. Find the IMA, AMA, and efficiency.
Ans. 80, 59, 74%

7.15. The screw of a certain press has a pitch of 0.20 cm. The diameter of the wheel to which a tangential turning force *F* is applied is 55 cm. If the efficiency is 40%, how large must *F* be to produce a force of 12 000 N in the press?   *Ans.*   35 N

7.16. The diameters of the two upper pulleys of a chain hoist (Fig. 7-6) are 18 cm and 16 cm. If the efficiency of the hoist is 45%, what force is required to lift a 400 kg crate?   *Ans.*   484 N

Chapter 8

Impulse and Momentum

LINEAR MOMENTUM of a body = (mass of body) $\times$ (velocity of body) = $m\mathbf{v}$.

Momentum is a vector quantity whose direction is that of the velocity. The units of momentum are a mass unit times a velocity unit, $\text{kg}\cdot\text{m/s}$ in the SI and $\text{slug}\cdot\text{ft/s}$ in the British system.

IMPULSE = (force) $\times$ (length of time the force acts) = $\mathbf{F}t$.

Impulse is a vector quantity whose direction is that of the force. Its units are $\text{N}\cdot\text{s}$ in the SI and $\text{lb}\cdot\text{s}$ in the British system.

AN IMPULSE CAUSES A CHANGE IN MOMENTUM: The change of momentum produced by an impulse is equal to the impulse in both magnitude and direction. Thus, if a constant force $\mathbf{F}$ acting for a time t on a body of mass m changes its velocity from an initial value $\mathbf{v}_0$ to a final value $\mathbf{v}_f$, then

$$\text{impulse} = \text{change in momentum}$$
$$\mathbf{F}t = m(\mathbf{v}_f - \mathbf{v}_0)$$

CONSERVATION OF LINEAR MOMENTUM: If the resultant external force on a system of objects is zero, then the vector sum of the momenta of the objects will remain constant.

IN COLLISIONS AND EXPLOSIONS, the vector sum of the momenta just before the event equals the vector sum of the momenta just after the event. The vector sum of the momenta of the objects involved does not change during the collision or explosion.

Thus, when two bodies of masses m_1 and m_2 collide,

$$\text{total momentum before impact} = \text{total momentum after impact}$$
$$m_1\mathbf{u}_1 + m_2\mathbf{u}_2 = m_1\mathbf{v}_1 + m_2\mathbf{v}_2$$

where $\mathbf{u}_1$ and $\mathbf{u}_2$ are the velocities before impact and $\mathbf{v}_1$ and $\mathbf{v}_2$ are the velocities after. Or, in component form,

$$m_1 u_{1x} + m_2 u_{2x} = m_1 v_{1x} + m_2 v_{2x}$$

and similarly for the y- and z-components.

A PERFECTLY ELASTIC COLLISION is one in which the sum of the translational KE's of the objects is not changed during the collision. In the case of two bodies, one has

$$\tfrac{1}{2}m_1 u_1^2 + \tfrac{1}{2}m_2 u_2^2 = \tfrac{1}{2}m_1 v_1^2 + \tfrac{1}{2}m_2 v_2^2$$

COEFFICIENT OF RESTITUTION: For any collision between two bodies in which the bodies move only along a single straight line, a *coefficient of restitution*, e, is defined. It is a pure number given by

$$e = \frac{v_2 - v_1}{u_1 - u_2}$$

where
$$u_1, u_2 = \text{velocities before impact}$$
$$v_1, v_2 = \text{velocities after impact}$$

60

Notice that $u_1 - u_2$ is the relative velocity of approach and $v_2 - v_1$ is the relative velocity of recession. For a perfectly elastic collision, $e = 1$. For inelastic collisions, $e < 1$. If the bodies stick together after collision, $e = 0$.

Solved Problems

8.1. An 8 g bullet is fired horizontally into a 9 kg block of wood and sticks in it. The block, which is free to move, has a velocity of 40 cm/s after impact. Find the initial velocity of the bullet.

Consider the system (block + bullet). The velocity, and hence the momentum, of the block before impact is zero. The momentum conservation law tells us that

momentum of system before impact = momentum of system after impact

(mass) × (velocity of bullet) + 0 = (mass) × (velocity of block + bullet)

$$(0.008 \text{ kg})v + 0 = (9.008 \text{ kg})(0.40 \text{ m/s})$$

where v is the velocity of the bullet. Solving gives $v = 450 \text{ m/s}$.

8.2. A 16 g mass is moving in the $+x$-direction at 30 cm/s while a 4 g mass is moving in the $-x$-direction at 50 cm/s. They collide head on and stick together. Find their velocity after collision.

Apply the law of conservation of momentum to the system consisting of the two masses.

momentum before impact = momentum after impact

$$(0.016 \text{ kg})(0.30 \text{ m/s}) + (0.004 \text{ kg})(-0.50 \text{ m/s}) = (0.020 \text{ kg})v$$

Notice that the 4 g mass has negative momentum. Solving gives $v = 0.14 \text{ m/s}$.

8.3. A 2 kg brick is moving at a speed of 6 m/s. How large a force, F, is needed if the brick is to be stopped in a time of 7×10^{-4} s?

Let us solve this by use of the impulse equation.

impulse on brick = change in momentum of brick

$$Ft = mv_f - mv_0$$

$$F(7 \times 10^{-4} \text{ s}) = 0 - (2 \text{ kg})(6 \text{ m/s})$$

from which $F = -1.71 \times 10^4 \text{ N}$. The minus sign indicates that the force opposes the motion.

8.4. As shown in Fig. 8-1, a 15 g bullet is fired horizontally into a 3 kg block of wood suspended by a long cord. The bullet sticks in the block. Compute the velocity of the bullet if the impact causes the block to swing 10 cm above its initial level.

Consider first the collision of block and bullet. During the collision, momentum is conserved, so

momentum just before = momentum just after

$$(0.015 \text{ kg})v + 0 = (3.015 \text{ kg})V$$

where v is the initial speed of the bullet and V is the speed of block and bullet just after collision.

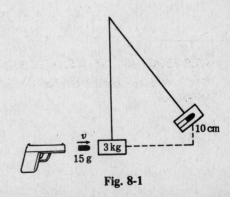

Fig. 8-1

We have two unknowns in this equation. To find another equation, we can use the fact that the block swings 10 cm high. Therefore, choosing GPE = 0 at the initial level of the block,

$$\text{KE just after collision} = \text{final GPE}$$

$$\tfrac{1}{2}(3.015 \text{ kg})V^2 = (3.015 \text{ kg})(9.8 \text{ m/s}^2)(0.10 \text{ m})$$

From this we find $V = 1.40$ m/s. Substituting this in the previous equation gives $v = 281$ m/s for the speed of the bullet.

Notice that we cannot write the equation $\tfrac{1}{2}mv^2 = (m + M)gh$, where $m = 0.015$ kg and $M = 3$ kg. Why not?

8.5. The nucleus of a certain atom has a mass of 3.8×10^{-25} kg and is at rest. The nucleus is radioactive and suddenly ejects from itself a particle of mass 6.6×10^{-27} kg and speed 1.5×10^7 m/s. Find the recoil speed of the nucleus left behind.

The momentum of the system is conserved during the explosion.

$$\text{momentum before} = \text{momentum after}$$

$$0 = (3.73 \times 10^{-25} \text{ kg})(v) + (6.6 \times 10^{-27} \text{ kg})(1.5 \times 10^7 \text{ m/s})$$

where the mass of the remaining nucleus is 3.73×10^{-25} kg and its recoil velocity is v. Solving gives

$$-v = \frac{(6.6 \times 10^{-27})(1.5 \times 10^7)}{3.73 \times 10^{-25}} = \frac{10.0 \times 10^{-20}}{3.73 \times 10^{-25}} = 2.7 \times 10^5 \text{ m/s}$$

8.6. A 500 lb gun fires a 2 lb projectile with a muzzle velocity of 1600 ft/s. (a) Determine the initial recoil velocity v of the gun. (b) If the recoil is against a constant resisting force of 400 lb, find the time taken to bring the gun to rest and the distance it recoils.

(a) Consider the system (gun + projectile). The momentum of the system before firing is zero. Therefore, since momentum is conserved,

$$\text{momentum before firing} = \text{momentum just after}$$

$$0 = \left(\frac{2}{32} \text{ slug}\right)(1600 \text{ ft/s}) + \left(\frac{500}{32} \text{ slug}\right)v$$

from which $v = -6.4$ ft/s.

(b) The loss in momentum during recoil is due to the impulse exerted on the gun by the 400 lb resisting force. Therefore

$$\text{impulse} = mv_f - mv_0$$

$$(-400 \text{ lb})t = 0 - \left(\frac{500}{32} \text{ slug}\right)(6.4 \text{ ft/s})$$

from which $t = 0.25$ s.

Since the resisting force is constant, the gun's recoil is uniformly decelerated. We may therefore write

$$\bar{v} = \tfrac{1}{2}(0 + 6.4) \text{ ft/s} = 3.2 \text{ ft/s}$$

Then $x = \bar{v}t$ gives the recoil distance as $x = (3.2 \text{ ft/s})(0.25 \text{ s}) = 0.80$ ft.

8.7. A 0.25 kg ball moving in the $+x$-direction at 13 m/s is hit by a bat. Its final velocity is 19 m/s in the $-x$-direction. The bat acts on the ball for 0.010 s. Find the average force F exerted on the ball by the bat.

We have $v_0 = 13$ m/s and $v_f = -19$ m/s. The impulse equation then gives

$$Ft = mv_f - mv_0$$

$$F(0.01 \text{ s}) = (0.25 \text{ kg})(-19 \text{ m/s}) - (0.25 \text{ kg})(13 \text{ m/s})$$

from which $F = -800$ N.

8.8. A 7500 kg truck traveling at 5 m/s east collides with a 1500 kg car moving at 20 m/s in a direction 30° south of west. After collision, the two vehicles remain tangled together. With what speed and in what direction does the wreckage begin to move?

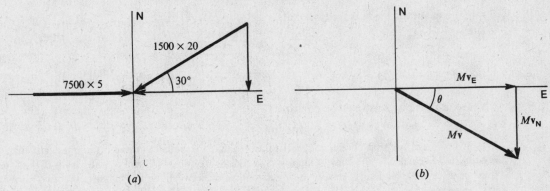

Fig. 8-2

The original momenta are shown in Fig. 8-2(a), while the final momentum, $M\mathbf{v}$, is shown in Fig. 8-2(b). Momentum must be conserved in both the north and east directions. Therefore

$$(\text{momentum before})_E = (\text{momentum after})_E$$

$$(7500 \text{ kg})(5 \text{ m/s}) - (1500 \text{ kg})[(20 \text{ m/s}) \cos 30°] = Mv_E$$

where $M = 7500 + 1500 = 9000$ kg, and v_E is the eastward component of the velocity of the wreckage. Similarly,

$$(\text{momentum before})_N = (\text{momentum after})_N$$

$$(7500 \text{ kg})(0) - (1500 \text{ kg})[(20 \text{ m/s}) \sin 30°] = Mv_N$$

The first equation gives $v_E = 1.28$ m/s and the second gives $v_N = -1.67$ m/s.
The resultant velocity is

$$v = \sqrt{(1.67)^2 + (1.28)^2} = 2.1 \text{ m/s}$$

The angle θ in Fig. 8-2(b) is given by

$$\theta = \arctan\left(\frac{1.67}{1.28}\right) = 53°$$

8.9. Two identical balls collide head on. The initial velocity of one is 0.75 m/s, while that of the other is -0.43 m/s. If the collision is perfectly elastic, what is the final velocity of each ball?

Since the collision is head-on, all motion takes place along a straight line. Let us call the mass of each ball m. Momentum is conserved in a collision and so we can write

$$\text{momentum before} = \text{momentum after}$$

$$m(0.75 \text{ m/s}) + m(-0.43 \text{ m/s}) = mv_1 + mv_2$$

where v_1 and v_2 are the final velocities. This equation simplifies to give

$$0.32 \text{ m/s} = v_1 + v_2 \tag{1}$$

Because the collision is assumed to be perfectly elastic, KE is also conserved.

$$\text{KE before} = \text{KE after}$$

$$\tfrac{1}{2}m(0.75 \text{ m/s})^2 + \tfrac{1}{2}m(0.43 \text{ m/s})^2 = \tfrac{1}{2}mv_1^2 + \tfrac{1}{2}mv_2^2$$

This equation can be simplified to give

$$0.747 = v_1^2 + v_2^2 \tag{2}$$

We can solve for v_2 in (1) to give $v_2 = 0.32 - v_1$ and substitute this in (2). This yields

$$0.747 = (0.32 - v_1)^2 + v_1^2$$

from which
$$2v_1^2 - 0.64\,v_1 - 0.645 = 0$$

Using the quadratic formula we find that

$$v_1 = \frac{0.64 \pm \sqrt{(0.64)^2 + 5.16}}{4} = 0.16 \pm 0.59$$

from which $v_1 = 0.75$ m/s or -0.43 m/s. Substitution back into (1) gives $v_2 = -0.43$ m/s or 0.75 m/s.

Two choices for answers are available:

$$(v_1 = 0.75 \text{ m/s}, \; v_2 = -0.43 \text{ m/s}) \qquad (v_1 = -0.43 \text{ m/s}, \; v_2 = 0.75 \text{ m/s})$$

We must discard the first choice because it implies that the balls continue on unchanged; that is to say, no collision occurred. The correct answer is therefore $v_1 = -0.43$ m/s and $v_2 = 0.75$ m/s, which tells us that in a perfectly elastic, head-on collision between equal masses, the two bodies simply exchange velocities.

Alternate (Better) Method

If we recall that $e = 1$ for a perfectly elastic, head-on collision, then

$$e = \frac{v_2 - v_1}{u_1 - u_2} \qquad \text{becomes} \qquad 1 = \frac{v_2 - v_1}{0.75 - (-0.43)}$$

which gives
$$v_2 - v_1 = 1.18 \text{ m/s} \tag{3}$$

Equations (1) and (3) determine v_1 and v_2 uniquely.

8.10. A 1 kg ball moving at 12 m/s collides head on with a 2 kg ball moving in the opposite direction at 24 m/s. Find the velocity of each after impact if (a) $e = 2/3$, (b) the balls stick together, (c) the collision is perfectly elastic.

In all three cases momentum is conserved and so we can write

$$\text{momentum before} = \text{momentum after}$$
$$(1 \text{ kg})(12 \text{ m/s}) + (2 \text{ kg})(-24 \text{ m/s}) = (1 \text{ kg})v_1 + (2 \text{ kg})v_2$$

which becomes

$$-36 \text{ m/s} = v_1 + 2v_2$$

(a) In this case $e = 2/3$ and so

$$e = \frac{v_2 - v_1}{u_1 - u_2} \qquad \text{becomes} \qquad \frac{2}{3} = \frac{v_2 - v_1}{12 - (-24)}$$

from which 24 m/s $= v_2 - v_1$. Combining this with the momentum equation found above gives $v_2 = -4$ m/s and $v_1 = -28$ m/s.

(b) In this case $v_1 = v_2 = v$ and so the momentum equation becomes

$$3v = -36 \text{ m/s} \qquad \text{or} \qquad v = -12 \text{ m/s}$$

(c) Here $e = 1$ and so

$$e = \frac{v_2 - v_1}{u_1 - u_2} \qquad \text{becomes} \qquad 1 = \frac{v_2 - v_1}{12 - (-24)}$$

from which $v_2 - v_1 = 36$ m/s. Adding this to the momentum equation gives $v_2 = 0$. Using this value for v_2 then gives $v_1 = -36$ m/s.

8.11. A ball is dropped from a height h above a tile floor and rebounds to a height of $0.65\,h$. Find the coefficient of restitution between ball and floor.

The initial and final velocities of the floor, u_1 and v_1, are zero. Therefore

$$e = \frac{v_2 - v_1}{u_1 - u_2} = -\frac{v_2}{u_2}$$

But from the interchange of GPE and KE we can write

$$mgh = \tfrac{1}{2} mu_2^2 \qquad mg(0.65\,h) = \tfrac{1}{2} mv_2^2$$

Therefore, taking *down* as positive, $u_2 = \sqrt{2gh}$ and $v_2 = -\sqrt{1.30\,gh}$. Substitution gives

$$e = \frac{\sqrt{1.30\,gh}}{\sqrt{2gh}} = \sqrt{\frac{1.30\,gh}{2gh}} = \sqrt{0.65} = 0.81$$

8.12. The two balls shown in Fig. 8-3 collide and bounce off each other as shown. (*a*) What is the final velocity of the 500 g ball if the 800 g ball has a speed of 15 cm/s after collision? (*b*) Is the collision perfectly elastic?

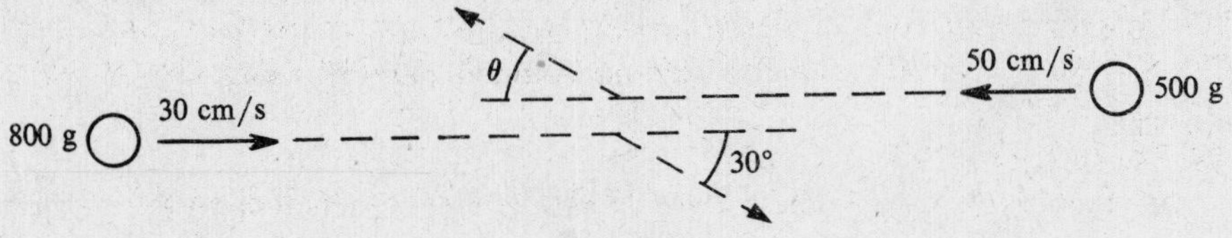

800 g — 30 cm/s — θ — 50 cm/s — 500 g — 30°

Fig. 8-3

(*a*) From the law of conservation of momentum,

$$(\text{momentum before})_x = (\text{momentum after})_x$$

$$(0.80 \text{ kg})(0.3 \text{ m/s}) - (0.50 \text{ kg})(0.5 \text{ m/s}) = (0.8 \text{ kg})[(0.15 \text{ m/s}) \cos 30°] + (0.5 \text{ kg})v_x$$

from which $v_x = -0.228$ m/s. Also,

$$(\text{momentum before})_y = (\text{momentum after})_y$$

$$0 = (0.8 \text{ kg})[(0.15 \text{ m/s}) \sin 30°] + (0.5 \text{ kg})v_y$$

from which $v_y = 0.120$ m/s. Then

$$v = \sqrt{v_x^2 + v_y^2} = \sqrt{(0.228)^2 + (0.120)^2} = 0.26 \text{ m/s}$$

Also, for the angle θ shown in Fig. 8-3,

$$\theta = \arctan\left(\frac{0.120}{0.228}\right) = 28°$$

(*b*) $$\text{total KE before} = \tfrac{1}{2}(0.8)(0.3)^2 + \tfrac{1}{2}(0.5)(0.5)^2 = 0.0985 \text{ J}$$

$$\text{total KE after} = \tfrac{1}{2}(0.8)(0.15)^2 + \tfrac{1}{2}(0.5)(0.26)^2 = 0.026 \text{ J}$$

As we see, KE is lost in the collision and so the collision is not perfectly elastic.

8.13. What force is exerted on a stationary flat plate held perpendicular to a jet of water as shown in Fig. 8-4? The horizontal speed of the water is 80 cm/s and 30 cm³ of the water hits the plate each second. Assume the water moves parallel to the plate after striking it. One cubic centimeter of water has a mass of one gram.

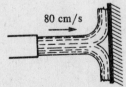

Fig. 8-4

The plate exerts an impulse on the water and changes its horizontal momentum. We can therefore write

$$(\text{impulse})_x = \text{change in } x\text{-directed momentum}$$

$$F_x t = (mv_x)_{\text{final}} - (mv_x)_{\text{initial}}$$

Let us take t to be 1 s so that m will be the mass that strikes in 1 s, namely 30 g. Then the above equation becomes

$$F_x(1 \text{ s}) = (0.030 \text{ kg})(0 \text{ m/s}) - (0.030 \text{ kg})(0.80 \text{ m/s})$$

from which $F_x = -0.024$ N. This is the force of the plate on the water. The law of action and reaction tells us that the jet exerts an equal but opposite force on the plate.

8.14. A rocket standing on its launch platform points straight upward. Its jet engines are activated and eject gas at a rate of 1500 kg/s. The molecules are expelled with a speed of 50 km/s. How much mass can the rocket initially have if it is slowly to rise because of the thrust of the engines?

Because the motion of the rocket itself is negligible in comparison to the speed of the expelled gas, we can assume the gas to be accelerated from rest to a speed of 50 km/s. The impulse required to provide this acceleration to a mass m of gas is

$$Ft = mv_f - mv_0 = m(50\,000 \text{ m/s}) - 0$$

from which

$$F = (50\,000 \text{ m/s}) \frac{m}{t}$$

But we are told that the mass ejected per second (m/t) is 1500 kg/s and so the force exerted on the expelled gas is

$$F = (50\,000 \text{ m/s})(1500 \text{ kg/s}) = 7.5 \times 10^7 \text{ N}$$

An equal but opposite reaction force acts on the rocket and this is the upward thrust on the rocket. The engines can therefore support a weight of 7.5×10^7 N and so the maximum mass the rocket could have would be

$$M_{\text{rocket}} = \frac{\text{weight}}{g} = \frac{7.5 \times 10^7 \text{ N}}{9.8 \text{ m/s}^2} = 7.65 \times 10^6 \text{ kg}$$

Supplementary Problems

8.15. A 40 000 kg freight car is coasting at a speed of 5.0 m/s along a straight track when it strikes a 30 000 kg stationary freight car and couples to it. What will be their combined speed after impact?
Ans. 2.86 m/s

8.16. An empty 15 000 kg coal car is coasting on a level track at 5 m/s. Suddenly 5000 kg of coal is dumped into it from directly above it. The coal initially has zero horizontal velocity. Find the final speed of the car. *Ans.* 3.75 m/s

8.17. Sand drops at a rate of 2000 kg/min from the bottom of a hopper onto a belt conveyor moving horizontally at 250 m/min. Determine the force needed to drive the conveyor, neglecting friction.
Ans. 139 N

8.18. Two bodies of masses 8 and 4 kg move along the x-axis in opposite directions with velocities of 11 m/s and -7 m/s respectively. They collide and stick together. Find their velocity just after collision.
Ans. 5 m/s

8.19. A 1200 lb gun mounted on wheels shoots an 8 lb projectile with a muzzle velocity of 1800 ft/s at an angle of 30° above the horizontal. Find the horizontal recoil velocity of the gun. *Ans.* 10.4 ft/s

8.20. (a) What minimum thrust must the jet engines of a 2×10^5 kg rocket have if the rocket is to be able to rise from the earth when aimed straight upward? (b) If the engines eject fuel at the rate of 20 kg/s, how fast must the gaseous fuel be moving as it leaves the engines? Neglect the small change in mass of the rocket due to the ejected fuel.   *Ans.* (a) 19.6×10^5 N; (b) 98 km/s

8.21. A 2 kg block of wood rests on a long tabletop. A 5 g bullet moving horizontally with a speed of 150 m/s is shot into the block and sticks in it. The block then slides 270 cm along the table and stops. (a) Find the speed of the block just after impact. (b) Find the friction force between block and table. *Ans.* (a) 0.374 m/s; (b) 0.052 N

8.22. A 2 kg block of wood rests on a tabletop. A 7 g bullet is shot straight up through a hole in the table beneath the block. The bullet lodges in the block and the block flies 25 cm above the tabletop. How fast was the bullet going initially?   *Ans.* 635 m/s

8.23. A 6000 kg truck traveling north at 5 m/s collides with a 4000 kg truck moving west at 15 m/s. If the two trucks remain locked together after impact, with what speed and in what direction do they move immediately after the collision?   *Ans.* 6.71 m/s at 26.6° north of west

8.24. What average resisting force must act on a 3 kg mass to reduce its speed from 65 to 15 cm/s in 0.2 s? *Ans.* 7.5 N

8.25. A 7 g bullet moving horizontally at 200 m/s strikes and passes through a 150 g tin can sitting on a post. Just after impact, the can has a horizontal speed of 180 cm/s. What was the bullet's speed after leaving the can?   *Ans.* 161 m/s

8.26. Two balls of equal mass, moving with speeds of 3 m/s, collide head on. Find the velocity of each after impact if (a) they are stuck together, (b) the collision is perfectly elastic, (c) the coefficient of restitution is 1/3.   *Ans.* (a) 0 m/s; (b) each rebounds at 3 m/s; (c) each rebounds at 1 m/s

8.27. A 9 lb ball moving at 10 ft/s collides head on with a stationary 1 lb ball. Determine the velocity of each after impact if (a) they stick together, (b) the collision is perfectly elastic, (c) the coefficient of restitution is 0.90.   *Ans.* (a) 9 ft/s; (b) 8 ft/s, 18 ft/s; (c) 8.1 ft/s, 17.1 ft/s

8.28. A ball is dropped onto a horizontal floor. It reaches a height of 144 cm on the first bounce and 81 cm on the second bounce. Find (a) the coefficient of restitution between the ball and floor and (b) the height it attains on the third bounce.   *Ans.* (a) 0.75; (b) 45.6 cm

8.29. Two identical balls undergo a collision at the origin of coordinates. Before collision their velocity components are ($u_x = 40$ cm/s, $u_y = 0$) and ($u_x = -30$ cm/s, $u_y = 20$ cm/s). After collision, the first ball is standing still. Find the velocity components of the second ball. *Ans.* $v_x = 10$ cm/s, $v_y = 20$ cm/s

8.30. Two identical balls traveling parallel to the x-axis have oppositely directed velocities of 30 cm/s. They collide perfectly elastically. After collision, one ball is moving at an angle of 30° above the $+x$-axis. Find its speed and the velocity of the other ball. *Ans.* 30 cm/s, 30 cm/s at 30° below the $-x$-axis (opposite to the first ball)

Chapter 9

Angular Motion

ANGULAR DISPLACEMENT (θ) is usually expressed in *radians*, in *degrees*, or in *revolutions*.

$$1 \text{ rev} = 360° = 2\pi \text{ rad} \qquad \text{and} \qquad 1 \text{ rad} = 57.3°$$

One radian is the angle subtended at the center of a circle by an arc equal in length to the radius of the circle. Thus,

$$\text{radian measure of an angle} = (\text{length of arc}) \div (\text{length of radius})$$

The radian measure of an angle is a pure number and has no unit in the true sense of the word.

ANGULAR VELOCITY (ω) of an object is the rate at which its angular coordinate, the angular displacement θ, changes with time. If θ changes from θ_0 to θ_f in a time t, then the *average angular speed* is

$$\bar{\omega} = \frac{\theta_f - \theta_0}{t}$$

The units of $\bar{\omega}$ are typically rad/s, deg/s, or rev/min (rpm)—always an angle measure divided by a time unit. Also

$$\omega \text{ (in rad/s)} = 2\pi f$$

where f is the *frequency of rotation* in rev/s.

ANGULAR ACCELERATION (α) of an object is the rate at which its angular velocity is changing with time. If the angular velocity changes uniformly from ω_0 to ω_f in a time t, then

$$\alpha = \frac{\omega_f - \omega_0}{t}$$

The units of α are typically rad/s^2, rev/min^2, etc.

EQUATIONS FOR UNIFORMLY ACCELERATED ANGULAR MOTION are exactly analogous to those for uniformly accelerated linear motion. In the usual notation we have:

Linear	Angular
$\bar{v} = \frac{1}{2}(v_0 + v_f)$	$\bar{\omega} = \frac{1}{2}(\omega_0 + \omega_f)$
$s = \bar{v}t$	$\theta = \bar{\omega}t$
$v_f = v_0 + at$	$\omega_f = \omega_0 + \alpha t$
$v_f^2 = v_0^2 + 2as$	$\omega_f^2 = \omega_0^2 + 2\alpha\theta$
$s = v_0 t + \frac{1}{2}at^2$	$\theta = \omega_0 t + \frac{1}{2}\alpha t^2$

Taken alone, the second of these equations is just the definition of average velocity, so it is valid whether the acceleration is constant or not.

RELATIONS BETWEEN ANGULAR AND TANGENTIAL QUANTITIES: When a wheel of radius r rotates on its axis, a point on the rim of the wheel is described in terms of the circumferential distance it has moved, s; its tangential speed, v; and its tangential acceleration, a. These quantities

68

are related to the angular quantities θ, ω, and α that describe the rotation of the wheel, through the relations

$$s = \theta r \qquad v = \omega r \qquad a = \alpha r$$

provided radian measure is used for θ, ω, and α. By simple reasoning, s can be shown to be the length of belt wound on the wheel or the distance the wheel would roll (without slipping) if free to do so. In such cases, v and a refer to the speed and acceleration of the belt or of the center of the wheel.

CENTRIPETAL ACCELERATION: A point mass m moving with constant speed v around a circle of radius r is undergoing acceleration. Although the magnitude of its velocity is not changing, the direction of the velocity is continually changing. This change in velocity gives rise to an acceleration a_c of the mass, directed toward the center of the circle. We call this acceleration the *centripetal acceleration*; its value is given by

$$a_c = \frac{(\text{tangential speed})^2}{\text{radius of circular path}} = \frac{v^2}{r}$$

where v is the speed of the mass around the perimeter of the circle.

Because $v = \omega r$, we also have $a_c = \omega^2 r$, where ω must be in radian measure.

THE CENTRIPETAL FORCE is the unbalanced force that must act on the mass m moving in its circular path of radius r to give it the centripetal acceleration v^2/r. From $F = ma$, we have

$$\text{required centripetal force} = \frac{mv^2}{r} = m\omega^2 r$$

It must be directed toward the center of the circular path.

Solved Problems

9.1. Express the following in terms of the three angular measures: (a) $28°$, (b) $\frac{1}{4}$ rev/s, (c) 2.18 rad/s^2.

(a)
$$28° = (28 \text{ deg})\left(\frac{1 \text{ rev}}{360 \text{ deg}}\right) = 0.078 \text{ rev}$$

$$= (28 \text{ deg})\left(\frac{2\pi \text{ rad}}{360 \text{ deg}}\right) = 0.49 \text{ rad}$$

(b)
$$\frac{1}{4} \frac{\text{rev}}{\text{s}} = \left(0.25 \frac{\text{rev}}{\text{s}}\right)\left(\frac{360 \text{ deg}}{1 \text{ rev}}\right) = 90 \frac{\text{deg}}{\text{s}}$$

$$= \left(0.25 \frac{\text{rev}}{\text{s}}\right)\left(\frac{2\pi \text{ rad}}{1 \text{ rev}}\right) = \frac{\pi}{2} \frac{\text{rad}}{\text{s}}$$

(c)
$$2.18 \frac{\text{rad}}{\text{s}^2} = \left(2.18 \frac{\text{rad}}{\text{s}^2}\right)\left(\frac{360 \text{ deg}}{2\pi \text{ rad}}\right) = 125 \frac{\text{deg}}{\text{s}^2}$$

$$= \left(2.18 \frac{\text{rad}}{\text{s}^2}\right)\left(\frac{1 \text{ rev}}{2\pi \text{ rad}}\right) = 0.35 \frac{\text{rev}}{\text{s}^2}$$

9.2. The bob of a pendulum 90 cm long swings through a 15 cm arc, as shown in Fig. 9-1. Find the angle θ, in radians and in degrees, through which it swings.

We recall that $s = r\theta$ applies only to angles in radian measure. Therefore

$$\theta \text{ in radians} = \frac{s \text{ (length of arc)}}{r \text{ (radius)}} = \frac{0.15 \text{ m}}{0.90 \text{ m}} = 0.167 \text{ rad}$$

$$\theta \text{ in degrees} = (0.167 \text{ rad})\left(\frac{360 \text{ deg}}{2\pi \text{ rad}}\right) = 9.55°$$

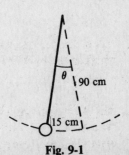

Fig. 9-1

9.3. A fan turns at a rate of 900 rpm (rev/min). (a) Find the angular speed of any point on one of the fan blades. (b) Find the tangential speed of the tip of the blade if the distance from the center to tip is 20 cm.

(a)
$$\omega = 900 \frac{\text{rev}}{\text{min}} = 15 \frac{\text{rev}}{\text{s}} = 94 \frac{\text{rad}}{\text{s}}$$

for all points of the fan blade.

(b) The tangential speed is ωr, where ω must be in radian measure. Therefore,

$$v = \omega r = \left(94 \frac{\text{rad}}{\text{s}}\right)(0.20 \text{ m}) = 18.8 \text{ m/s}$$

Notice that rad, not a true unit, does not carry through the equations properly.

9.4. A belt passes over a wheel of radius 25 cm, as shown in Fig. 9-2. If a point on the belt has a speed of 5.0 m/s, how fast is the wheel turning?

$$\omega = \frac{v}{r} = \frac{5 \text{ m/s}}{0.25 \text{ m}} = 20 \frac{\text{rad}}{\text{s}} = 3.2 \frac{\text{rev}}{\text{s}}$$

Notice that we must supply the angle measure radians. We know it to be radians because $v = \omega r$ applies only to angles in radian measure.

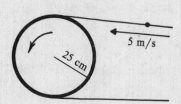

Fig. 9-2

9.5. A wheel of 40 cm radius rotates on a stationary axle. It is uniformly speeded up from rest to a speed of 900 rpm in a time of 20 s. (a) Find the angular acceleration of the wheel and (b) the tangential acceleration of a point on its rim.

(a) Because the acceleration is constant, we can use the definition $\alpha = (\omega_f - \omega_0)/t$ to give

$$\alpha = \frac{\dfrac{900 \text{ rev}}{60 \text{ s}} - 0 \dfrac{\text{rev}}{\text{s}}}{20 \text{ s}} = 0.75 \text{ rev/s}^2$$

(b) In order to use $a = r\alpha$, we must have α in radian measure.

$$0.75 \frac{\text{rev}}{\text{s}^2} = \left(0.75 \frac{\text{rev}}{\text{s}^2}\right)\left(\frac{2\pi \text{ rad}}{1 \text{ rev}}\right) = 4.7 \frac{\text{rad}}{\text{s}^2}$$

Then
$$a = r\alpha = (0.40 \text{ m})\left(4.7 \frac{\text{rad}}{\text{s}^2}\right) = 1.88 \frac{\text{m}}{\text{s}^2}$$

9.6. A pulley of 5 cm radius on a motor is turning at 30 rev/s and slows down uniformly to 20 rev/s in 2 s. Calculate (a) the angular acceleration of the motor, (b) the number of revolutions it makes in this time, and (c) the length of belt it winds in this time.

(a)
$$\alpha = \frac{\omega_f - \omega_0}{t} = \frac{(20 - 30) \text{ rev/s}}{2 \text{ s}} = -5 \text{ rev/s}^2$$

(b)
$$\theta = \bar{\omega}t = \tfrac{1}{2}(\omega_f + \omega_0)t = \tfrac{1}{2}(50 \text{ rev/s})(2 \text{ s}) = 50 \text{ rev}$$

(c) To use $s = \theta r$, we must express θ in radian measure: $\theta = 50 \text{ rev} = 314 \text{ rad}$.

$$s = \theta r = (314 \text{ rad})(0.05 \text{ m}) = 15.7 \text{ m}$$

9.7. A car has wheels of 30 cm radius. It starts from rest and accelerates uniformly to a speed of 15 m/s in a time of 8 s. Find the angular acceleration of its wheels and the number of rotations one makes in this time.

We know that $a = (v_f - v_0)/t$ and so

$$a = \frac{15 \text{ m/s}}{8 \text{ s}} = 1.875 \text{ m/s}^2$$

Then $a = r\alpha$ gives

$$\alpha = \frac{a}{r} = \frac{1.875 \text{ m/s}^2}{0.30 \text{ m}} = 6.2 \text{ rad/s}^2$$

Notice that we must introduce the proper angular measure, radians.

Now we can use $\theta = \omega_0 t + \frac{1}{2}\alpha t^2$ to find

$$\theta = 0 + \tfrac{1}{2}(6.2 \text{ rad/s}^2)(8 \text{ s})^2 = 200 \text{ rad}$$

$$= (200 \text{ rad})\left(\frac{1 \text{ rev}}{2\pi \text{ rad}}\right) = 32 \text{ rev}$$

9.8. The spin-drier of a washing machine revolving at 900 rpm slows down uniformly to 300 rpm while making 50 revolutions.   Find (a) the angular acceleration and (b) the time required to turn through these 50 revolutions.

We easily find that   900 rev/min = 15 rev/s   and   300 rev/min = 5 rev/s.

(a) From $\omega_f^2 = \omega_0^2 + 2\alpha\theta$,   we have

$$\alpha = \frac{\omega_f^2 - \omega_0^2}{2\theta} = \frac{(15^2 - 5^2) \text{ (rev/s)}^2}{(2)(50 \text{ rev})} = 2 \text{ rev/s}^2$$

(b) Because $\bar{\omega} = \frac{1}{2}(\omega_0 + \omega_f) = 10 \text{ rev/s}$, $\theta = \bar{\omega}t$ gives

$$t = \frac{\theta}{\bar{\omega}} = \frac{50 \text{ rev}}{10 \text{ rev/s}} = 5 \text{ s}$$

9.9. A 200 g object is tied to the end of a cord and whirled in a horizontal circle of radius 1.20 m at 3 rev/s.   Assume that the cord is horizontal, i.e. that gravity can be neglected.   Determine (a) the acceleration of the object and (b) the tension in the cord.

(a) The object is not accelerating tangentially to the circle but is undergoing a radial, or centripetal, acceleration, given by

$$a_c = \frac{v^2}{r} = \omega^2 r$$

where ω must be in radian measure.   Since $\omega = 3 \text{ rev/s} = 6\pi \text{ rad/s}$,

$$a_c = (6\pi \text{ rad/s})^2 (1.20 \text{ m}) = 426 \text{ m/s}^2$$

(b) In order to cause the acceleration found in (a), the cord must pull on the 0.200 kg mass with a centripetal force given by

$$F_c = ma_c = (0.20 \text{ kg})(426 \text{ m/s}^2) = 85 \text{ N}$$

This is the tension in the cord.

9.10. Repeat Problem 9.9 for a 2 lb object at the end of a 4 ft string.

(a) $$a_c = \omega^2 r = (6\pi \text{ rad/s})^2(4 \text{ ft}) = 1420 \text{ ft/s}^2$$

(b) $$F_c = \left(\frac{\text{weight}}{g}\right)a_c = \left(\frac{2 \text{ lb}}{32 \text{ ft/s}^2}\right)(1420 \text{ ft/s}^2) = 88 \text{ lb}$$

9.11. What is the maximum speed at which a car can round a curve of 25 m radius on a level road if the coefficient of static friction between the tires and road is 0.80?

The radial force required to keep the car in the curved path (the centripetal force) is supplied by the force of friction between the tires and the road.   If the mass of the car is m, then the maximum friction (and centripetal) force is $0.80\,mg$; this arises when the car is on the verge of skidding sideways. Therefore, the maximum speed is given by

$$\frac{mv^2}{r} = 0.80\,mg$$

$$v = \sqrt{0.80\,gr} = \sqrt{(0.80)(9.8 \text{ m/s}^2)(25 \text{ m})} = 14 \text{ m/s}$$

9.12. As shown in Fig. 9-3, a ball B is fastened to one end of a 24 cm string, and the other end is held fixed at point O. The ball whirls in the horizontal circle shown. Find the speed of the ball in its circular path if the string makes an angle of 30° to the vertical.

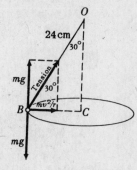

The only forces acting on the ball are the ball's weight, mg, and the tension in the cord, T. The tension must do two things: (1) balance the weight of the ball by means of its vertical component, $T\cos 30°$; (2) supply the required centripetal force by means of its horizontal component, $T\sin 30°$. Therefore we can write

$$T\cos 30° = mg \qquad T\sin 30° = \frac{mv^2}{r}$$

Solving for T in the first equation and substituting it in the second gives

$$\frac{mg\sin 30°}{\cos 30°} = \frac{mv^2}{r} \qquad \text{or} \qquad v = \sqrt{rg(0.577)}$$

However $r = \overline{BC} = (0.24\,\text{m})\sin 30° = 0.12\,\text{m}$ and $g = 9.8\,\text{m/s}^2$, from which $v = 0.82\,\text{m/s}$.

Fig. 9-3

9.13. As shown in Fig. 9-4, a 20 g bead slides from rest at A along a frictionless wire. If h is 25 cm and $R = 5$ cm, how large a force must the wire exert on the bead when it is at point B?

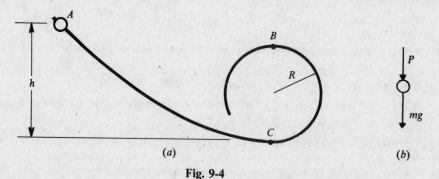

(a)   (b)

Fig. 9-4

Let us first find the speed of the bead at point B. It has fallen through a distance $h - 2R$ and so its loss in GPE is $mg(h - 2R)$. This must equal its KE at point B:

$$\tfrac{1}{2}mv^2 = mg(h - 2R)$$

where v is the speed of the bead at point B. Hence,

$$v = \sqrt{2g(h - 2R)} = \sqrt{2(9.8\ \text{m/s}^2)(0.15\ \text{m})} = 1.715\ \text{m/s}$$

As shown in Fig. 9-4(b), two forces act on the bead when it is at B: (1) the weight of the bead, mg; (2) the (assumed downward) push, P, of the wire on the bead. Together, these two forces must supply the required centripetal force, mv^2/R, if the bead is to follow the circular path. We therefore write

$$mg + P = \frac{mv^2}{R}$$

or

$$P = \frac{mv^2}{R} - mg = (0.02\ \text{kg})\left[\left(\frac{1.715^2}{0.05} - 9.8\right)\ \text{m/s}^2\right] = 0.98\ \text{N}$$

The wire must exert a 0.98 N downward force on the bead to hold it in a circular path.

9.14. As shown in Fig. 9-5, a 0.9 kg body attached to a cord is whirled in a vertical circle of radius 2.5 m. (a) What minimum speed, v_t, must it have at the top of the circle so as not to depart from the circular path? (b) Under the condition (a), what speed, v_b, will the object have at the bottom of the circle? (c) Find the tension T_b in the cord when the body is at the bottom of the circle and moving with the critical speed v_b.

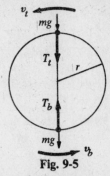

Fig. 9-5

(a) As shown in Fig. 9-5, two radial forces act on the object at the top: (1) its weight mg, and (2) the tension T_t. The resultant of these two forces must supply the required centripetal force:

$$\frac{mv^2}{r} = mg + T_t$$

For a given r, v will be smallest when $T_t = 0$. Thus,

$$\frac{mv_t^2}{r} = mg \qquad \text{or} \qquad v_t = \sqrt{rg}$$

Using $r = 2.5$ m and $g = 9.8$ m/s^2 gives $v_t = 4.95$ m/s as the speed at the top.

(b) In traveling from bottom to top, the body rises a distance $2r$. Therefore, with $v_t = 4.95$ m/s as the speed at the top and with v_b as the speed at the bottom,

$$\text{KE at bottom} = \text{KE at top} + \text{GPE at top}$$

$$\tfrac{1}{2}mv_b^2 = \tfrac{1}{2}mv_t^2 + mg(2r)$$

where we have chosen the bottom of the circle as the zero level for GPE. Notice that m cancels. Using $v_t = 4.95$ m/s, $r = 2.5$ m, and $g = 9.8$ m/s^2 gives $v_b = 11.1$ m/s.

(c) When the object is at the bottom of its path, we see from Fig. 9-5 that the unbalanced radial force on it is $T_b - mg$. This force supplies the required centripetal force:

$$T_b - mg = \frac{mv_b^2}{r}$$

Using $m = 0.9$ kg, $g = 9.8$ m/s^2, $v_b = 11.1$ m/s, and $r = 2.5$ m gives

$$T_b = m\left(g + \frac{v_b^2}{r}\right) = 53 \text{ N}$$

9.15. A curve of radius 30 m is to be banked so that a car may make the turn at a speed of 13 m/s without depending on friction. What must be the slope of the curve?

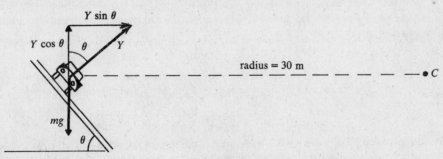

Fig. 9-6

The situation is shown in Fig. 9-6 if friction is absent. Only two forces act upon the car: (1) the weight of the car, mg, and (2) the normal force Y exerted by the pavement on the car.

The force Y must do two things: (1) its vertical component, $Y \cos \theta$, must balance the car's weight; (2) its horizontal component, $Y \sin \theta$, must supply the required centripetal force. We can therefore write

$$Y \cos \theta = mg \qquad Y \sin \theta = \frac{mv^2}{r}$$

Dividing the second equation by the first causes Y and m to cancel, and gives

$$\tan \theta = \frac{v^2}{gr} = \frac{(13 \text{ m/s})^2}{(9.8 \text{ m/s}^2)(30 \text{ m})} = 0.575$$

From this we find that θ, the banking angle, must be 30°.

Supplementary Problems

9.16. Convert (*a*) 50 rev to radians, (*b*) 48π rad to revolutions, (*c*) 72 rps to rad/s, (*d*) 1500 rpm to rad/s, (*e*) 22 rad/s to rpm, (*f*) 2 rad/s to deg/s.
Ans. (*a*) 314 rad; (*b*) 24 rev; (*c*) 452 rad/s; (*d*) 157 rad/s; (*e*) 210 rev/min; (*f*) 114.6 deg/s

9.17. Express the angular speed 40 deg/s in (*a*) rev/s, (*b*) rev/min, and (*c*) rad/s.
Ans. (*a*) 0.111 rev/s; (*b*) 6.67 rev/min; (*c*) 0.698 rad/s

9.18. A flywheel turns at 480 rpm. Compute the angular speed at any point on the wheel and the tangential speed 30 cm from the center. *Ans.* 50 rad/s, 15.1 m/s

9.19. It is desired that the outer edge of a grinding wheel 9 cm in radius move at a rate of 6 m/s. (*a*) Determine the angular speed of the wheel. (*b*) What length of thread could be wound on the rim of the wheel in 3 s when turning at this rate? *Ans.* (*a*) 66.7 rad/s = 10.6 rev/s; (*b*) 18 m

9.20. Through how many radians does a point on the earth's surface move in 6 h as a result of the earth's rotation? What is the speed of a point on the equator? Take the radius of the earth to be 6370 km. *Ans.* 1.57 rad, 463 m/s

9.21. A wheel 2.5 ft in radius turning at 120 rpm increases its speed to 660 rpm in 9 s. Find (*a*) the constant angular acceleration in rev/s^2 and in rad/s^2, (*b*) the tangential acceleration of a point on its rim.
Ans. (*a*) 1 rev/s^2 = 6.28 rad/s^2; (*b*) 15.7 ft/s^2

9.22. The angular speed of a disk decreases uniformly from 12 to 4 rad/s in 16 s. Compute the angular acceleration and the number of revolutions made in this time. *Ans.* -0.5 rad/s^2, 20.4 rev

9.23. A car wheel 30 cm in radius is turning at a rate of 8 rev/s when the car begins to slow uniformly to rest in a time of 14 s. Find the number of revolutions made by the wheel and the distance the car goes in the 14 s. *Ans.* 56 rev, 106 m

9.24. A wheel revolving at 6 rev/s has an angular acceleration of 4 rad/s^2. Find the number of turns the wheel must make to acquire an angular speed of 26 rev/s and the time required.
Ans. 502 rev, 31.4 s

9.25. A string wound on the rim of a wheel 20 cm in diameter is pulled out at a rate of 75 cm/s. Through how many revolutions will the wheel have turned by the time that 9.0 m of string has been unwound? How long does it take? *Ans.* 14.3 rev, 12 s

9.26. A mass of 1.5 kg moves in a circle of radius 25 cm at 2 rev/s. Calculate (*a*) the tangential velocity, (*b*) the acceleration, (*c*) the required centripetal force for the motion.
Ans. (*a*) 3.14 m/s; (*b*) 39.4 m/s^2 radially inward; (*c*) 59 N

9.27. A 20 lb body moves at 4 rev/s in a circular path of 5 ft radius. Calculate (*a*) its tangential speed, (*b*) its acceleration, (*c*) the required centripetal force for the motion.
Ans. (*a*) 126 ft/s; (*b*) 3160 ft/s^2 radially inward; (*c*) 1960 lb

9.28. (*a*) Compute the radial acceleration of a point at the equator of the earth. (*b*) Repeat for the north pole of the earth. Take the radius of the earth to be 6370 km. *Ans.* (*a*) 0.0337 m/s^2; (*b*) zero

9.29. A car going at 5 m/s tries to round a corner in a circular arc of 8 m radius. The roadway is flat. How large must the coefficient of friction be between wheels and roadway if the car is not to skid? *Ans.* 0.32

9.30. A box rests at a point 2.0 m from the axis of a horizontal circular platform. The coefficient of static friction between box and platform is 0.25. As the rate of rotation of the platform is slowly increased from zero, at what angular speed will the box first slide? *Ans.* 1.107 rad/s

9.31. A stone rests in a pail that is moved in a vertical circle of radius 60 cm. What is the least speed the stone must have as it rounds the top of the circle if it is to remain in contact with the pail? *Ans.* 2.4 m/s

9.32. A pendulum 80 cm long is pulled aside, so that its bob is raised 20 cm from its lowest position, and is then released. As the 50 g bob moves through its lowest position, (a) what is its speed and (b) what is the tension in the pendulum cord? *Ans.* (a) 1.98 m/s; (b) 0.735 N

9.33. Refer back to Fig. 9-4. How large must h be (in terms of R) if the frictionless wire is to exert no force on the bead as it passes by point B? Assume the bead is released from rest at A. *Ans.* 2.5 R

9.34. If, in Fig. 9-4 and in Problem 9.33, $h = 2.5\,R$, how large a force will the 50 g bead exert on the wire as it passes point C? *Ans.* 2.94 N

9.35. A satellite orbits the earth at a height of 200 km in a circle of radius 6570 km. Find the speed of the satellite and the time taken to complete one revolution. Assume the earth's mass is 6.0×10^{24} kg. (*Hint*: The gravitational force provides the centripetal force.) *Ans.* 7800 m/s, 88 min

9.36. A roller coaster is just barely moving as it goes over the top of one hill. It rolls nearly frictionless down the hill and then up over a lower hill that has a radius of curvature of 15 m. How much higher must the first hill be than the second if the passengers are to exert no force on the seat as they pass over the top of the lower hill? *Ans.* 7.5 m

9.37. The human body can safely stand an acceleration 9 times that due to gravity. With what minimum radius of curvature may a pilot safely turn the plane upward at the end of a dive if the plane's speed is 770 km/h? *Ans.* 519 m

9.38. A 60 kg glider pilot traveling in a glider at 40 m/s wishes to turn an inside vertical loop such that he exerts a force of 350 N on the seat when the glider is at the top of the loop. What must be the radius of the loop under these conditions? (*Hint*: Gravity and the seat exert forces on the pilot.) *Ans.* 102 m

9.39. Suppose the earth is a perfect sphere with $R = 6370$ km. If a person weighs exactly 600.0 N at the north pole, how much will the person weigh at the equator? (*Hint*: The upward push of the scale on the person is what the scale will read and is what we are calling the weight in this case.) *Ans.* 597.9 N

Chapter 10

Rigid Body Rotation

THE TORQUE (OR MOMENT) due to a force about an axis was defined in Chapter 3.

THE MOMENT OF INERTIA (I) of a body is a measure of the rotational inertia of the body. If an object that is free to rotate about an axis is difficult to set into rotation, its moment of inertia about that axis is large. An object with small I has little rotational inertia.

If a body is considered to be made up of tiny masses $m_1, m_2, m_3, \ldots$ at respective distances $r_1, r_2, r_3, \ldots$ from an axis, its moment of inertia about that axis is

$$I = m_1 r_1^2 + m_2 r_2^2 + m_3 r_3^2 + \cdots = \sum m_i r_i^2$$

The units of I are kg·m² or slug·ft².

It is convenient to define a *radius of gyration* (k) for an object about an axis by the relation

$$I = Mk^2$$

where M is the total mass of the object. Hence k is the distance from the axis that a point mass M must be placed if the mass is to have the same I as the actual body does.

TORQUE AND ANGULAR ACCELERATION: An unbalanced torque τ, acting on a body of moment of inertia I, produces in it an angular acceleration α given by

$$\tau = I\alpha$$

Here, τ, I and α are all computed with respect to the same axis. As for units:

$$\tau\ (\text{N·m}) = I(\text{kg·m}^2) \times \alpha\ (\text{rad/s}^2)$$

$$\tau\ (\text{lb·ft}) = I(\text{slug·ft}^2) \times \alpha(\text{rad/s}^2)$$

Note that α *must be in rad/s²*.

KINETIC ENERGY OF ROTATION (KE_r) of a mass whose moment of inertia about an axis is I, and which is rotating about this axis with angular velocity ω, is

$$\text{KE}_r = \tfrac{1}{2} I\omega^2$$

where the energy is in J (or ft·lb) and ω *must be in rad/s*.

COMBINED ROTATION AND TRANSLATION: The KE of a rolling ball or other object is the sum of (1) its rotational KE about an *axis through its center of mass* (Chapter 6) and (2) the translational KE of an equivalent point mass moving with the center of mass. In symbols,

$$\text{KE}_{\text{total}} = \tfrac{1}{2} I\omega^2 + \tfrac{1}{2} Mv^2$$

Note however the restriction as to the axis that must be used.

WORK (W) done on a rotating body during an angular displacement θ by a constant torque τ is given by

$$W = \tau\theta$$

where W is in J (or ft·lb) and θ *must be in radians*.

POWER (P) transmitted to a body by a torque is given by

$$P = \tau\omega$$

where τ is the applied torque and ω is the angular velocity in the same rotation direction as the torque. Radian measure must be used for ω.

ANGULAR MOMENTUM, a vector quantity, has magnitude $I\omega$ and is directed along the axis of rotation. If the net torque on a body is zero, its angular momentum will remain unchanged in both magnitude and direction. This is the *law of conservation of angular momentum*.

ANGULAR IMPULSE has magnitude τt, where t is the time for which the constant torque τ acts on the object. In analogy to the linear case, an angular impulse τt on a body causes a change in angular momentum of the body given by

$$\tau t = I\omega_f - I\omega_0$$

PARALLEL-AXIS THEOREM: The moment of inertia I of a body about an axis parallel to an axis through the center of mass is

$$I = I_c + Mh^2$$

where I_c = moment of inertia about an axis through the mass center
          M = total mass of the body
          h = perpendicular distance between the two parallel axes

ANALOGOUS LINEAR AND ANGULAR QUANTITIES:

Linear displacement	s	$\longleftrightarrow$	Angular displacement	θ
Linear speed	v	$\longleftrightarrow$	Angular speed	ω
Linear acceleration	a	$\longleftrightarrow$	Angular acceleration	α
Mass (inertia)	m	$\longleftrightarrow$	Moment of inertia	I
Force	F	$\longleftrightarrow$	Torque	τ
Linear momentum	mv	$\longleftrightarrow$	Angular momentum	$I\omega$
Linear impulse	Ft	$\longleftrightarrow$	Angular impulse	τt

If, in the equations for linear motion, we replace the linear quantities by the corresponding angular quantities, we get the corresponding equations for angular motion. Thus:

Linear: $F = ma$ $\mathrm{KE} = \frac{1}{2}mv^2$ work $= Fs$ power $= Fv$

Angular: $\tau = I\alpha$ $\mathrm{KE} = \frac{1}{2}I\omega^2$ work $= \tau\theta$ power $= \tau\omega$

In these equations, θ, ω, and α must be expressed in radian measure.

MOMENTS OF INERTIA OF SYMMETRICAL BODIES: The following are expressions for the moments of inertia of various symmetrical bodies about an axis through the center of mass.

$I = Mr^2$ for a small mass M at a distance r from the axis of rotation. Also for a thin ring, hollow cylinder, or hoop, about its central axis, from which all the mass particles are at a distance r.

$I = \frac{1}{2}Mr^2$ for a uniform solid cylinder or disk, of any length and of mass M and radius r, about an axis through its center and perpendicular to the flat faces.

$I = \dfrac{1}{12}Ml^2$ for a thin uniform rod, of mass M and length l, about a transverse axis through its center.

$I = \dfrac{1}{12}M(l^2 + b^2)$ for a uniform rectangular body, of mass M, length l, width b, and arbitrary depth, about an axis through its center and perpendicular to the bl-faces.

$I = \dfrac{2}{5}Mr^2$ for a uniform solid sphere, of mass M and radius r, about any diameter.

Solved Problems

10.1. A wheel of radius 30 cm has forces applied to it as shown in Fig. 10-1. Find the torque produced by the force of (*a*) 4 N, (*b*) 9 N, (*c*) 7 N, (*d*) 6 N.

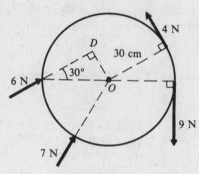

Fig. 10-1

In each case, $\tau = (\text{force}) \times (\text{lever arm})$, where the lever arm is the length of the perpendicular dropped from O to the line of the force.

(*a*) $\tau_4 = (4\ \text{N})(0.30\ \text{m}) = 1.20\ \text{N} \cdot \text{m}$ counterclockwise

(*b*) $\tau_9 = (9\ \text{N})(0.30\ \text{m}) = 2.70\ \text{N} \cdot \text{m}$ clockwise

(*c*) $\tau_7 = (7\ \text{N})(0\ \text{m}) = 0\ \text{N} \cdot \text{m}$

(*d*) $\tau_6 = (6\ \text{N})(\overline{OD}) = (6\ \text{N})(0.15\ \text{m}) = 0.90\ \text{N} \cdot \text{m}$ clockwise

10.2. A wheel, of mass 6 kg and radius of gyration 40 cm, is rotating at 300 rpm. Find its moment of inertia and its rotational KE.

$$I = Mk^2 = (6\ \text{kg})(0.40\ \text{m})^2 = 0.96\ \text{kg} \cdot \text{m}^2$$

The rotational KE is $\frac{1}{2}I\omega^2$, where ω must be in rad/s.

$$\omega = \left(300\ \frac{\text{rev}}{\text{min}}\right)\left(\frac{1\ \text{min}}{60\ \text{s}}\right)\left(\frac{2\pi\ \text{rad}}{1\ \text{rev}}\right) = 31.4\ \text{rad/s}$$

so

$$\text{KE}_r = \tfrac{1}{2}I\omega^2 = \tfrac{1}{2}(0.96\ \text{kg} \cdot \text{m}^2)(31.4\ \text{rad/s})^2 = 473\ \text{J}$$

10.3. An airplane propeller has a mass of 70 kg and a radius of gyration of 75 cm. Find its moment of inertia. How large an unbalanced torque is needed to give it an angular acceleration of 4 rev/s^2?

$$I = Mk^2 = (70\ \text{kg})(0.75\ \text{m})^2 = 39\ \text{kg} \cdot \text{m}^2$$

To use $\tau = I\alpha$, we must have α in rad/s^2.

$$\alpha = \left(4\ \frac{\text{rev}}{\text{s}^2}\right)\left(2\pi\ \frac{\text{rad}}{\text{rev}}\right) = 8\pi\ \text{rad/s}^2$$

Then

$$\tau = I\alpha = (39\ \text{kg} \cdot \text{m}^2)(8\pi\ \text{rad/s}^2) = 990\ \text{N} \cdot \text{m}$$

10.4. Repeat Problem 10.3 if the propeller weighs 144 lb and has $k = 2$ ft.

$$I = Mk^2 = (144/32)\ \text{slugs} \times 4\ \text{ft}^2 = 18\ \text{slug} \cdot \text{ft}^2$$

and

$$\tau = I\alpha = (18\ \text{slug} \cdot \text{ft}^2)(8\pi\ \text{rad/s}^2) = 450\ \text{lb} \cdot \text{ft}$$

10.5. An 8 kg uniform solid disk has a radius of 20 cm and spins on an axis through its center and perpendicular to its plane. Find its moment of inertia and its radius of gyration.

The moment of inertia of such a disk is

$$I = \tfrac{1}{2}Mr^2 = \tfrac{1}{2}(8 \text{ kg})(0.20 \text{ m})^2 = 0.16 \text{ kg} \cdot \text{m}^2$$

Because $I = Mk^2$, we have

$$k = \sqrt{I/M} = \sqrt{(0.16 \text{ kg} \cdot \text{m}^2)/(8 \text{ kg})} = 0.14 \text{ m}$$

10.6. As shown in Fig. 10-2, a constant force of 40 N is applied tangentially to the rim of a wheel 20 cm in radius. The wheel has a moment of inertia of 30 kg·m². Find (a) the angular acceleration, (b) the angular speed after 4 s from rest, (c) the number of revolutions made in that 4 s. (d) Show that the work done on the wheel in the 4 s is equal to the KE$_r$ of the wheel after 4 s.

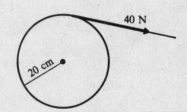

Fig. 10-2

(a) Using $\tau = I\alpha$, we have

$$(40 \text{ N})(0.20 \text{ m}) = (30 \text{ kg} \cdot \text{m}^2)\alpha$$

from which $\alpha = 0.267 \text{ rad/s}^2$. Notice that α will be in radian measure.

(b) Use $\omega_f = \omega_0 + \alpha t$ to find

$$\omega_f = 0 + (0.267 \text{ rad/s}^2)(4 \text{ s}) = 1.07 \text{ rad/s}$$

(c) Because $\theta = \bar{\omega}t = \tfrac{1}{2}(\omega_f + \omega_0)t$, we have

$$\theta = \tfrac{1}{2}(1.07 \text{ rad/s})(4 \text{ s}) = 2.14 \text{ rad} = 0.34 \text{ rev}$$

(d) We know that work = (torque) × (θ) and so

$$\text{work} = (40 \text{ N} \times 0.20 \text{ m})(2.14 \text{ rad}) = 17.1 \text{ J}$$

Notice that radian measure must be used. The final KE$_r$ is $\tfrac{1}{2}I\omega_f^2$ and so

$$\text{KE}_r = \tfrac{1}{2}(30 \text{ kg} \cdot \text{m}^2)(1.07 \text{ rad/s})^2 = 17.1 \text{ J}$$

The work done equals KE$_r$.

10.7. The wheel on a grinder is a uniform 0.9 kg disk of 8 cm radius. It coasts uniformly to rest from a speed of 1400 rpm in a time of 35 s. How large a friction torque slows its motion?

We will first find α from a motion problem; then we will use $\tau = I\alpha$ to find τ. We know that

$$\omega_0 = 1400 \text{ rev/min} = 23.3 \text{ rev/s} = 146 \text{ rad/s}$$

and $\omega_f = 0$. Therefore,

$$\alpha = \frac{\omega_f - \omega_0}{t} = \frac{-146 \text{ rad/s}}{35 \text{ s}} = -4.2 \text{ rad/s}^2$$

We also need I. For a uniform disk,

$$I = \tfrac{1}{2}Mr^2 = \tfrac{1}{2}(0.9 \text{ kg})(0.08 \text{ m})^2 = 0.0029 \text{ kg} \cdot \text{m}^2$$

Then $\tau = I\alpha = (0.0029 \text{ kg} \cdot \text{m}^2)(-4.2 \text{ rad/s}^2) = -0.0121 \text{ N} \cdot \text{m}$

10.8. Rework Problem 10.7 using the relation between work and energy.

The wheel originally had KE$_r$, but, as the wheel slowed, this energy was lost doing friction work. We therefore write

$$\text{initial KE}_r = \text{work done against friction torque}$$

$$\tfrac{1}{2}I\omega_0^2 = \tau\theta$$

To find θ, we note that

$$\theta = \bar{\omega}t = \tfrac{1}{2}(\omega_0 + \omega_f)t = \tfrac{1}{2}(146 \text{ rad/s})(35 \text{ s}) = 2550 \text{ rad}$$

From Problem 10.7, $I = 0.0029 \text{ kg} \cdot \text{m}^2$ and so the work-energy equation is

$$\tfrac{1}{2}(0.0029 \text{ kg} \cdot \text{m}^2)(146 \text{ rad/s})^2 = \tau(2550 \text{ rad})$$

N

from which $\tau = 0.0121 \text{ N} \cdot \text{m}$.

10.9.   A flywheel has a moment of inertia of $8.9 \text{ slug} \cdot \text{ft}^2$. What constant unbalanced torque is required to increase its speed from 2 rev/s to 5 rev/s in 6 revolutions?

Given

$$\theta = 6 \text{ rev} = 12\pi \text{ rad} \qquad \omega_0 = 2 \text{ rev/s} = 4\pi \text{ rad/s} \qquad \omega_f = 5 \text{ rev/s} = 10\pi \text{ rad/s}$$

we can write

$$\text{work done on wheel} = \text{change in KE}_r \text{ of wheel}$$

$$\tau\theta = \tfrac{1}{2}I\omega_f^2 - \tfrac{1}{2}I\omega_0^2$$

$$\tau(12\pi \text{ rad}) = \tfrac{1}{2}(8.9 \text{ slug} \cdot ft^2)\big[(100\pi^2 - 16\pi^2)(\text{rad/s})^2\big]$$

which gives $\tau = 98 \text{ lb} \cdot \text{ft}$.   Notice in all of these problems that radians and seconds must be used.

10.10.   As shown in Fig. 10-3, a mass $m = 400 \text{ g}$ hangs from the rim of a wheel of radius $r = 15 \text{ cm}$.   When released from rest, the mass falls 2.0 m in 6.5 s.   Find the moment of inertia of the wheel.

We will write $\tau = I\alpha$ for the wheel and $F = ma$ for the mass. But first we find a from a motion problem, using $y = v_0 t + \tfrac{1}{2}at^2$:

$$2.0 \text{ m} = 0 + \tfrac{1}{2}a(6.5 \text{ s})^2$$

which gives $a = 0.095 \text{ m/s}^2$.   Then, from $a = \alpha r$,

$$\alpha = \frac{a}{r} = \frac{0.095 \text{ m/s}^2}{0.15 \text{ m}} = 0.63 \text{ rad/s}^2$$

The unbalanced force on the mass m is $mg - T$ and so $F = ma$ becomes

$$mg - T = ma$$

$$(0.40 \text{ kg})(9.8 \text{ m/s}^2) - T = (0.40 \text{ kg})(0.095 \text{ m/s}^2)$$

from which $T = 3.88 \text{ N}$.

Now we write $\tau = I\alpha$ for the wheel:

$$(T)(r) = I\alpha \qquad \text{or} \qquad (3.88 \text{ N})(0.15 \text{ m}) = I(0.63 \text{ rad/s}^2)$$

from which $I = 0.92 \text{ kg} \cdot \text{m}^2$.

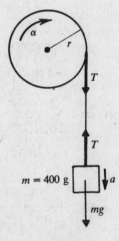

Fig. 10-3

10.11.   Repeat Problem 10.10 using energy considerations.

Originally the mass m had $\text{GPE} = mgh$,   where $h = 2 \text{ m}$.   It loses all this GPE, and an equal amount of KE results.   Part of this KE is translational KE of the mass, and the rest is KE$_r$ of the wheel.

$$\text{original GPE} = \text{final KE of } m + \text{final KE}_r \text{ of wheel}$$

$$mgh = \tfrac{1}{2}mv_f^2 + \tfrac{1}{2}I\omega_f^2$$

To find v_f, we note that $v_0 = 0$, $y = 2 \text{ m}$,   and $t = 6.5 \text{ s}$. (Observe that $a \neq g$ for the mass, because it does not fall freely.)   Then

$$\bar{v} = \frac{y}{t} = \frac{2 \text{ m}}{6.5 \text{ s}} = 0.308 \text{ m/s}$$

and $\bar{v} = \tfrac{1}{2}(v_0 + v_f)$ with $v_0 = 0$ gives

$$v_f = 2\bar{v} = 0.616 \text{ m/s}$$

Moreover, $v = \omega r$ gives

$$\omega_f = \frac{v_f}{r} = \frac{0.616 \text{ m/s}}{0.15 \text{ m}} = 4.1 \text{ rad/s}$$

Substitution in the energy equation gives

$$(0.4 \text{ kg})(9.8 \text{ m/s}^2)(2 \text{ m}) = \tfrac{1}{2}(0.4 \text{ kg})(0.62 \text{ m/s})^2 + \tfrac{1}{2}I(4.1 \text{ rad/s})^2$$

from which $I = 0.92 \text{ kg} \cdot \text{m}^2$.

10.12. Refer to Fig. 10-4. The moment of inertia of the pulley system is $I = 1.70 \text{ kg} \cdot \text{m}^2$, while $r_1 = 50$ cm and $r_2 = 20$ cm. Find the angular acceleration of the pulley system and the tensions T_1 and T_2.

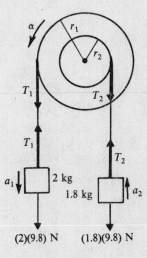

Note at the beginning that $a = \alpha r$ gives $a_1 = (0.50 \text{ m})\alpha$ and $a_2 = (0.20 \text{ m})\alpha$. We shall write $F = ma$ for both masses and also $\tau = I\alpha$ for the wheel, taking the direction of motion to be the positive direction:

$$(2)(9.8) \text{ N} - T_1 = 2a_1 \qquad\qquad 19.6 \text{ N} - T_1 = 1.0\,\alpha$$

$$T_2 - (1.8)(9.8) \text{ N} = 1.8\,a_2 \qquad \text{or} \qquad T_2 - 17.6 \text{ N} = 0.36\,\alpha$$

$$(T_1)(r_1) - (T_2)(r_2) = I\alpha \qquad\qquad 0.5\,T_1 - 0.2\,T_2 = 1.70\,\alpha$$

These three equations have three unknowns. Solve for T_1 in the first equation and substitute it in the third to obtain

$$9.8 - 0.5\,\alpha - 0.2\,T_2 = 1.70\,\alpha$$

Solve this equation for T_2 and substitute in the second equation to obtain

$$-11\alpha + 49 - 17.6 = 0.36\,\alpha$$

from which $\alpha = 2.76 \text{ rad/s}^2$.

We can now go back to the first equation to find $T_1 = 16.8$ N, and to the second to find $T_2 = 18.6$ N.

Fig. 10-4

10.13. A motor runs at 20 rev/s and supplies a torque of 75 N $\cdot$ m. What horsepower is it delivering?

Using $\omega = 20 \text{ rev/s} = 40\pi \text{ rad/s}$,

$$\text{power} = \tau\omega = (75 \text{ N} \cdot \text{m})(40\pi \text{ rad/s}) = 9400 \text{ W} = 12.6 \text{ hp}$$

10.14. The driving wheel of a belt drive attached to an electric motor has a diameter of 38 cm and makes 1200 rpm. The tension in the belt is 130 N on the slack side and 600 N on the tight side. Find the horsepower transmitted to the wheel by the belt.

We make use of power $= \tau\omega$. In this case two torques, due to the two parts of the belt, act on the wheel. We have

$$\omega = (1200 \text{ rev/min}) = 20 \text{ rev/s} = 40\pi \text{ rad/s}$$

and so

$$\text{power} = (600 - 130)(0.19) \text{ N} \cdot \text{m} \times 40\pi \text{ rad/s} = 11\,200 \text{ W} = 15.0 \text{ hp}$$

10.15. A 0.75 hp motor acts for 8 s on an initially nonrotating wheel having a moment of inertia 2.0 kg $\cdot$ m^2. Find the angular speed developed in the wheel, assuming no losses.

$$\text{work done by motor in 8 s} = \text{KE of wheel after 8 s}$$

$$(\text{power}) \times (\text{time}) = \tfrac{1}{2}I\omega^2$$

$$(0.75 \text{ hp})(746 \text{ W/hp}) \times (8 \text{ s}) = \tfrac{1}{2}(2 \text{ kg} \cdot \text{m}^2)\omega^2$$

from which $\omega = 67 \text{ rad/s}$.

10.16. As shown in Fig. 10-5, a uniform solid sphere rolls on a horizontal surface at 20 m/s. It then rolls up the incline shown. If friction losses are negligible, what will be the value of h where the ball stops?

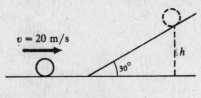

Fig. 10-5

The rotational and translational KE of the ball at the bottom will be changed to GPE when the sphere stops. We therefore write

$$\left(\tfrac{1}{2}Mv^2 + \tfrac{1}{2}I\omega^2\right)_{\text{start}} = (Mgh)_{\text{end}}$$

But for a solid sphere, $I = (2/5)Mr^2$. Also, $\omega = v/r$. The equation becomes

$$\tfrac{1}{2}Mv^2 + \tfrac{1}{2}\left(\tfrac{2}{5}\right)(Mr^2)\left(\tfrac{v}{r}\right)^2 = Mgh \qquad \text{or} \qquad \tfrac{1}{2}v^2 + \tfrac{1}{5}v^2 = (9.8 \text{ m/s}^2)h$$

Using $v = 20$ m/s gives $h = 28.6$ m. Notice that the answer does not depend upon the mass of the ball or the angle of the incline.

10.17. As a solid disk rolls over the top of a hill on a track, its speed is 80 cm/s. If friction losses are negligible, how fast is the disk moving when it is 18 cm below the top?

At the top, the disk has translational and rotational KE, plus its GPE relative to the point 18 cm lower. At the final point, the GPE has been transformed to more KE of rotation and translation. We therefore write, with $h = 18$ cm,

$$(KE_t + KE_r)_{\text{start}} + Mgh = (KE_t + KE_r)_{\text{end}}$$

$$\tfrac{1}{2}Mv_0^2 + \tfrac{1}{2}I\omega_0^2 + Mgh = \tfrac{1}{2}Mv_f^2 + \tfrac{1}{2}I\omega_f^2$$

For a solid disk, $I = \tfrac{1}{2}Mr^2$. Also, $\omega = v/r$. Substituting these values and simplifying gives

$$\tfrac{1}{2}v_0^2 + \tfrac{1}{4}v_0^2 + gh = \tfrac{1}{2}v_f^2 + \tfrac{1}{4}v_f^2$$

But $v_0 = 0.80$ m/s and $h = 0.18$ m. Substitution gives $v_f = 1.73$ m/s.

10.18. Starting from rest, a hoop of 20 inch radius rolls down a hill to a point 50 ft below its starting point. How fast is it rotating at that point?

$$\text{GPE at start} = (KE_r + KE_t) \text{ at end}$$

$$Mgh = \tfrac{1}{2}I\omega^2 + \tfrac{1}{2}Mv^2$$

But $I = Mr^2$ for a hoop and $v = \omega r$. The equation becomes

$$Mgh = \tfrac{1}{2}M\omega^2 r^2 + \tfrac{1}{2}M\omega^2 r^2$$

from which

$$\omega = \sqrt{\frac{gh}{r^2}} = \sqrt{\frac{(32 \text{ ft/s}^2)(50 \text{ ft})}{\left(\frac{20}{12} \text{ ft}\right)^2}} = 24 \text{ rad/s}$$

10.19. Find the radius of gyration of a solid sphere, of diameter 6 cm, rotating about a diameter as axis.

For a solid sphere rotating about a diameter, $I = (2/5)Mr^2$. Then

$$I = Mk^2 = \tfrac{2}{5}Mr^2 \qquad \text{or} \qquad k = r\sqrt{\tfrac{2}{5}} = 1.90 \text{ cm}$$

10.20. As shown in Fig. 10-6, three masses (0.5 kg each) are fastened to a very light rod pivoted at one end. Find the moment of inertia for the rotation axis shown. Assume that the masses are much smaller geometrically than shown.

$$I = m_1 r_1^2 + m_2 r_2^2 + \cdots = m(0.10 \text{ m})^2 + m(0.22 \text{ m})^2 + m(0.42 \text{ m})^2$$

$$= (0.235 \text{ m}^2)m = (0.235 \text{ m}^2)(0.5 \text{ kg}) = 0.117 \text{ kg} \cdot \text{m}^2$$

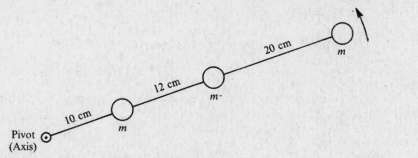

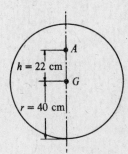

Fig. 10-6   Fig. 10-7

10.21. The uniform circular disk shown in Fig. 10-7 has mass 6.5 kg and diameter 80 cm. Compute its moment of inertia about (a) an axis perpendicular to the page through G, (b) an axis perpendicular to the page through A.

(a) $$I_G = \tfrac{1}{2} Mr^2 = \tfrac{1}{2}(6.5 \text{ kg})(0.40 \text{ m})^2 = 0.52 \text{ kg} \cdot \text{m}^2$$

(b) Use the result of (a) and the parallel-axis theorem.

$$I_A = I_G + Mh^2 = 0.52 \text{ kg} \cdot \text{m}^2 + (6.5 \text{ kg})(0.22 \text{ m})^2 = 0.83 \text{ kg} \cdot \text{m}^2$$

10.22. As shown in Fig. 10-8, a thin, uniform rod AB of mass M and length l is hinged at end A to the level floor. It originally stands vertically. If allowed to fall to the floor as shown, with what angular speed will it strike the floor?

The moment of inertia about a transverse axis through end A is

$$I_A = I_G + Mh^2 = \frac{1}{12} Ml^2 + M\left(\frac{l}{2}\right)^2 = \frac{Ml^2}{3}$$

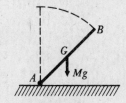

When the rod falls down, the center of mass, G, falls a distance $l/2$. We can write

GPE lost by rod = KE_r gained by rod

$$Mg\left(\frac{l}{2}\right) = \frac{1}{2}\left(\frac{Ml^2}{3}\right)\omega^2$$

Fig. 10-8

from which $\omega = \sqrt{3g/l}$.

10.23. A boy stands on a freely rotating platform, as shown in Fig. 10-9. With his arms extended, his rotation speed is 0.25 rev/s. But when he draws them in, his speed is 0.80 rev/s. Find the ratio of his moment of inertia in the first case to that in the second.

Because there is no net torque on the system (why?), the law of conservation of angular momentum tells us that

angular momentum before = angular momentum after

$$I_0 \omega_0 = I_f \omega_f$$

Fig. 10-9

Or, since we desire I_0/I_f,

$$\frac{I_0}{I_f} = \frac{\omega_f}{\omega_0} = \frac{0.80 \text{ rev/s}}{0.25 \text{ rev/s}} = 3.2$$

10.24. A disk of moment of inertia I_1 is rotating freely with angular speed ω_1 when a second, nonrotating, disk with moment of inertia I_2 is dropped on it (Fig. 10-10). The two then rotate as a unit. Find the final angular speed.

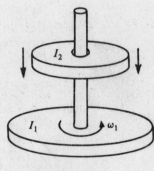

From the law of conservation of angular momentum,

angular momentum before = angular momentum after

$$I_1\omega_1 + I_2(0) = I_1\omega + I_2\omega$$

Solving gives

$$\omega = \frac{I_1\omega_1}{I_1 + I_2}$$

Fig. 10-10

Supplementary Problems

10.25. (a) A force of 200 N acts tangentially on the rim of a wheel 25 cm in radius. Find the torque. (b) Repeat if the force makes an angle of 40° to a spoke of the wheel.
Ans. (a) 50 N·m; (b) 32 N·m

10.26. A certain 8 kg wheel has a radius of gyration of 25 cm. (a) What is its moment of inertia? (b) How large a torque is required to give it an angular acceleration of 3 rad/s²?
Ans. (a) 0.50 kg·m²; (b) 1.5 N·m

10.27.   Determine the constant torque that must be applied to a 50 kg flywheel, of radius of gyration 40 cm, to give it an angular speed of 300 rpm in 10 s.   *Ans.*   25 N · m

10.28.   An 80 lb wheel of 2 ft radius of gyration is rotating at 360 rpm.   The retarding frictional torque is 4 lb · ft.   Compute the time it will take the wheel to coast to rest.   *Ans.*   94 s

10.29.   Compute the rotational KE of a 25 kg wheel rotating at 6 rev/s if the radius of gyration of the wheel is 22 cm.   *Ans.*   860 J

10.30.   A cord 3 m long is coiled around the axle of a wheel.   The cord is pulled with a constant force of 40 N. When the cord leaves the axle, the wheel is rotating at 2 rev/s.   Determine the moment of inertia of the wheel and axle.   Neglect friction.   (*Hint*: Easiest solution is by energy method.)   *Ans.*   1.52 kg · m^2

10.31.   A 500 g wheel that has a moment of inertia of 0.015 kg · m^2 is initially turning at 30 rev/s.   It coasts to rest after 163 rev.   How large is the torque that slowed it?   *Ans.*   0.26 N · m

10.32.   When 100 J of work is done upon a flywheel, its angular speed increases from 60 rev/min to 180 rev/min.   What is its moment of inertia?   *Ans.*   0.63 kg · m^2

10.33.   A 72 lb wheel, radius of gyration 9 inches, is to be given a speed of 10 rev/s in 25 revolutions from rest. Find the constant unbalanced torque required.   *Ans.*   15.8 lb · ft

10.34.   An electric motor runs at 900 rpm and delivers 2 hp.   How much torque does it deliver?
Ans.   15.8 N · m

10.35.   The driving side of a belt has a tension of 1600 N and the slack side has 500 N tension.   The belt turns a pulley 40 cm in radius at a rate of 300 rpm.   This pulley drives a dynamo having 90% efficiency. How many kilowatts are being delivered by the dynamo?   *Ans.*   12.4 kW

10.36.   A 25 kg wheel has a radius of 40 cm and turns freely on a horizontal axis.   The radius of gyration of the wheel is 30 cm.   A 1.2 kg mass hangs at the end of a cord that is wound around the rim of the wheel. This mass falls and causes the wheel to rotate.   Find the acceleration of the falling mass and the tension in the cord.   *Ans.*   0.77 m/s^2, 10.8 N

10.37.   A wheel and axle, having a total moment of inertia 0.002 kg · m^2, is caused to rotate about a horizontal axis by means of an 800 g mass attached to a cord wrapped around the axle.   The radius of the axle is 2 cm.   Starting from rest, how far must the mass fall in order to give the wheel a speed of 3 rev/s? *Ans.*   5.25 cm

10.38.   A 20 kg solid disk ($I = \frac{1}{2} Mr^2$) rolls on a horizontal surface at a rate of 4 m/s.   Compute its total KE. *Ans.*   240 J

10.39.   A 6 kg bowling ball ($I = 2Mr^2/5$) starts from rest and rolls down a gradual slope until it reaches a point 80 cm lower than its starting point.   How fast is it then moving?   Ignore friction losses.
Ans.   3.35 m/s

10.40.   A tiny solid ball ($I = 2Mr^2/5$) rolls without slipping on the inside surface of a hemisphere as shown in Fig. 10-11.   (The ball is much smaller than shown.)   If the ball is released at *A*, how fast is it moving as it passes (*a*) point *B*?   (*b*) point *C*?   *Ans.*   (*a*) 2.65 m/s;   (*b*) 2.32 m/s

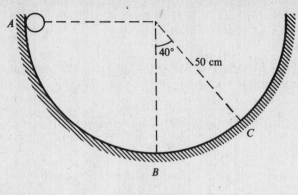

Fig. 10-11

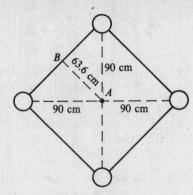

Fig. 10-12

10.41. Compute the radius of gyration of a solid disk, diameter 24 cm, about an axis through its center of mass and perpendicular to its face.   *Ans.*   8.49 cm

10.42. In Fig. 10-12 are shown four masses, each 200 g, held at the corners of a square by a very light frame. What is the moment of inertia of the system about an axis perpendicular to the page (*a*) through *A*? (*b*) through *B*?   *Ans.*   (*a*) 0.648 kg·m^2;   (*b*) 0.972 kg·m^2

10.43. Determine the moment of inertia (*a*) of a vertical thin hoop, mass 2 kg and radius 9 cm, about a horizontal, parallel axis at its rim; (*b*) of a solid sphere, mass 2 kg and radius 5 cm, about an axis tangent to the sphere.
Ans.   (*a*) I = $Mr^2 + Mr^2$ = 0.0324 kg·m^2;   (*b*) I = $(2/5)Mr^2 + Mr^2$ = 7×10^{-3} kg·m^2

10.44. The rod *OA* in Fig. 10-13 is a meterstick. It is hinged at *O* so that it can turn in a vertical plane. It is held horizontally and then released. Compute the angular speed of the rod and the linear speed of its free end as it passes through the position shown in the figure. (*Hint*: Show that $I = ml^2/3$.)
Ans.   5.05 rad/s, 5.05 m/s

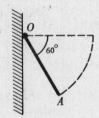

Fig. 10-13

10.45. Suppose that a satellite ship orbits the moon in an elliptical orbit. At its closest point to the moon it has a speed v_c and a radius r_c from the center of the moon. At its farthest point, it has a speed v_f and a radius r_f. Find the ratio v_c/v_f. (*Hint*: The moment of inertia of the ship about an axis through the moon's center is mr^2, where *m* is the mass of the satellite.)   *Ans.* r_f/r_c

10.46. A large horizontal disk is rotating on a vertical axis through its center; for the disk, $I = 4000$ kg·m^2. The disk is coasting at a rate of 0.150 rev/s when a 90 kg person drops onto the disk from an overhanging tree limb. The person lands and remains at a distance of 3 m from the axis of rotation. What will be the rotation speed after the person has landed?   *Ans.*   0.125 rev/s

Chapter 11

Simple Harmonic Motion and Springs

THE PERIOD (T) of a vibrating system is the time required for the system to complete one full cycle of vibration.   It is the total time for the combined back and forth motion.

THE FREQUENCY (f) is the number of vibrations made per unit time.   Because T is the time for one vibration, $f = 1/T$.   One cycle/s is one *hertz* (Hz).

THE GRAPH OF A VIBRATORY MOTION is shown in Fig. 11-1.   The motion shown there is the up-and-down motion of a mass at the end of a spring.   One complete cycle is from *a* to *b*, or from *c* to *d*, or from *e* to *f*.   The time taken for one cycle is T, the period.

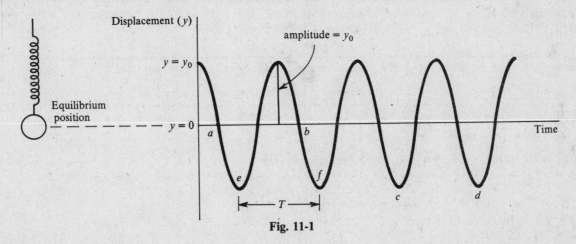

Fig. 11-1

THE DISPLACEMENT (x or y) is the distance of the vibrating object from its equilibrium position (normal rest position), i.e. from the center of its vibration path.   The maximum displacement is called the *amplitude* (see Fig. 11-1).

A RESTORING FORCE is necessary if vibration is to occur.   It is a force that is always directed so as to push or pull the system back to its equilibrium (normal rest) position.   For a mass at the end of a spring, the stretched spring pulls the mass back toward the equilibrium position, while the compressed spring pushes the mass back toward the equilibrium position.

HOOKE'S LAW is obeyed by a system if the magnitude of the restoring force is proportional to the magnitude of the displacement (or distortion).

SIMPLE HARMONIC MOTION (SHM) is the vibratory motion of a system that obeys Hooke's law. The motion illustrated in Fig. 11-1 is SHM.   Because of its resemblance to a sine or cosine curve, SHM is frequently called *sinusoidal motion*.

A HOOKEAN SPRING obeys Hooke's law. When such a spring is stretched a distance x (for a compressed spring, x is negative), the restoring force exerted by the spring is given by

$$F = -kx$$

where k is a positive constant called the *spring constant*. The units of k are N/m or lb/ft and it is a measure of the stiffness of the spring. Most springs obey Hooke's law for small distortions.

It is sometimes useful to express Hooke's law in terms of F_{ext}, the external force needed to stretch the spring a given amount x. This force is the negative of the restoring force and so

$$F_{ext} = kx$$

THE POTENTIAL ENERGY stored in a Hookean spring (SPE) when distorted a distance x is $\frac{1}{2}kx^2$. If the amplitude of motion is x_0 for a mass at the end of a spring, then the energy of the vibrating system is $\frac{1}{2}kx_0^2$ at all times. However, this energy is completely stored in the spring only when $x = \pm x_0$, i.e. when the mass has its maximum displacement.

ENERGY INTERCHANGE between kinetic and potential energy occurs constantly in a vibrating system. When the system passes through its equilibrium position, KE = maximum and PE = 0. When the system has its maximum displacement, then KE = 0 and PE = maximum. From the law of conservation of energy, in the absence of friction-type losses,

$$KE + PE = \text{constant}$$

For a mass m at the end of a spring (whose own mass is negligible), this becomes

$$\tfrac{1}{2}mv^2 + \tfrac{1}{2}kx^2 = \tfrac{1}{2}kx_0^2$$

where x_0 is the amplitude of the motion.

SPEED IN SHM is given by the above energy equation as

$$|v| = \sqrt{(x_0^2 - x^2)\frac{k}{m}}$$

ACCELERATION IN SHM is given by Hooke's law, $F = -kx$, and $F = ma$. Equating these two expressions for F, we have

$$a = -\frac{k}{m}x$$

The minus sign is present because the direction of **a** (and **F**) is always opposite to the direction of the displacement **x**.

REFERENCE CIRCLE: Suppose that a point P moves with constant speed v_0 around a circle, as shown in Fig. 11-2. This circle is called the *reference circle* for SHM. Point A is the projection of point P on the x-axis, the line of the horizontal diameter of the circle. The motion of point A back and forth about point O as center is SHM. The amplitude of the motion is x_0, the radius of the circle. The time taken for P to go around the circle once is the period of the motion, T. The velocity of point A is

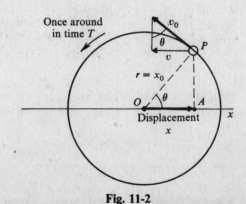

Fig. 11-2

$$v = -v_0\sin\theta$$

PERIOD IN SHM: The period T of an SHM is the time taken for point P to go around the reference circle in Fig. 11-2. Therefore

$$T = \frac{2\pi r}{v_0} = \frac{2\pi x_0}{v_0}$$

But v_0 is the maximum speed of point A in Fig. 11-2, i.e. v_0 is the value of $|v|$ in SHM when $x = 0$.

$$|v| = \sqrt{(x_0^2 - x^2)(k/m)} \qquad \text{gives} \qquad v_0 = x_0\sqrt{k/m}$$

This then gives the period of SHM to be

$$T = 2\pi\sqrt{m/k}$$

for a Hookean spring system.

ACCELERATION IN TERMS OF T: Eliminating the quantity k/m between the two equations $a = -(k/m)x$ and $T = 2\pi\sqrt{m/k}$, we find

$$a = -\frac{4\pi^2}{T^2}x$$

THE SIMPLE PENDULUM very nearly undergoes SHM if its angle of swing is not too large. The period of vibration for a pendulum of length L at a location where the gravitational acceleration is g is given by

$$T = 2\pi\sqrt{L/g}$$

ANGULAR SHM is periodic, oscillating angular motion. It occurs in systems that obey Hooke's law in the following form: the restoring torque is proportional to the angular distortion (or displacement). A typical example would be a horizontal disk suspended by a vertical wire through its center. The disk will undergo rotational oscillation about the wire as axis.

TORSION CONSTANT (K): Suppose that a disk (or some other object) suspended by a vertical wire or rod is twisted through a small angle θ, with the wire or rod as axis. Then, provided the system obeys Hooke's law, the wire or rod exerts a restoring torque τ given by

$$\tau = -K\theta$$

The positive constant K is called the *torsion constant*. The potential energy stored in a torsion spring is $\frac{1}{2}K\theta^2$.

A TORSION PENDULUM consists of an object suspended by a rod or wire whose line passes through the mass center of the object. Rotational oscillations occur with the wire as axis. If I is the moment of inertia of the object about this axis, then the period of oscillation is

$$T = 2\pi\sqrt{I/K}$$

The energy equation for the system is

$$\tfrac{1}{2}I\omega^2 + \tfrac{1}{2}K\theta^2 = \tfrac{1}{2}K\theta_0^2$$

where θ_0 is the angular amplitude. The angular acceleration is given by

$$\alpha = -\frac{K}{I}\theta$$

where the minus sign shows that the acceleration is directed so as to decrease the distortion.

Solved Problems

11.1. For the motion shown in Fig. 11-3, what are the amplitude, period, and frequency?

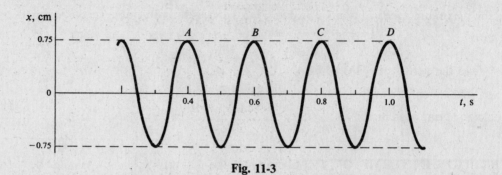

Fig. 11-3

The amplitude is the maximum displacement from the equilibrium position and so it is 0.75 cm.
The period is the time for one complete cycle, the time from A to B, for example. Therefore the period is 0.20 s. The frequency is given by

$$f = \frac{1}{T} = \frac{1}{0.20 \text{ s}} = 5 \text{ cycles/s} = 5 \text{ Hz}$$

11.2. A spring makes 12 vibrations in 40 s. Find the period and frequency of the vibration.

$$T = \frac{\text{elapsed time}}{\text{vibrations made}} = \frac{40 \text{ s}}{12} = 3.3 \text{ s}$$

$$f = \frac{\text{vibrations made}}{\text{elapsed time}} = \frac{12}{40 \text{ s}} = 0.30 \text{ Hz}$$

11.3. A 50 g mass vibrates in SHM at the end of a spring. The amplitude of the motion is 12 cm and the period is 1.70 s. Find: (a) the frequency, (b) the spring constant, (c) the maximum speed of the mass, (d) the maximum acceleration of the mass, (e) the speed when the displacement is 6 cm, (f) the acceleration when $x = 6$ cm.

(a)
$$f = \frac{1}{T} = \frac{1}{1.70 \text{ s}} = 0.588 \text{ Hz}$$

(b) Since $T = 2\pi\sqrt{m/k}$,

$$k = \frac{4\pi^2 m}{T^2} = \frac{4\pi^2(0.050 \text{ kg})}{(1.70 \text{ s})^2} = 0.68 \text{ N/m}$$

(c)
$$v_0 = x_0\sqrt{k/m} = (0.12 \text{ m})\sqrt{(0.68 \text{ N/m})/(0.050 \text{ kg})} = 0.44 \text{ m/s}$$

(d) From $a = -(k/m)x$ it is seen that a has maximum magnitude when x has maximum magnitude, i.e. at the endpoints $x = \pm x_0$. Thus,

$$a_0 = \frac{k}{m}x_0 = \frac{0.68 \text{ N/m}}{0.050 \text{ kg}}(0.12 \text{ m}) = 1.63 \text{ m/s}^2$$

(e) From $|v| = \sqrt{(x_0^2 - x^2)(k/m)}$,

$$|v| = \sqrt{(0.12^2 - 0.06^2)\frac{0.68}{0.050}} = 0.38 \text{ m/s}$$

(f)
$$a = -\frac{k}{m}x = -\frac{0.68 \text{ N/m}}{0.050 \text{ kg}}(0.06 \text{ m}) = -0.82 \text{ m/s}^2$$

11.4. A 50 g mass hangs at the end of a Hookean spring. When 20 g more is added to the end of the spring, it stretches 7.0 cm more. (*a*) Find the spring constant. (*b*) If the 20 g is now removed, what will be the period of the motion?

(*a*) Under the weight of the 50 g mass, $F_{ext\,1} = kx_1$, where x_1 is the original stretching of the spring. When 20 g more is added, the force becomes $F_{ext\,1} + F_{ext\,2} = k(x_1 + x_2)$, where $F_{ext\,2}$ is the weight of 20 g and x_2 is the stretching it causes. Subtracting the two force equations gives

$$F_{ext\,2} = kx_2$$

(Note that this is the same as $F_{ext} = kx$, where F_{ext} is the additional stretching force and x is the amount of stretch due to it. Hence we could have ignored the fact that the spring had the 50 g mass at its end to begin with.) Solving for k,

$$k = \frac{F_{ext\,2}}{x_2} = \frac{(0.020 \text{ kg})(9.8 \text{ m/s}^2)}{0.07 \text{ m}} = 2.8 \text{ N/m}$$

(*b*)
$$T = 2\pi\sqrt{\frac{m}{k}} = 2\pi\sqrt{\frac{0.050 \text{ kg}}{2.8 \text{ N/m}}} = 0.84 \text{ s}$$

11.5. A 3 lb weight hangs at the end of a spring which has $k = 25$ lb/ft. If the weight is displaced slightly and released, with what frequency will it vibrate?

Using $T = 2\pi\sqrt{m/k}$, with

$$m = \frac{3 \text{ lb}}{32.2 \text{ ft/s}^2} = 0.093 \text{ slug}$$

gives

$$T = 2\pi\sqrt{\frac{0.093 \text{ slug}}{25 \text{ lb/ft}}} = 0.38 \text{ s} \qquad \text{and} \qquad f = \frac{1}{0.38 \text{ s}} = 2.61 \text{ Hz}$$

11.6. As shown in Fig. 11-4, a long, light piece of spring steel is clamped at its lower end and a 2 kg ball is fastened to its top end. A force of 8 N is required to displace the ball 20 cm to one side as shown. Assume the system to undergo SHM when released. Find (*a*) the force constant of the spring and (*b*) the period with which the ball will vibrate back and forth.

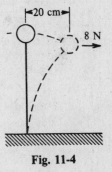

Fig. 11-4

(*a*)
$$k = \frac{\text{external force } F_{ext}}{\text{displacement } x} = \frac{8 \text{ N}}{0.20 \text{ m}} = 40 \text{ N/m}$$

(*b*)
$$T = 2\pi\sqrt{\frac{m}{k}} = 2\pi\sqrt{\frac{2 \text{ kg}}{40 \text{ N/m}}} = 1.40 \text{ s}$$

11.7. When a mass m is hung on a spring, the spring stretches 6 cm. Determine its period of vibration if it is then pulled down a little and released.

$$k = \frac{F_{ext}}{x} = \frac{mg}{0.06 \text{ m}}$$

$$T = 2\pi\sqrt{\frac{m}{k}} = 2\pi\sqrt{\frac{0.06 \text{ m}}{g}} = 0.49 \text{ s}$$

11.8. Two identical springs each have $k = 20$ N/m. A 0.3 kg mass is connected to them as shown in Fig. 11-5(*a*) and (*b*). Find the period of motion for each system. Ignore friction forces.

(*a*) Consider what happens when the mass is given a displacement $x > 0$. One spring will be stretched x and the other will be compressed x. They will each exert a force of magnitude

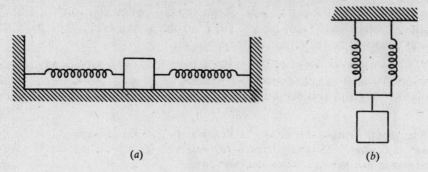

(a)   (b)

Fig. 11-5

(20 N/m)x on the mass in the direction opposite to the displacement. Hence the total restoring force is

$$F = -(20 \text{ N/m})x - (20 \text{ N/m})x = -(40 \text{ N/m})x$$

Comparing with $F = -kx$ tells us that the system has a spring constant of $k = 40$ N/m. Hence

$$T = 2\pi\sqrt{\frac{m}{k}} = 2\pi\sqrt{\frac{0.3 \text{ kg}}{40 \text{ N/m}}} = 0.54 \text{ s}$$

(b) When the mass is displaced a distance y downward, each spring is stretched a distance y. The net restoring force on the mass is then

$$F = -(20 \text{ N/m})y - (20 \text{ N/m})y = -(40 \text{ N/m})y$$

Comparison with $F = -ky$ shows k to be 40 N/m, the same as in (a). Hence the period in this case is also 0.54 s.

11.9. In a certain engine, a piston undergoes vertical SHM with amplitude 7 cm. A washer rests on top of the piston. As the motor is slowly speeded up, at what frequency will the washer no longer stay in contact with the piston?

The maximum downward acceleration of the washer will be that for free fall, g. If the piston accelerates downward faster than this, the washer will lose contact.

In SHM, the acceleration is given in terms of the displacement and the period by

$$a = -\frac{4\pi^2}{T^2}x$$

With the upward direction chosen as positive, the largest downward (most negative) acceleration occurs for $x = +x_0 = 0.07$ m, i.e.

$$a_0 = \frac{4\pi^2}{T^2}(0.07 \text{ m})$$

The washer will separate from the piston when a_0 first becomes equal to g. Therefore, the critical period of vibration, T_c, is given by

$$\frac{4\pi^2}{T_c^2}(0.07 \text{ m}) = g \qquad \text{or} \qquad T_c = 2\pi\sqrt{\frac{0.07 \text{ m}}{g}} = 0.53 \text{ s}$$

This corresponds to a frequency $f_c = 1/T_c = 1.88$ Hz. The washer will separate from the piston if the piston's frequency exceeds 1.88 vib/s.

11.10. A 40 lb electric motor is mounted on four vertical springs, each having a spring constant of 15 lb/in. Find the period with which the motor vibrates vertically.

As in Problem 11.8, we may replace the springs by an equivalent single spring. Its force constant will be 4(15 lb/in) or 60 lb/in. Then

$$T = 2\pi\sqrt{m/k} = 2\pi\sqrt{\left(\frac{40 \text{ lb}}{32.2 \text{ ft/s}^2}\right) \Big/ \left(60 \frac{\text{lb}}{\text{in}} \times \frac{12 \text{ in}}{\text{ft}}\right)} = 0.26 \text{ s}$$

11.11. Nine kilograms of mercury is poured into a glass U-tube, as shown in Fig. 11-6. The tube's inner diameter is 1.2 cm and the mercury oscillates freely up and down about its position of equilibrium ($x = 0$). Compute (a) the effective spring constant for the oscillation, and (b) the period of oscillation. One m^3 of mercury has a mass $\mu = 13\,600$ kg. Ignore frictional and surface tension effects.

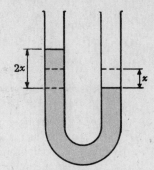

Fig. 11-6

(a) When the mercury is displaced x meters from its equilibrium position as shown, the restoring force is the weight of the unbalanced column of mercury with height $2x$. We have

$$\text{weight} = (\text{volume})(\text{weight of 1 m}^3 \text{ of mercury}) = \left[(\pi r^2)(2x)\right](\mu g)$$

Therefore Hooke's law, in the form $F = -(2\pi r^2 \mu g)x$, applies, from which we see that the effective spring constant for the system is

$$k = 2\pi r^2 \mu g = 2\pi(0.006 \text{ m})^2(13.6 \times 10^3 \text{ kg})(9.8 \text{ m/s}^2) = 30 \text{ N/m}$$

(b) The period of the vibration is

$$T = 2\pi\sqrt{\frac{m}{k}} = 2\pi\sqrt{\frac{9 \text{ kg}}{30 \text{ N/m}}} = 3.4 \text{ s}$$

11.12. Compute the acceleration due to gravity at a place where a simple pendulum 150.3 cm long makes 100.00 vibrations in 246.7 s.

$$T = \frac{246.7 \text{ s}}{100} = 2.467 \text{ s}$$

Squaring $T = 2\pi\sqrt{L/g}$ and solving for g,

$$g = (4\pi^2/T^2)L = 9.75 \text{ m/s}^2$$

11.13. The 200 g mass shown in Fig. 11-7 is pushed to the left against the spring and compresses the spring 15 cm from its equilibrium position. The system is then released and the mass shoots to the right. If friction can be ignored, how fast will the mass be moving as it shoots away? Assume the mass of the spring to be very small.

When the spring is compressed, energy is stored in it. This energy is $\frac{1}{2}kx_0^2$, in which $x_0 = 0.15$ m. After release, this energy will be given to the mass as KE. When the spring passes through the equilibrium position, all the SPE will be changed to KE. (Since the mass of the spring is small, its KE can be ignored.) Therefore

original SPE = final KE of mass

$$\frac{1}{2}kx_0^2 = \frac{1}{2}mv^2$$

$$\frac{1}{2}(400 \text{ N/m})(0.15 \text{ m})^2 = \frac{1}{2}(0.200 \text{ kg})v^2$$

from which $v = 6.7$ m/s.

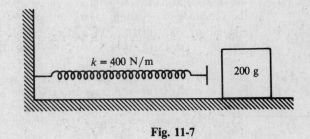

Fig. 11-7

11.14. Suppose that, in Fig. 11-7, the 200 g mass initially moves to the left at a speed of 8.0 m/s. It strikes the spring and becomes attached to it. (a) How far does it compress the spring? (b) The system then oscillates back and forth. What is the amplitude of the oscillation? Ignore friction and the small mass of the spring.

(a) Because the spring can be considered massless, all of the KE of the mass will go into compressing the spring. We can therefore write

$$\text{original KE of mass} = \text{final SPE}$$

$$\tfrac{1}{2}mv_0^2 = \tfrac{1}{2}kx_0^2$$

where $v_0 = 8$ m/s and x_0 is the maximum compression of the spring. Using $m = 0.200$ kg and $k = 400$ N/m, the above relation gives $x_0 = 0.179$ m.

(b) The spring compresses 0.179 m from its equilibrium position. At that point, all of the energy of the system is SPE. As the spring pushes the mass back toward the right, the mass moves through the equilibrium position. The mass stops at a point to the right of the equilibrium position where the energy is again all SPE. Since no losses occurred, the same energy must be stored in the stretched spring as in the compressed spring. Therefore, it will be stretched $x_0 = 0.179$ m from the equilibrium point. The amplitude of oscillation is therefore 0.179 m.

11.15. As shown in Fig. 11-8, a uniform bar is suspended in a horizontal position by a vertical wire attached to its center. When a torque of 5 N·m is applied to the bar as shown, the bar moves through an angle of 12°. If the bar is then released, it oscillates as a torsion pendulum with a period of 0.5 s. Determine its moment of inertia.

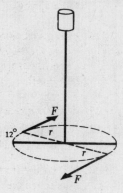

For a torsion pendulum, $T = 2\pi\sqrt{I/K}$. In our case

$$K = \frac{\text{external torque}}{\text{resulting angular displacement}} = \frac{5 \text{ N} \cdot \text{m}}{(12 \text{ deg})\left(\dfrac{2\pi \text{ rad}}{360 \text{ deg}}\right)} = 23.9 \text{ N} \cdot \text{m/rad}$$

We then have, after squaring the equation for T and rearranging,

$$I = \left(\frac{T}{2\pi}\right)^2 K = \left(\frac{0.5 \text{ s}}{2\pi}\right)^2 (23.9 \text{ N} \cdot \text{m}) = 0.151 \text{ kg} \cdot \text{m}^2$$

Fig. 11-8

Supplementary Problems

11.16. A pendulum is timed as it swings back and forth. The clock is started when the bob is at the left end of its swing. When the bob returns to the left end for the 90th return, the clock reads 60.00 s. What is the period of vibration? the frequency? *Ans.* 0.667 s, 1.50 Hz

11.17. A 300 g mass at the end of a Hookean spring vibrates up and down in such a way that it is 2 cm above the tabletop at its lowest point and 16 cm above at its highest point. Its period is 4 s. Determine (a) the amplitude of vibration, (b) the spring constant, (c) the speed and acceleration of the mass when it is 9 cm above the tabletop, (d) the speed and acceleration of the mass when it is 12 cm above the tabletop. *Ans.* (a) 7 cm; (b) 0.74 N/m; (c) 0.110 m/s, zero; (d) 0.099 m/s, 0.074 m/s²

11.18. A coiled Hookean spring is stretched 10 cm when a 1.5 kg mass is hung from it. Suppose a 4 kg mass hangs from the spring and is set into vibration with an amplitude of 12 cm. Find (a) the force constant of the spring, (b) the maximum restoring force acting on the vibrating body, (c) the period of vibration, (d) the maximum speed and the maximum acceleration of the vibrating object, (e) the speed and acceleration when the displacement is 9 cm. *Ans.* (a) 147 N/m; (b) 17.6 N; (c) 1.04 s; (d) 0.73 m/s, 4.4 m/s²; (e) 0.48 m/s, 3.3 m/s²

11.19. A 2.5 kg mass undergoes SHM and makes 3 vibrations each second. Compute the acceleration and the restoring force acting on the body when its displacement from the equilibrium position is 5 cm. *Ans.* 17.8 m/s², 44 N

11.20. A 0.70 lb object undergoes SHM at the end of a spring. If the object vibrates with a frequency of 3 Hz, what is the spring constant of the spring? What is the acceleration of the object when it is 4 inches from the equilibrium position? *Ans.* 7.7 lb/ft, 118 ft/s²

11.21. A 150 g mass, when hung on a spiral spring, stretches it 40 cm. Assuming the spring to be Hookean, what is the period of vibration if the mass is pulled down a little and released? *Ans.* 1.27 s

11.22. A 9 lb block hangs from a spiral spring of such stiffness that an added weight of 1 lb would stretch it 2 inches farther. With only the 9 lb hanging from it, what will be its frequency of vibration? *Ans.* 0.74 Hz

11.23. A certain Hookean spring is stretched 20 cm when a given mass is hung from it. What is the frequency of vibration of the mass if pulled down a little and released? *Ans.* 1.11 Hz

11.24. A 300 g mass executes SHM with a period of 2.4 s at the end of a spring. Find the period of oscillation of a 133 g mass attached to the same spring. *Ans.* 1.60 s

11.25. With a 50 g mass at its end, a spring undergoes SHM with a frequency of 0.70 vib/s. How much work is done in stretching the spring 15 cm from its unstretched length? How much energy is then stored in the spring? *Ans.* 0.0109 J, 0.0109 J

11.26. In a situation similar to that shown in Fig. 11-7, a mass is pressed back against a light spring for which $k = 400$ N/m. The mass compresses the spring 8 cm and is then released. After sliding 55 cm along the flat table from the point of release, the mass comes to rest. How large a friction force opposed its motion? *Ans.* 2.33 N

11.27. A 500 g mass is attached to the end of an initially unstretched vertical spring for which $k = 30$ N/m. The mass is then released, so that it falls and stretches the spring. How far will it fall before stopping? (*Hint*: The GPE lost by the mass must appear as SPE.) *Ans.* 33 cm

11.28. Find the frequency of vibration on Mars for a simple pendulum that is 50 cm long. Objects weigh 0.40 as much on Mars as on the earth. *Ans.* 0.45 Hz

11.29. A "seconds pendulum" beats seconds; that is, it takes 1 s for half a cycle. (*a*) What is the length of a simple "seconds pendulum" at a place where $g = 9.80$ m/s^2? (*b*) What is the length there of a pendulum for which $T = 1$ s? *Ans.* (*a*) 99.3 cm; (*b*) 24.8 cm

11.30. Show that the natural period of vertical oscillation of a mass hung on a Hookean spring is the same as the period of a simple pendulum whose length is equal to the elongation the mass causes when hung on the spring.

11.31. The torsion constant for a rod 80 cm long is determined by finding that a torque of 0.0250 N·m applied to the end of the rod twists the end through 30° about the rod as axis. If a 4 kg uniform sphere is fastened to the end of the rod, with what period will the sphere oscillate about the line of the rod as axis? The line passes through the center of the sphere and the radius of the sphere is 6 cm. *Ans.* 2.18 s

11.32. A certain torsion pendulum consists of a horizontal metal disk ($I = 0.0245$ kg·m^2) suspended at its center by a vertical wire. The pendulum oscillates with a period of 4.0 s. (*a*) What is the torsion constant of the wire? (*b*) What is the maximum angular speed of the disk if the amplitude of oscillation is 30°? *Ans.* (*a*) 0.0604 N·m/rad; (*b*) 0.82 rad/s

11.33. A uniform metal disk has a mass of 2.16 kg and a radius of 9.0 cm. Hanging horizontal by a vertical wire attached to its center, the disk oscillates as a torsion pendulum. The wire can be twisted through an angle of 10° by a torque of 0.81 N·m applied to its end. What is the period of oscillation of this torsion pendulum? *Ans.* 0.27 s

11.34. To determine the moment of inertia of a certain wheel, the wheel is suspended with its plane horizontal by a vertical rod attached at its center. The rod and wheel act as a torsion pendulum. A horizontal force of 25 N, applied 10 cm from the center of the wheel and perpendicular to a radius, causes the wheel to rotate through 18°. When released, the wheel oscillates (with the rod as axis) with a period of 0.20 s. What is the moment of inertia of the wheel about the rod as axis? *Ans.* 0.0081 kg·m^2

Chapter 12

Density; Elasticity

THE MASS DENSITY (ρ) of a material is the mass per unit volume of the material.

$$\rho = \frac{\text{mass of body}}{\text{volume of body}} = \frac{m}{V}$$

The SI unit for mass density is kg/m^3. Also used are g/cm^3 and, much less frequently, $slug/ft^3$.

$$1000 \text{ kg/m}^3 = 1 \text{ g/cm}^3 = 1.94 \text{ slug/ft}^3$$

The density of water is close to 1000 kg/m^3.

THE WEIGHT DENSITY (D) of a material is the weight per unit volume of the material.

$$D = \frac{\text{weight of body}}{\text{volume of body}} = \frac{mg}{V} = \rho g$$

The units of D are N/m^3, lb/ft^3, or lb/in^3. Water has a weight density close to 62.4 lb/ft^3.

THE SPECIFIC GRAVITY (sp gr) of a substance is the ratio of the density of the substance to the density of some standard substance. The standard is usually water (at $4°$ C) for liquids and solids, while for gases, it is usually air.

$$\text{sp gr} = \frac{\rho}{\rho_{\text{standard}}} = \frac{D}{D_{\text{standard}}}$$

Since sp gr is a dimensionless ratio, it has the same value for all units systems.

ELASTICITY is that property by virtue of which a body returns to its original size and shape when the forces that deformed it are removed.

STRESS is a measure of the strength of the agent that is causing a deformation. Precisely, if a force F is applied to a surface of area A, then

$$\text{stress} = \frac{\text{force}}{\text{area of surface on which force acts}} = \frac{F}{A}$$

Its units are N/m^2, N/cm^2, lb/ft^2, lb/in^2, and so on.

STRAIN is the fractional deformation resulting from a stress. It is measured by the ratio of the change in some dimension of a body to the original dimension in which the change occurred.

$$\text{strain} = \frac{\text{change in dimension}}{\text{original dimension}}$$

Strain has no units because it is a ratio of like quantities. The exact definition of strain for various situations is given later.

HOOKE'S LAW (Chapter 11) can be stated in terms of stress and strain. If the system obeys Hooke's law, then stress $\propto$ strain. We then define a constant, called the *modulus of elasticity*, by the relation

$$\text{modulus of elasticity} = \frac{\text{stress}}{\text{strain}}$$

The modulus has the same units as stress. A large modulus means that a large stress is required to produce a given strain.

ELASTIC LIMIT is the smallest value of the stress required to produce a permanent distortion in a body. When a stress in excess of this limit is applied, the body will not return exactly to its original state after the stress is removed.

YOUNG'S MODULUS (or *tensile modulus*) describes the length elasticity of a material. Suppose a wire or rod of original length L and cross-sectional area A elongates an amount ΔL under a stretching force F applied to its end. Then

$$\text{tensile stress} = \frac{F}{A} \qquad \text{tensile strain} = \frac{\Delta L}{L}$$

and

$$\text{Young's modulus} = Y = \frac{\text{stress}}{\text{strain}} = \frac{F/A}{\Delta L/L} = \frac{FL}{A\,\Delta L}$$

Commonly its units are N/m^2 or lb/ft^2. The value of Y depends only on the material of the wire or rod and not on its dimensions.

BULK MODULUS (B) describes the volume elasticity of a material. Suppose that a uniformly distributed compressive force acts on the surface of an object and is directed perpendicular to the surface at all points. Then if F is the force acting on and perpendicular to an area A, we define

$$\text{pressure on } A = p = \frac{F}{A}$$

The SI unit for pressure is the *pascal* (Pa), where $1\ Pa = 1\ N/m^2$. Other units used for pressure are lb/in^2 and lb/ft^2.

Suppose that the pressure on an object of original volume V is increased by an amount Δp. The pressure increase causes a volume change ΔV, where ΔV will be negative. We then define

$$\text{volume stress} = \Delta p \qquad \text{volume strain} = -\frac{\Delta V}{V}$$

$$\text{bulk modulus} = B = \frac{\text{stress}}{\text{strain}} = -\frac{\Delta p}{\Delta V/V} = -\frac{V\,\Delta p}{\Delta V}$$

The minus sign is used so as to cancel the minus numerical value of ΔV and thereby make B a positive number. The bulk modulus has the units of pressure.

The reciprocal of the bulk modulus is called the *compressibility*, k, of the substance.

SHEAR MODULUS (S) describes the shape elasticity of a material. Suppose, as shown in Fig. 12-1, that equal and opposite tangential forces F act on a rectangular block. These *shearing forces* distort the block as indicated, but its volume remains unchanged. We define

$$\text{shearing stress} = \frac{\text{tangential force acting}}{\text{area of surface being sheared}} = \frac{F}{A}$$

$$\text{shearing strain} = \frac{\text{distance sheared}}{\text{distance between surfaces}} = \frac{\Delta L}{L}$$

Then

$$\text{shear modulus} = S = \frac{\text{stress}}{\text{strain}} = \frac{F/A}{\Delta L/L} = \frac{FL}{A\,\Delta L}$$

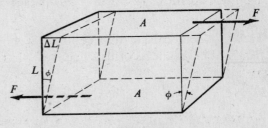

Fig. 12-1

Since ΔL is usually very small, the ratio $\Delta L/L$ is equal approximately to the shear angle ϕ in radians. In that case

$$S = \frac{F}{A\phi}$$

Solved Problems

12.1. Find the density and specific gravity of gasoline if 51 g occupies 75 cm³.

$$\text{density} = \frac{\text{mass}}{\text{volume}} = \frac{0.051 \text{ kg}}{75 \times 10^{-6} \text{ m}^3} = 680 \text{ kg/m}^3$$

$$\text{sp gr} = \frac{\text{density of gasoline}}{\text{density of water}} = \frac{680 \text{ kg/m}^3}{1000 \text{ kg/m}^3} = 0.68$$

or

$$\text{sp gr} = \frac{\text{mass of 75 cm}^3 \text{ gasoline}}{\text{mass of 75 cm}^3 \text{ water}} = \frac{51 \text{ g}}{75 \text{ g}} = 0.68$$

12.2. What volume does 300 g of mercury occupy? The density of mercury is 13 600 kg/m³.

From $\rho = m/V$,

$$V = \frac{m}{\rho} = \frac{0.30 \text{ kg}}{13\,600 \text{ kg/m}^3} = 2.21 \times 10^{-5} \text{ m}^3 = 22.1 \text{ cm}^3$$

12.3. The specific gravity of cast iron is 7.20. (*a*) Find its density in kg/m³ and the mass of 60 cm³ of cast iron. (*b*) Find its weight density in lb/ft³ and the weight of 20 ft³ of cast iron.

We make use of

$$\text{sp gr} = \frac{\text{density of substance}}{\text{density of water}} \qquad \text{and} \qquad \rho = \frac{m}{V}$$

(*a*) density of iron = (sp gr of iron)(density of water) = $(7.20)(1000 \text{ kg/m}^3)$ = 7200 kg/m³

mass of 60 cm³ of iron = $\rho V = (7200 \text{ kg/m}^3)(60 \times 10^{-6} \text{ m}^3)$ = 0.432 kg

(*b*) weight density of iron = (sp gr of iron)(weight density of water)

$$= (7.20)(62.4 \text{ lb/ft}^3) = 449 \text{ lb/ft}^3$$

weight of 20 ft³ of iron = (weight density of iron)(V) = $(449 \text{ lb/ft}^3)(20 \text{ ft}^3)$ = 8980 lb

12.4. A certain calibrated flask "weighs" 25.0 g when empty, 75.0 g when filled with water, and 88.0 g when filled with glycerin. Find the specific gravity of glycerin.

Let us first point out that "weights" in grams or kilograms are not really weights at all. We often obtain them by use of scales or balances, hence the terminology. But, in fact, these are masses. Hence the data tell us that the flask has a mass of 25 g, the mass of the water in it is   75 − 25 = 50 g,   and the mass of the glycerin in it is   88 − 25 = 63 g.

$$\text{sp gr} = \frac{\text{mass of glycerin}}{\text{mass of equal volume of water}} = \frac{63 \text{ g}}{50 \text{ g}} = 1.26$$

12.5. A solid cube of aluminum is 2.00 cm on each edge. The density of aluminum is 2700 kg/m³. Find the mass of the cube.

$$\text{mass of cube} = \rho V = (2700 \text{ kg/m}^3)(0.02 \text{ m})^3 = 0.0216 \text{ kg} = 21.6 \text{ g}$$

12.6. What is the mass of one liter (1000 cm³) of cottonseed oil of density 926 kg/m³? How much does it weigh?

$$m = \rho V = (926 \text{ kg/m}^3)(1000 \times 10^{-6} \text{ m}^3) = 0.926 \text{ kg}$$

$$\text{weight} = mg = (0.926 \text{ kg})(9.8 \text{ m/s}^2) = 9.1 \text{ N}$$

12.7. An electrolytic tin-plating process gives a tin coating that is 7.5×10^{-5} cm thick. How large an area can be coated in this way by use of 1 kg of tin. The density of tin is 7.3 g/cm^3.

The volume of the plating is equal to its area A times its thickness.

$$V = A(7.5 \times 10^{-5}\ cm) = (7.5 \times 10^{-7}\ m)A$$

But the volume of 1 kg of tin is given by $\rho = m/V$ to be

$$V \text{ for 1 kg} = \frac{m}{\rho} = \frac{1\ kg}{7300\ kg/m^3} = 1.37 \times 10^{-4}\ m^3$$

Equating these two expressions for the volume gives

$$(7.5 \times 10^{-7}\ m)A = 1.37 \times 10^{-4}\ m^3$$

from which $A = 183\ m^2$ is the area that can be coated.

12.8. A thin sheet of gold foil has an area of 3.12 cm^2 and "weighs" 6.5 milligrams. How thick is the sheet? ρ for gold is 19 300 kg/m^3.

One milligram is 10^{-6} kg, so the mass of the sheet is 6.5×10^{-6} kg. Its volume is

$$V = (\text{area}) \times (\text{thickness}) = (3.12 \times 10^{-4}\ m^2)t$$

where t is the thickness of the sheet. We equate this expression for the volume to m/ρ:

$$(3.12 \times 10^{-4}\ m^2)t = \frac{6.5 \times 10^{-6}\ kg}{19\ 300\ kg/m^3}$$

from which $t = 1.08 \times 10^{-6}\ m = 1.08\ \mu m$.

12.9. The mass of a liter of milk is 1.032 kg. The butterfat that it contains has a density of 865 kg/m^3 when pure, and it constitutes 4% of the milk by volume. What is the density of the fat-free skimmed milk?

$$\text{volume of fat in 1000 cm}^3 \text{ of milk} = 4\% \times 1000\ cm^3 = 40\ cm^3$$

$$\text{mass of 40 cm}^3 \text{ fat} = V\rho = (40 \times 10^{-6}\ m^3)(865\ kg/m^3) = 0.0346\ kg$$

$$\text{density of skimmed milk} = \frac{\text{mass}}{\text{volume}} = \frac{(1.032 - 0.0346)\ kg}{(1000 - 40) \times 10^{-6}\ m^3} = 1039\ kg/m^3$$

12.10. A metal wire 75 cm long and 0.130 cm in diameter stretches 0.035 cm when a load of 8.0 kg is hung on its end. Find the stress, strain, and Young's modulus for the material of the wire.

$$\text{stress} = \frac{F}{A} = \frac{(8.0\ kg)(9.8\ m/s^2)}{\pi(6.5 \times 10^{-4}\ m)^2} = 5.91 \times 10^7\ N/m^2$$

$$\text{strain} = \frac{\Delta L}{L} = \frac{0.035\ cm}{75\ cm} = 4.67 \times 10^{-4}$$

$$Y = \frac{\text{stress}}{\text{strain}} = \frac{5.91 \times 10^7\ N/m^2}{4.67 \times 10^{-4}} = 1.27 \times 10^{11}\ N/m^2$$

12.11. A solid cylindrical steel column is 4.0 m long and 9.0 cm in diameter. What will be its decrease in length when carrying a load of 80 000 kg? $Y = 1.9 \times 10^{11}\ N/m^2$.

$$\text{cross-sectional area of column} = \pi r^2 = \pi(0.045\ m)^2 = 6.36 \times 10^{-3}\ m^2$$

From $Y = (F/A)/(\Delta L/L)$ we have

$$\Delta L = \frac{FL}{AY} = \frac{[(8 \times 10^4)(9.8)\ N](4.0\ m)}{(6.36 \times 10^{-3}\ m^2)(1.9 \times 10^{11}\ N/m^2)} = 2.6 \times 10^{-3}\ m = 2.6\ mm$$

12.12. Atmospheric pressure is about 1.01×10^5 Pa or 14.7 lb/in². How large a force does the atmosphere exert on a 2 cm² area on the top of your head? Express your answer in both newtons and pounds.

Because $p = F/A$, where F is perpendicular to A, we have $F = pA$. Assuming that 2 cm² of your head is flat (nearly correct) and that the force due to the atmosphere is perpendicular to the surface (as it is), we have

$$F = pA = (1.01 \times 10^5 \text{ N/m}^2)(2 \times 10^{-4} \text{ m}^2) = 20 \text{ N}$$

and

$$F = (14.7 \text{ lb/in}^2)\left(\frac{1 \text{ in}}{2.54 \text{ cm}}\right)^2 (2 \text{ cm}^2) = 4.6 \text{ lb}$$

12.13. A 60 kg woman stands on a light, cubical box that is 5 cm on each edge. The box sits on the floor. What pressure does the box exert on the floor?

$$p = \frac{F}{A} = \frac{(60)(9.8) \text{ N}}{(5 \times 10^{-2} \text{ m})^2} = 2.4 \times 10^5 \text{ N/m}^2 = 240 \text{ kPa}$$

12.14. The bulk modulus of mercury is 4×10^6 lb/in². Compute the volume contraction of 100 in³ of mercury when subjected to a pressure of 200 lb/in².

From $B = -\Delta p/(\Delta V/V)$,

$$\Delta V = -\frac{V \Delta p}{B} = -\frac{(100 \text{ in}^3)(200 \text{ lb/in}^2)}{4 \times 10^6 \text{ lb/in}^2} = -0.005 \text{ in}^3$$

12.15. A box-shaped piece of gelatin dessert has a top area of 15 cm² and a height of 3 cm. When a shearing force of 0.50 N is applied to the upper surface, the upper surface displaces 4 mm relative to the bottom surface. What are the shearing stress, shearing strain, and the shear modulus for the gelatin?

$$\text{shear stress} = \frac{\text{tangential force}}{\text{area of face}} = \frac{0.50 \text{ N}}{15 \times 10^{-4} \text{ m}^2} = 333 \text{ N/m}^2$$

$$\text{shear strain} = \frac{\text{displacement}}{\text{height}} = \frac{0.4 \text{ cm}}{3 \text{ cm}} = 0.133$$

$$\text{shear modulus } S = \frac{\text{stress}}{\text{strain}} = \frac{333 \text{ N/m}^2}{0.133} = 2500 \text{ N/m}^2$$

12.16. A 15 kg ball, of radius 4 cm, is suspended from a point 2.94 m above the floor by an iron wire of unstretched length 2.85 m. The diameter of the wire is 0.090 cm and its Young's modulus is 1.8×10^{11} N/m². If the ball is set swinging so that its center passes through the lowest point at 5 m/s, by how much does the bottom of the ball clear the floor?

Call the tension in the wire T when swinging through the lowest point. Since T must supply the centripetal force as well as balance the weight,

$$T = mg + \frac{mv^2}{r} = m\left(9.8 + \frac{25}{r}\right)$$

all in proper SI units. This is complicated, because r is the distance from the pivot to the center of the ball when the wire is stretched and so it is $r_0 + \Delta r$, where r_0 is the unstretched length of the pendulum,

$$r_0 = 2.85 \text{ m} + 0.04 \text{ m} = 2.89 \text{ m}$$

and where Δr is as yet unknown. However, the unstretched distance from the pivot to the bottom of the ball is

$$2.85 \text{ m} + 0.08 \text{ m} = 2.93 \text{ m}$$

and so the maximum possible value for Δr is

$$2.94 \text{ m} - 2.93 \text{ m} = 0.01 \text{ m}$$

We will therefore incur no more than a 1/3 percent error in r by using $r = r_0 = 2.89$ m. This gives $T = 277$ N.

Under this tension, the wire stretches by

$$\Delta L = \frac{FL}{AY} = \frac{(277 \text{ N})(2.85 \text{ m})}{\pi(4.5 \times 10^{-4} \text{ m})^2(1.8 \times 10^{11} \text{ N/m}^2)} = 6.9 \times 10^{-3} \text{ mm}$$

Hence the ball misses by

$$2.94 \text{ m} - (2.85 + 0.0069 + 0.08) \text{ m} = 0.0031 \text{ m} = 3.1 \text{ mm}$$

To check the approximation we have made, we could use $r = 2.90$ m, its maximum possible value. Then we find that $\Delta L = 6.9$ mm, showing that the approximation has caused negligible error.

12.17. A vertical wire 5 m long and of 0.0088 cm^2 cross-sectional area has $Y = 2.0 \times 10^{11}$ N/m^2. A 2 kg object is fastened to its end and stretches the wire elastically. If the object is now pulled down a little and released, the object undergoes vertical SHM. Find the period of its vibration.

The force constant of the wire acting as a vertical spring is given by $k = F/\Delta L$, where ΔL is the deformation produced by the force (weight) F. But, from $F/A = Y(\Delta L/L)$,

$$k = \frac{F}{\Delta L} = \frac{AY}{L} = \frac{(8.8 \times 10^{-7} \text{ m}^2)(2 \times 10^{11} \text{ N/m}^2)}{5 \text{ m}} = 35\,000 \text{ N/m}$$

Then for the period we have

$$T = 2\pi\sqrt{m/k} = 2\pi\sqrt{\frac{2 \text{ kg}}{35\,000 \text{ N/m}}} = 0.047 \text{ s}$$

Supplementary Problems

12.18. Find the density and specific gravity of ethyl alcohol if 63.3 g occupies 80.0 cm^3. *Ans.* 791 kg/m^3, 0.791

12.19. Determine the volume of 200 g of carbon tetrachloride, for which sp gr = 1.60. *Ans.* 125 cm^3

12.20. The density of aluminum is 2.70 g/cm^3. What volume does 2.0 kg occupy? *Ans.* 740 cm^3

12.21. Determine the mass of an aluminum cube that is 5 cm on each edge. The density of aluminum is 2700 kg/m^3. *Ans.* 0.338 kg

12.22. A drum holds 200 lb of water or 132 lb of gasoline. Determine for the gasoline (a) sp gr, (b) ρ in kg/m^3, (c) D in lb/ft^3. *Ans.* (a) 0.66; (b) 660 kg/m^3; (c) 41.2 lb/ft^3

12.23. Air has a density of 1.29 kg/m^3 under standard conditions. What is the mass of air in a room with dimensions 10 m × 8 m × 3 m? *Ans.* 310 kg

12.24. What is the density of the material in the nucleus of the hydrogen atom? The nucleus can be considered to be a sphere of radius 1.20×10^{-15} m and its mass is 1.67×10^{-27} kg. The volume of a sphere is $(4/3)\pi r^3$ *Ans.* 2.3×10^{17} kg/m^3

12.25. To determine the inner radius of a uniform capillary tube, the tube is filled with mercury. A column of mercury 2.375 cm long is found to "weigh" 0.242 g. What is the inner radius, r, of the tube? The density of mercury is 13 600 kg/m³ and the volume of a right circular cylinder is $\pi r^2 h$. *Ans.* 0.49 mm

12.26. Battery acid has sp gr = 1.285 and is 38% sulfuric acid by weight. What mass of sulfuric acid is contained in a liter of battery acid?   *Ans.* 488 g

12.27. A thin, semitransparent, film of gold ($\rho = 19\,300$ kg/m³) has an area of 14.5 cm² and a mass of 1.93 mg. (*a*) What is the volume of 1.93 mg of gold? (*b*) What is the thickness of the film in angstroms, where $1\,\text{Å} = 10^{-10}$ m? (*c*) Gold atoms have a diameter of about 5 Å. How many atoms thick is the film? *Ans.* (*a*) 1×10^{-10} m³; (*b*) 690 Å; (*c*) 138 atoms thick

12.28. In an unhealthy, dusty cement mill, there were 2.6×10^9 dust particles (sp gr = 3.0) per cubic meter of air. Assuming the particles to be spheres of 2 micrometer diameter, calculate the mass of dust (*a*) in a $20\ \text{m} \times 15\ \text{m} \times 8\ \text{m}$ room, (*b*) inhaled in each average breath of 400 cm³ volume. One micrometer $= 1\ \mu\text{m} = 10^{-6}$ m. *Ans.* (*a*) 78 g; (*b*) 13.1 μg

12.29. An iron rod 4 m long and 0.5 cm² in cross section stretches 1 mm when a mass of 225 kg is hung from its lower end. Compute Young's modulus for the iron.   *Ans.* 1.76×10^{11} N/m²

12.30. A load of 100 lb is applied to the lower end of a steel rod 3 ft long and 0.20 inches in diameter. How much will the rod stretch? $Y = 3.3 \times 10^7$ lb/in² for steel.   *Ans.* 2.9×10^{-4} ft $= 3.5 \times 10^{-3}$ in

12.31. A platform is suspended by four wires at its corners. The wires are 3 m long and have a diameter of 2.0 mm. Young's modulus for the material of the wires is 1.8×10^{11} N/m². How far will the platform drop (due to elongation of the wires) if a 50 kg load is placed at the center of the platform? *Ans.* 0.65 mm

12.32. Determine the fractional change in volume as the pressure of the atmosphere (1×10^5 Pa) around a metal block is reduced to zero by placing the block in vacuum. The bulk modulus for the metal is 1.25×10^{11} N/m².   *Ans.* 8×10^{-7}

12.33. Compute the volume change of a solid copper cube, 40 mm on each edge, when subjected to a pressure of 2×10^7 Pa. The bulk modulus for copper is 1.25×10^{11} N/m².   *Ans.* -10 mm³

12.34. The compressibility of water is 5×10^{-10} m²/N. Find the decrease in volume of 100 cm³ of water when subjected to a pressure of 15 MPa.   *Ans.* 0.75 cm³

12.35. Two parallel and opposite forces, each 4000 N, are applied tangentially to the upper and lower faces of a cubical metal block 25 cm on a side. Find the angle of shear and the displacement of the upper surface relative to the lower surface. The shear modulus for the metal is 8×10^{10} N/m². *Ans.* 8.0×10^{-7} rad, 2×10^{-7} m

12.36. A 60 kg motor sits on four cylindrical rubber blocks. Each cylinder has a height of 3 cm and a cross-sectional area of 15 cm². The shear modulus for this rubber is 2×10^6 N/m². (*a*) If a sideways force of 300 N is applied to the motor, how far will it move sideways? (*b*) With what frequency will the motor vibrate back and forth sideways if disturbed?   *Ans.* (*a*) 0.075 cm; (*b*) 13 Hz

Chapter 13

Fluids at Rest

THE AVERAGE PRESSURE on a surface of area A is force divided by area, where it is stipulated that the force must be perpendicular (or normal) to the area.

$$\text{average pressure } p = \frac{\text{force } F \text{ acting normal to an area}}{\text{area } A \text{ over which the force is distributed}}$$

The units of pressure are N/m^2, lb/ft^2, lb/in^2, and so on. The SI name given to the N/m^2 is the pascal (Pa), where $1 \text{ Pa} = 1 \text{ N}/m^2$.

STANDARD ATMOSPHERIC PRESSURE is 1.01×10^5 Pa and this is equivalent to $14.7 \text{ lb}/in^2$. Other units used for pressure are

$$1 \text{ atmosphere (atm)} = 1.013 \times 10^5 \text{ Pa}$$

$$1 \text{ torr} = 1 \text{ mm of mercury (mmHg)} = 133.32 \text{ Pa}$$

HYDROSTATIC PRESSURE due to a column of fluid of height h and mass density ρ (or weight density D) is

$$p = h\rho g = hD$$

PASCAL'S PRINCIPLE: When the pressure on any part of a confined fluid (liquid or gas) is changed, the pressure on every other part of the fluid is also changed by the same amount.

ARCHIMEDES' PRINCIPLE: A body wholly or partly immersed in a fluid is buoyed up by a force equal to the weight of the fluid it displaces. The buoyant force can be considered to act vertically upward through the center of gravity of the displaced fluid.

$$\text{BF} = \text{buoyant force} = \text{weight of displaced fluid}$$

Solved Problems

13.1. An 80 kg metal cylinder 2 m long and with each end of area 25 cm² stands vertically on one end. What pressure does the cylinder exert on the floor?

$$p = \frac{\text{normal force}}{\text{area}} = \frac{(80 \text{ kg})(9.8 \text{ m/s}^2)}{25 \times 10^{-4} \text{ m}^2} = 3.14 \times 10^5 \text{ N/m}^2 = 314 \text{ kPa}$$

13.2. Atmospheric pressure is about 1×10^5 Pa. How large a force does the air in a room exert on the inside of a window pane that is 40 cm × 80 cm?

The atmosphere exerts a force normal to any surface placed in it. Consequently, the force on the window pane is perpendicular to the pane and is given by

$$F = pA = (1 \times 10^5 \text{ N/m}^2)(0.40 \times 0.80 \text{ m}^2) = 3.2 \times 10^4 \text{ N}$$

Of course, a nearly equal force due to the atmosphere on the outside keeps the window from breaking.

13.3. Find the pressure due to the fluid at a depth of 76 cm in still (*a*) water ($\rho = 1.00$ g/cm³), (*b*) mercury ($\rho = 13.6$ g/cm³).

(*a*) $\qquad\qquad p = h\rho g = (0.76 \text{ m})(1000 \text{ kg/m}^3)(9.8 \text{ m/s}^2) = 7450 \text{ N/m}^2 = 7.45 \text{ kPa}$

(*b*) $\qquad\qquad p = h\rho g = (0.76 \text{ m})(13\,600 \text{ kg/m}^3)(9.8 \text{ m/s}^2) = 1.01 \times 10^5 \text{ N/m}^2 \approx 1 \text{ atm}$

13.4. What is the pressure due to the water at a depth of 50 ft in a lake? The weight density of water is 62.4 lb/ft³.

$$p = h\rho g = hD = (50 \text{ ft})(62.4 \text{ lb/ft}^3) = 3120 \text{ lb/ft}^2$$

13.5. When a submarine dives to a depth of 120 m, to how large a total pressure is its exterior surface subjected? The density of seawater is about 1.03 g/cm³.

p = atmospheric pressure + pressure of water

$\qquad = 1 \times 10^5 \text{ N/m}^2 + h\rho g = 1 \times 10^5 \text{ N/m}^2 + (120 \text{ m})(1030 \text{ kg/m}^3)(9.8 \text{ m/s}^2)$

$\qquad = 1 \times 10^5 \text{ N/m}^2 + 12.1 \times 10^5 \text{ N/m}^2 = 13.1 \times 10^5 \text{ N/m}^2 = 1.31 \text{ MPa}$

13.6. How high would water rise in the pipes of a building if the water pressure gauge shows the pressure at the ground floor to be 270 kPa (about 40 lb/in²)?

Water pressure gauges read the excess pressure due to the water, that is, the difference between the pressure in the water and the pressure of the atmosphere. The water pressure at the bottom of the highest column that can be supported is 270 kPa. Therefore, $p = h\rho g$ gives

$$h = \frac{p}{\rho g} = \frac{2.7 \times 10^5 \text{ N/m}^2}{(1000 \text{ kg/m}^3)(9.8 \text{ m/s}^2)} = 27.6 \text{ m}$$

13.7. As shown in Fig. 13-1, a weighted piston holds compressed gas in a tank. The piston and its weights have a mass of 20 kg. The cross-sectional area of the piston is 8 cm². What is the total pressure of the gas in the tank? What would an ordinary pressure gauge on the tank read?

The *total* pressure in the tank will be the pressure of the atmosphere (about 1.0×10^5 Pa) plus the pressure due to the piston and weights.

$$p = 1.0 \times 10^5 \text{ N/m}^2 + \frac{(20)(9.8) \text{ N}}{8 \times 10^{-4} \text{ m}^2}$$

$$= 1.0 \times 10^5 \text{ N/m}^2 + 2.45 \times 10^5 \text{ N/m}^2 = 3.45 \times 10^5 \text{ N/m}^2 = 345 \text{ kPa}$$

A pressure gauge on the tank would read the difference between the pressure inside and outside the tank.

$$\text{gauge reading} = 2.45 \times 10^5 \text{ N/m}^2 = 245 \text{ kPa}$$

It reads the pressure due to the piston and weights.

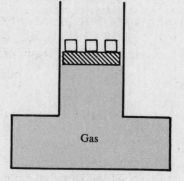

Fig. 13-1

13.8. A reservoir dam holds an 8 km² lake behind it. Just behind the dam, the lake is 12 m deep. What is the water pressure (*a*) at the base of the dam? (*b*) at a point 3 m down from the surface?

The area of the lake behind the dam has no effect on the pressure against the dam. At any point, $p = h\rho g$.

(*a*) $\qquad\qquad p = (12 \text{ m})(1000 \text{ kg/m}^3)(9.8 \text{ m/s}^2) = 118\,000 \text{ Pa}$

(*b*) $\qquad\qquad p = (3 \text{ m})(1000 \text{ kg/m}^3)(9.8 \text{ m/s}^2) = 29\,400 \text{ Pa}$

13.9. In a hydraulic press such as the one shown in Fig. 13-2, the large piston has cross-sectional area $A_1 = 200$ cm^2 and the small piston has cross-sectional area $A_2 = 5$ cm^2. If a force of 250 N is applied to the small piston, find the force F_1 on the large piston.

By Pascal's principle,

$$\text{pressure under large piston} = \text{pressure under small piston}$$

$$\frac{F_1}{A_1} = \frac{F_2}{A_2}$$

$$F_1 = \frac{A_1}{A_2} F_2 = \frac{200}{5} (250 \text{ N}) = 10\,000 \text{ N}$$

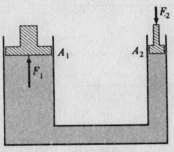

Fig. 13-2

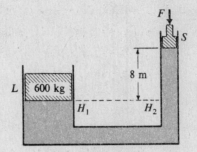

Fig. 13-3

13.10. For the system shown in Fig. 13-3, the cylinder on the left, at L, has a mass of 600 kg and a cross-sectional area of 800 cm^2. The piston on the right, at S, has cross-sectional area 25 cm^2 and negligible weight. If the apparatus is filled with oil ($\rho = 0.78$ g/cm^3), find the force F required to hold the system in equilibrium as shown.

The pressures at points H_1 and H_2 are equal since they are at the same level in a single connected fluid. Therefore,

$$\text{pressure at } A_L = \text{pressure at } A_S$$

$$\left(\begin{array}{c} \text{pressure due to} \\ \text{left piston} \end{array} \right) = \left(\begin{array}{c} \text{pressure due to } F \\ \text{and right piston} \end{array} \right) + \left(\text{pressure due to 8 m of oil} \right)$$

$$\frac{(600)(9.8) \text{ N}}{0.08 \text{ m}^2} = \frac{F}{25 \times 10^{-4} \text{ m}^2} + (8 \text{ m})(780 \text{ kg/m}^3)(9.8 \text{ m/s}^2)$$

Solving for F gives it to be 31 N.

13.11. A barrel will rupture when the gauge pressure within it reaches 50 lb/in^2. It is attached to the lower end of a vertical pipe, with the pipe and barrel filled with oil ($D = 47$ lb/ft^3). How long can the pipe be if the barrel is not to rupture?

From $p = hD$, we have

$$h = \frac{p}{D} = \frac{(50 \text{ lb/in}^2)(144 \text{ in}^2/\text{ft}^2)}{47 \text{ lb/ft}^3} = 153 \text{ ft}$$

13.12. A vertical test tube has 2 cm of oil ($\rho = 0.80$ g/cm^3) floating on 8 cm of water. What is the pressure at the bottom of the tube due to the fluid in it?

$$p = h_1\rho_1 g + h_2\rho_2 g = (0.02 \text{ m})(800 \text{ kg/m}^3)(9.8 \text{ m/s}^2) + (0.08 \text{ m})(1000 \text{ kg/m}^3)(9.8 \text{ m/s}^2)$$
$$= 941 \text{ Pa}$$

13.13. As shown in Fig. 13-4, a column of water 40 cm high supports a 31 cm column of an unknown fluid. What is the density of the unknown fluid?

The pressures at point A due to the two fluids must be equal (or the one with the higher pressure would push the lower-pressure fluid away). Therefore,

$$\text{pressure due to water} = \text{pressure due to unknown fluid}$$

$$h_1 \rho_1 g = h_2 \rho_2 g$$

from which

$$\rho_2 = \frac{h_1}{h_2} \rho_1 = \frac{40}{31} (1000 \text{ kg/m}^3) = 1290 \text{ kg/m}^3$$

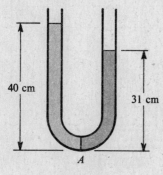

Fig. 13-4

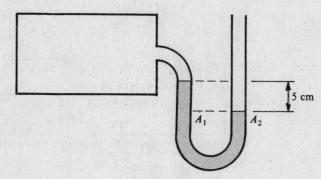

Fig. 13-5

13.14. The U-tube device connected to the tank in Fig. 13-5 is called a *manometer*. As you see, the mercury in the tube stands higher in one side than the other. What is the pressure in the tank if atmospheric pressure is 76 cm of mercury? The density of mercury is 13.6 g/cm³.

$$\text{pressure at } A_1 = \text{pressure at } A_2$$

$$(p \text{ in tank}) + (p \text{ due to 5 cm mercury}) = (p \text{ due to atmosphere})$$

$$p + (0.05 \text{ m})(13\,600 \text{ kg/m}^3)(9.8 \text{ m/s}^2) = (0.76 \text{ m})(13\,600 \text{ kg/m}^3)(9.8 \text{ m/s}^2)$$

from which $p = 95$ kPa.

Or, more simply perhaps, we could note that the pressure in the tank is 5 cm of mercury *lower* than atmospheric. So the pressure is 71 cm of mercury, which is 94.6 kPa.

13.15. A block of aluminum "weighs" 25 g in air. (*a*) What is its volume? (*b*) What will be the tension in a string that suspends the block when the block is totally submerged in water? The density of aluminum is 2700 kg/m³.

(*a*) Because $\rho = m/V$, we have

$$V = m/\rho = \frac{0.025 \text{ kg}}{2700 \text{ kg/m}^3} = 9.26 \times 10^{-6} \text{ m}^3 = 9.26 \text{ cm}^3$$

(*b*) The block displaces 9.26×10^{-6} m³ of water when submerged, so the buoyant force on it is

$$\text{BF} = \text{weight of displaced water} = (\text{volume})(\rho \text{ of water}) g$$

$$= (9.26 \times 10^{-6} \text{ m}^3)(1000 \text{ kg/m}^3)(9.8 \text{ m/s}^2) = 0.091 \text{ N}$$

The tension in the supporting cord plus the buoyant force must equal the weight of the block if it is to be in equilibrium (see Fig. 13-6). Therefore

$$T + \text{BF} = mg$$

$$T = mg - \text{BF} = (0.025 \text{ kg})(9.8 \text{ m/s}^2) - 0.091 \text{ N} = 0.154 \text{ N}$$

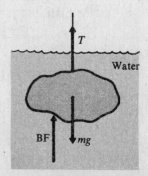

Fig. 13-6

13.16. A piece of alloy "weighs" 86 g in air and 73 g when immersed in water. Find its volume and its density.

Figure 13-6 shows the situation when the object is in water. From the figure,

$$BF + T = mg$$
$$BF = (0.086)(9.8) \text{ N} - (0.073)(9.8) \text{ N} = (0.013)(9.8) \text{ N}$$

But BF must be equal to the weight of the displaced water.

$$BF = \text{weight of water} = (\text{mass of water})(g)$$
$$= (\text{volume of water})(\text{density of water})(g)$$

or

$$(0.013)(9.8) \text{ N} = V(1000 \text{ kg/m}^3)(9.8 \text{ m/s}^2)$$

from which $V = 1.30 \times 10^{-5} \text{ m}^3$. This is also the volume of the piece of alloy. Therefore,

$$\rho \text{ of alloy} = \frac{\text{mass}}{\text{volume}} = \frac{0.086 \text{ kg}}{1.30 \times 10^{-5} \text{ m}^3} = 6600 \text{ kg/m}^3$$

13.17. A tank partly filled with water rests on a scale, which registers 436.1 lb. Determine the scale reading when a steel object of volume 0.0545 ft³ is suspended from a cord and submerged in the water. No water escapes from the tank.

The BF upward on the steel object is given by

$$\text{upward BF} = \text{weight of displaced water} = V \times (D \text{ for water})$$
$$= (0.0545 \text{ ft}^3)(62.4 \text{ lb/ft}^3) = 3.4 \text{ lb}$$

By Newton's third law, an equal but opposite reaction force of 3.4 lb acts downward on the tank. Hence the scale reading is $(436.1 + 3.4) \text{ lb} = 439.5 \text{ lb}$.

13.18. A solid aluminum cylinder, $\rho = 2700 \text{ kg/m}^3$, "weighs" 67 g in air and 45 g when immersed in turpentine. Determine the density of turpentine.

The BF acting on the cylinder when immersed is

$$BF = (0.067 - 0.045)(9.8) \text{ N} = (0.022)(9.8) \text{ N}$$

This is also the weight of the displaced turpentine.

The volume of the cylinder is, from $\rho = m/V$,

$$V \text{ of cylinder} = \frac{m}{\rho} = \frac{0.067 \text{ kg}}{2700 \text{ kg/m}^3} = 2.48 \times 10^{-5} \text{ m}^3$$

This is also the volume of the displaced turpentine. We therefore have for the turpentine

$$\rho = \frac{\text{mass}}{\text{volume}} = \frac{(\text{weight})/g}{\text{volume}} = \frac{(0.022)(9.8)/(9.8)}{2.48 \times 10^{-5}} \frac{\text{kg}}{\text{m}^3} = 887 \text{ kg/m}^3$$

13.19. A glass stopper "weighs" 2.50 g in air, and 1.50 g in water, and 0.70 g in sulfuric acid. What is the density of the acid? Its specific gravity?

The BF on it in water is $(0.0025 - 0.0015)(9.8) \text{ N}$. This is the weight of the displaced water. Since $\rho g = (\text{weight})/V$, we have

$$\text{volume of stopper} = \text{volume of displaced water} = (\text{weight})/\rho g$$
$$= \frac{(0.0010)(9.8) \text{ N}}{(1000 \text{ kg/m}^3)(9.8 \text{ m/s}^2)} = 1 \times 10^{-6} \text{ m}^3$$

The buoyant force in acid is

$$[(2.50 - 0.70) \times 10^{-3}](9.8) \text{ N} = (0.0018)(9.8) \text{ N}$$

But this is equal to the weight of displaced acid, mg. Since $\rho = m/V$, and since $m = 0.0018 \text{ kg}$ and $V = 1 \times 10^{-6} \text{ m}^3$, we have

$$\rho \text{ of acid} = \frac{0.0018 \text{ kg}}{1 \times 10^{-6} \text{ m}^3} = 1800 \text{ kg/m}^3$$

Then, for the acid,

$$\text{sp gr} = \frac{\rho \text{ of acid}}{\rho \text{ of water}} = \frac{1800}{1000} = 1.8$$

Alternate Method

$$\text{weight of displaced water} = \left[(2.50 - 1.50) \times 10^{-3}\right](9.8) \text{ N}$$

$$\text{weight of displaced acid} = \left[(2.50 - 0.70) \times 10^{-3}\right](9.8) \text{ N}$$

$$\text{sp gr of acid} = \frac{\text{weight of displaced acid}}{\text{weight of equal volume of displaced water}} = \frac{1.80}{1.00} = 1.8$$

Then, since   sp gr of acid $= (\rho \text{ of acid})/(\rho \text{ of water})$,

$$\rho \text{ of acid} = (\text{sp gr of acid})(\rho \text{ of water}) = (1.8)(1000 \text{ kg/m}^3) = 1800 \text{ kg/m}^3$$

13.20. A block of oak weighs 20 lb in air. A lead sinker weighs 30 lb in water. The sinker is attached to the wood and both together weigh 22 lb in water. Find the specific gravity of the wood.

$$\text{sp gr of wood} = \frac{\text{weight of wood in air}}{\text{weight of equal volume of water}}$$

$$= \frac{\text{weight of wood in air}}{\text{loss of weight in water}} = \frac{20 \text{ lb}}{(30 + 20 - 22) \text{ lb}} = 0.71$$

13.21. The density of ice is 917 kg/m^3. What fraction of the volume of a piece of ice will be above water when floating in fresh water?

The piece of ice will float in the water, since its density is less than 1000 kg/m^3, the density of water. As it does,

$$\text{BF} = \text{weight of displaced water} = \text{weight of piece of ice}$$

But the weight of the ice is $\rho_{\text{ice}} gV$, where V is the volume of the piece. In addition, the weight of the displaced water is $\rho_{\text{water}} gV'$, where V' is the volume of the displaced water. Substituting in the above equation gives

$$\rho_{\text{ice}} gV = \rho_{\text{water}} gV'$$

$$V' = \frac{\rho_{\text{ice}}}{\rho_{\text{water}}} V = \frac{917}{1000} V = 0.917\, V$$

The fraction of the volume floating is then

$$\frac{V - V'}{V} = \frac{V - 0.917\, V}{V} = 1 - 0.917 = 0.083$$

13.22. A rectangular box open at the top and "weighing" 60 kg has base dimensions 1.0 m by 0.80 m and depth 0.50 m. (a) How deep will it sink in fresh water? (b) What weight, w_b, of ballast will cause it to sink to a depth of 30 cm?

(a)   Assuming that the box floats,

$$\text{BF} = \text{weight of displaced water} = \text{weight of box}$$

$$(1000 \text{ kg/m}^3)(9.8 \text{ m/s}^2)(1 \text{ m} \times 0.8 \text{ m} \times y) = (60 \text{ kg})(9.8 \text{ m/s}^2)$$

where y is the depth the box sinks. Solving gives $y = 0.075$ m. Because this is smaller than 0.50 m, our assumption is shown to be correct.

(b) $\text{BF} = \text{weight of box} + \text{weight of ballast}$

But the BF is equal to the weight of the displaced water. Therefore, the above equation becomes

$$(1000 \text{ kg/m}^3)(9.8 \text{ m/s}^2)(1 \text{ m} \times 0.8 \text{ m} \times 0.3 \text{ m}) = (60)(9.8) \text{ N} + w_b$$

from which $w_b = 1760$ N. So the ballast must have a mass of $(1760/9.8)$ kg $= 180$ kg.

13.23. A foam plastic ($\rho = 0.58$ g/cm³) is to be used as a life preserver. What volume of plastic must be used if it is to keep 20% (by volume) of an 80 kg man above water in a lake? The average density of the man is 1.04 g/cm³.

At equilibrium we have

$$\text{BF on man} + \text{BF on plastic} = \text{weight of man} + \text{weight of plastic}$$

$$(\rho_w)(0.8\,V_m)g + \rho_w V_p g = \rho_m V_m g + \rho_p V_p g$$

or

$$(\rho_w - \rho_p)V_p = (\rho_m - 0.8\,\rho_w)V_m$$

where subscripts m, w, and p refer to man, water, and plastic respectively.

But $\rho_m V_m = 80$ kg and so $V_m = (80/1040)$ m³. Substitution gives

$$\left[(1000 - 580)\ \text{kg/m}^3\right]V_p = \left[(1040 - 800)\ \text{kg/m}^3\right]\left[(80/1040)\ \text{m}^3\right]$$

from which $V_p = 0.044$ m³.

13.24. A piece of pure gold ($\rho = 19.3$ g/cm³) is suspected to have a hollow center. It "weighs" 38.25 g in air and 36.22 g in water. How large is the central hole in the gold?

From $\rho = m/V$,

$$\text{volume of 38.25 g of pure gold} = \frac{0.03825\ \text{kg}}{19\,300\ \text{kg/m}^3} = 1.982 \times 10^{-6}\ \text{m}^3$$

$$\text{volume of displaced water} = \frac{(38.25 - 36.22) \times 10^{-3}\ \text{kg}}{1000\ \text{kg/m}^3} = 2.030 \times 10^{-6}\ \text{m}^3$$

$$\text{volume of hole} = (2.030 - 1.982)\ \text{cm}^3 = 0.048\ \text{cm}^3$$

Supplementary Problems

13.25. A 60 kg performer balances on a cane. The end of the cane in contact with the floor has an area of 0.92 cm². Find the pressure exerted on the floor by the cane. Neglect the weight of the cane. *Ans.* 6.4 MPa

13.26. A solid cylinder sustains a force of 5000 N at an angle of 20° to the normal to a wall, as shown in Fig. 13-7. Find the pressure exerted on the wall at the contact area of the cylinder. The contact area between cylinder and wall is 8 cm². *Ans.* 5.9 MPa

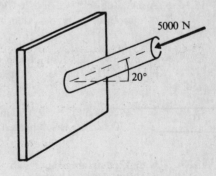

5000 N

20°

Fig. 13-7

13.27. A certain town receives its water directly from a water tower. If the top of the water in the tower is 26 m above the water faucet in a house, what should be the water pressure at the faucet? (Neglect the effects of other water users.) Repeat if the tower is 80 ft above the faucet. *Ans.* 255 kPa, 5000 lb/ft² = 35 lb/in²

13.28. At a height of 10 km (33 000 ft) above sea level, atmospheric pressure is about 210 mm of mercury. What is the resultant normal force on a 600 cm² window of an airplane flying at this height? Assume hydrostatic conditions and a pressure inside the plane of 760 mm of mercury. The density of mercury is 13 600 kg/m³. *Ans.* 4400 N

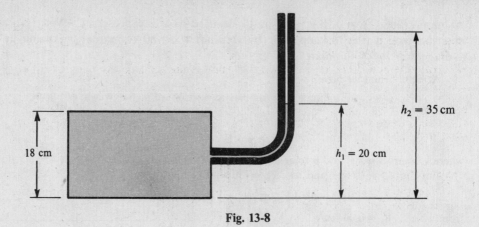

Fig. 13-8

13.29. A narrow tube is sealed onto a tank as shown in Fig. 13-8. The base of the tank has an area of 80 cm². (a) Find the force on the bottom of the tank due to the oil when the tank and capillary are filled with oil ($\rho = 0.72$ g/cm³) to the height h_1. (b) Repeat if filled to h_2.
Ans. (a) 11.3 N downward; (b) 19.8 N downward.

13.30. Repeat Problem 13.29 but now find the force on the top wall of the tank due to the oil.
Ans. (a) 1.13 N upward; (b) 9.6 N upward

13.31. Compute the pressure required for a water supply system that will raise water 150 ft vertically.
Ans. 9360 lb/ft² = 65 lb/in²

13.32. The area of a piston of a force pump is 8 cm². What force must be applied to the piston to raise oil ($\rho = 0.78$ g/cm³) to a height of 6.0 m? Assume the upper end of the oil is open to the atmosphere.
Ans. 37 N

13.33. The diameter of the large piston of a hydraulic press is 20 cm, and the area of the small piston is 0.50 cm². If a force of 400 N is applied to the small piston, (a) what is the resulting force exerted by the large piston? (b) What is the increase in pressure underneath the small piston? (c) underneath the large piston? *Ans.* (a) 2.5×10^5 N; (b) 8 MPa; (c) 8 MPa

13.34. A metal object "weighs" 26.03 g in air and 21.48 g when totally immersed in water. What is the volume of the object? Its mass density? *Ans.* 4.55 cm³, 5720 kg/m³

13.35. A solid piece of aluminum ($\rho = 2.70$ g/cm³) "weighs" 8.35 g in air. If it is now submerged in a vat of oil ($\rho = 0.75$ g/cm³) by a thread, what will be the tension in the thread? *Ans.* 0.059 N

13.36. A beaker contains oil of density 0.80 g/cm³. By means of a thread, a 1.6 cm cube of aluminum ($\rho = 2.70$ g/cm³) is submerged in the oil. Find the tension in the thread. *Ans.* 0.076 N

13.37. A tank containing oil of sp gr = 0.80 rests on a scale and weighs 320.2 lb. By means of a wire, a 6 inch cube of aluminum, sp gr = 2.70, is submerged in the oil. Find (a) the tension in the wire and (b) the scale reading if none of the oil overflows. *Ans.* (a) 14.8 lb; (b) 326.4 lb

13.38. Downward forces of 45 N and 15 N respectively are required to keep a plastic block totally immersed in water and in oil. If the volume of the block is 8000 cm³, find the density of the oil.
Ans. 620 kg/m³

13.39. Determine the unbalanced force acting on an iron ball ($r = 1.5$ cm, $\rho = 7.8$ g/cm^3) when just released while totally immersed in (a) water and (b) mercury ($\rho = 13.6$ g/cm^3). What will be the initial acceleration of the ball in each case?
Ans.   (a) 0.94 N down, 8.6 m/s^2 down;   (b) 0.80 N up, 7.3 m/s^2 up

13.40. A 2 cm cube of metal is suspended by a thread attached to a scale.   The cube appears to "weigh" 47.3 g when submerged in water.   What will it appear to "weigh" when submerged in glycerin,   sp gr = 1.26? (Hint: Find ρ too.)   Ans. 45.2 g

13.41. A balloon and its gondola has a total (empty) mass of 200 kg.   When filled, the balloon contains 900 m^3 of helium, which has a density of 0.183 kg/m^3.   Find the added load that the balloon can lift.   Density of air is 1.29 kg/m^3.   Ans. 7800 N

13.42. A certain piece of metal "weighs" 5.00 g in air, 3.00 g in water, and 3.24 g in benzene.   Determine the mass density of the metal and of the benzene.   Ans. 2500 kg/m^3, 880 kg/m^3

13.43. An alloy spring which may be either bronze (sp gr 8.8) or brass (sp gr 8.4) "weighs" 1.26 g in air and 1.11 g in water.   Which is it?   Ans. brass

13.44. What fraction of the volume of a piece of quartz ($\rho = 2.65$ g/cm^3) will be submerged when floating in a container of mercury ($\rho = 13.6$ g/cm^3)?   Ans. 0.195

13.45. A solid weighs 10 lb in air and 6 lb in a liquid whose specific gravity is 0.80.   What is the specific gravity of the solid?   Ans. 2.0

13.46. A cube of wood floating in water supports a 200 g mass resting on the center of its top face.   When the mass is removed, the cube rises 2 cm.   Determine the volume of the cube.   Ans. 1000 cm^3

13.47. A cork "weighs" 5 g in air.   A sinker "weighs" 86 g in water.   The cork is attached to the sinker and both together "weigh" 71 g when under water.   What is the density of the cork?   Ans. 250 kg/m^3

13.48. A glass of water has a 10 cm^3 ice cube floating in it.   The glass is filled to the brim with cold water.   By the time the ice cube has completely melted, how much water will have flowed out of the glass?   The sp gr of ice is 0.92.   Ans. none

13.49. A glass tube is bent into the form of a U.   A 50 cm height of olive oil in one arm is found to balance 46 cm of water in the other.   What is the density of the olive oil?   Ans. 920 kg/m^3

13.50. On a day when the pressure of the atmosphere is 1.000×10^5 Pa, a chemist distills a liquid under slightly reduced pressure.   The pressure within the distillation chamber is read by an oil-filled manometer (density of oil = 0.78 g/cm^3).   The difference in heights on the two sides of the manometer is 27 cm. What is the pressure in the distillation chamber?   Ans. 0.979×10^5 Pa

Chapter 14

Fluids in Motion

THE FLUID FLOW OR DISCHARGE (Q): When a fluid that fills a pipe flows through the pipe with an average velocity v, the *flow* or *discharge*, Q, is

$$Q = Av$$

where A is the cross-sectional area of the pipe. The units of Q are m^3/s in the SI and ft^3/s in the British system.

Sometimes Q is called the *rate of flow* or the *discharge rate*.

EQUATION OF CONTINUITY: Suppose an *incompressible* (constant-density) fluid fills a pipe and flows through it. Suppose further that the cross-sectional area of the pipe is A_1 at one point and A_2 at another. Since the flow through A_1 must equal the flow through A_2, one has

$$Q = A_1 v_1 = A_2 v_2 = \text{constant}$$

where v_1 and v_2 are the average fluid velocities over A_1 and A_2 respectively.

VISCOSITY (η) OF A FLUID is a measure of how difficult it is to cause the fluid to flow. A very viscous fluid, such as tar, has a high viscosity. Suppose that a nonelastic fluid is sheared between two plates, as shown in Fig. 14-1. If the velocity v of the upper plate is not too large, the fluid shears in the way indicated. The viscosity η is related to the force F required to produce the velocity v by

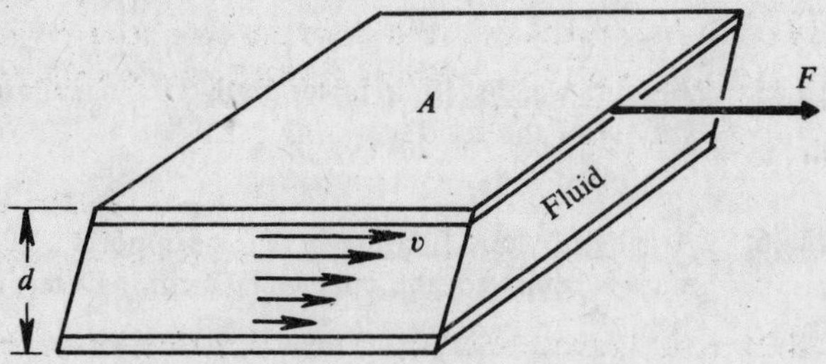

$$F = \frac{vA}{d}\,\eta$$

Fig. 14-1

where A is the area of either plate and d is the distance between the plates. The SI unit for η is $N \cdot s/m^2$ or $kg/m \cdot s$. Other units used are

$$1 \text{ poiseuille (Pl)} = 1 \text{ N} \cdot s/m^2 = 1 \text{ kg/m} \cdot s$$

$$1 \text{ poise (P)} = 0.1 \text{ kg/m} \cdot s$$

$$1 \text{ centipoise (cP)} = 10^{-3} \text{ kg/m} \cdot s$$

The British unit for η is $lb \cdot s/ft^2$.

POISEUILLE'S LAW: The fluid flow through a cylindrical pipe of length L and cross-sectional radius r is given by

$$Q = \frac{\pi r^4 (p_1 - p_2)}{8\eta L}$$

where $p_1 - p_2$ is the pressure difference between the two ends of the pipe.

WORK done by a piston in forcing a volume V of fluid into a cylinder against an opposing pressure p is given by pV.

BERNOULLI'S EQUATION for steady flow of a continuous stream of fluid: Consider two different points along the stream path.   Let point 1 be at a height h_1, and let v_1, ρ_1, and p_1 be the fluid velocity, density, and pressure at that point.   Similarly define h_2, v_2, ρ_2, and p_2 for point 2.   Then, provided the fluid is incompressible and has negligible viscosity,

$$p_1 + \tfrac{1}{2}\rho v_1^2 + h_1 \rho g = p_2 + \tfrac{1}{2}\rho v_2^2 + h_2 \rho g$$

where $\rho_1 = \rho_2 = \rho$ and g is the acceleration due to gravity.

TORRICELLI'S THEOREM: Suppose that a tank contains liquid open to the atmosphere at its top. If an orifice (opening) exists in the tank at a distance h below the top of the liquid, then the *velocity of outflow* from the orifice is $\sqrt{2gh}$, provided the liquid obeys Bernoulli's equation and the top of the liquid may be regarded as motionless.

Solved Problems

14.1.   Oil flows through a pipe 8 cm in diameter at an average velocity of 4 m/s.   What is the flow Q in m^3/s and m^3/h?

$$Q = Av = \pi(0.04 \text{ m})^2(4 \text{ m/s}) = 0.020 \text{ m}^3/\text{s}$$
$$= (0.020 \text{ m}^3/\text{s})(3600 \text{ s/h}) = 72 \text{ m}^3/\text{h}$$

14.2.   It is measured that 250 cm^3 of fluid flows out of a tube whose inner diameter is 7 mm in a time of 41 s.   What is the average velocity of the fluid in the tube?

From $Q = Av$,

$$v = \frac{Q}{A} = \frac{(250 \times 10^{-6} \text{ m}^3)/(41 \text{ s})}{\pi(0.0035 \text{ m})^2} = 0.158 \text{ m/s}$$

14.3.   A 14 cm inner diameter (i.d.) water main furnishes water (through intermediate pipes) to a 1.00 cm i.d. faucet pipe.   If the average velocity in the faucet pipe is 3 cm/s, what will be the average velocity it causes in the water main?

The two flows are equal.   From the continuity equation, we have

$$Q = A_1 v_1 = A_2 v_2$$

Letting 1 be the faucet and 2 be the water main,

$$v_2 = v_1 \frac{A_1}{A_2} = v_1 \frac{\pi r_1^2}{\pi r_2^2} = (3 \text{ cm/s})\left(\frac{1}{14}\right)^2 = 0.0153 \text{ cm/s}$$

14.4.   How much water will flow in 30 s through 200 mm of capillary tube 1.5 mm i.d., if the pressure differential across the tube is 5 centimeters of mercury?   The viscosity of water is 0.801 cP and ρ for mercury is 13 600 kg/m^3.

We shall make use of Poiseuille's law with

$$p_1 - p_2 = \rho g h = (13\,600 \text{ kg/m}^3)(9.8 \text{ m/s}^2)(0.05 \text{ m}) = 6660 \text{ N/m}^2$$
$$\eta = (0.801 \text{ cP})\left(10^{-3} \frac{\text{kg/m}\cdot\text{s}}{\text{cP}}\right) = 8.01 \times 10^{-4} \text{ kg/m}\cdot\text{s}$$

Then

$$Q = \frac{\pi r^4(p_1 - p_2)}{8\eta L} = \frac{\pi(7.5 \times 10^{-4} \text{ m})^4(6660 \text{ N/m}^2)}{8(8.01 \times 10^{-4} \text{ kg/m}\cdot\text{s})(0.2 \text{ m})} = 5.2 \times 10^{-6} \text{ m}^3/\text{s} = 5.2 \text{ cm}^3/\text{s}$$

In 30 s, the quantity that would flow out of the tube is $(5.2 \text{ cm}^3/\text{s})(30 \text{ s}) = 155 \text{ cm}^3$.

14.5. An artery in a certain man has been reduced to half its original inside diameter by deposits on the inner artery wall. By what factor will the blood flow through the artery be reduced if the pressure differential across the artery has remained unchanged?

From Poiseuille's law, $Q \propto r^4$. Therefore,

$$\frac{Q \text{ final}}{Q \text{ original}} = \left(\frac{r \text{ final}}{r \text{ original}} \right)^4 = \left(\frac{1}{2} \right)^4 = 0.0625$$

14.6. Under the same pressure differential, compare the flow of water through a pipe to the flow of SAE No. 10 oil. η for water is 0.801 cP; η for the oil is 200 cP.

From Poiseuille's law, $Q \propto 1/\eta$. Therefore,

$$\frac{Q \text{ of water}}{Q \text{ of oil}} = \frac{200 \text{ cP}}{0.80 \text{ cP}} = 250$$

so the flow of water is 250 times as large as that of the oil under the same pressure differential.

14.7. What volume of water will escape per minute from an open-top tank through an opening 3 cm in diameter that is 5 m below the water level in the tank? See Fig. 14-2.

We can use Bernoulli's equation, with *1* being at the top level and *2* being at the orifice. Then $p_1 = p_2$ and $h_1 = 5$ m, $h_2 = 0$.

$$p_1 + \tfrac{1}{2}\rho v_1^2 + h_1 \rho g = p_2 + \tfrac{1}{2}\rho v_2^2 + h_2 \rho g$$

$$\tfrac{1}{2}\rho v_1^2 + h_1 \rho g = \tfrac{1}{2}\rho v_2^2 + h_2 \rho g$$

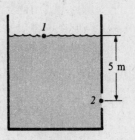

If the tank is large, v_1 can be approximated as zero. Then, solving for v_2, we obtain Torricelli's equation:

$$v_2 = \sqrt{2g(h_1 - h_2)} = \sqrt{2(9.8 \text{ m/s}^2)(5 \text{ m})} = 9.9 \text{ m/s}$$

and the flow is given by

Fig. 14-2

$$Q = v_2 A_2 = (9.9 \text{ m/s})\pi(1.5 \times 10^{-2} \text{ m})^2 = 7 \times 10^{-3} \text{ m}^3/\text{s} = 0.42 \text{ m}^3/\text{min}$$

14.8. A gauge reads the water pressure inside a water tank to be 500 kPa, when the tank springs a leak at the position shown in Fig. 14-3. What is the velocity of escape of the water through the hole?

Use Bernoulli's equation, with $p_1 - p_2 = 5 \times 10^5 \text{ N/m}^2$, $h_1 = h_2$, and approximating $v_1 = 0$. Then

$$(p_1 - p_2) + (h_1 - h_2)\rho g = \tfrac{1}{2}\rho v_2^2$$

whence

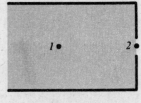

$$v_2 = \sqrt{\frac{2(p_1 - p_2)}{\rho}} = \sqrt{\frac{2(5 \times 10^5 \text{ N/m}^2)}{1000 \text{ kg/m}^3}} = 32 \text{ m/s}$$

Fig. 14-3

14.9. Water flows at the rate of 30 cm³/s through an opening at the bottom of a tank in which the water is 4 m deep. Calculate the rate of escape of the water if an added pressure of 50 kPa is applied to the top of the water.

From Bernoulli's equation in the case that v_1 is essentially zero,

$$(p_1 - p_2) + (h_1 - h_2)\rho g = \tfrac{1}{2}\rho v_2^2$$

We can write this twice, before the pressure is added and after.

$$(p_1 - p_2)_{\text{before}} + (h_1 - h_2)\rho g = \tfrac{1}{2}\rho (v_2^2)_{\text{before}}$$

$$(p_1 - p_2)_{\text{before}} + 5 \times 10^4 \text{ N/m}^2 + (h_1 - h_2)\rho g = \tfrac{1}{2}\rho (v_2^2)_{\text{after}}$$

If the opening and the top of the tank were originally at atmospheric pressure,

$$(p_1 - p_2)_{\text{before}} = 0$$

Then, division of the second equation by the first gives

$$\frac{(v_2^2)_{\text{after}}}{(v_2^2)_{\text{before}}} = \frac{5 \times 10^4 \text{ N/m}^2 + (h_1 - h_2)\rho g}{(h_1 - h_2)\rho g}$$

But $(h_1 - h_2)\rho g = (4 \text{ m})(1000 \text{ kg/m}^3)(9.8 \text{ m/s}^2) = 3.9 \times 10^4 \text{ N/m}^2$

Therefore,

$$\frac{(v_2)_{\text{after}}}{(v_2)_{\text{before}}} = \sqrt{\frac{8.9 \times 10^4 \text{ N/m}^2}{3.9 \times 10^4 \text{ N/m}^2}} = 1.51$$

Since $Q = Av$,   this can be written as

$$\frac{Q_{\text{after}}}{Q_{\text{before}}} = 1.51 \qquad \text{or} \qquad Q_{\text{after}} = (30 \text{ cm}^3/\text{s})(1.51) = 45 \text{ cm}^3/\text{s}$$

14.10. (a) How much work W is done by a pump in raising 5 m^3 of water 20 m and forcing it into a main at a pressure of 150 kPa?   (b) Repeat if 100 ft^3 of water is raised 40 ft and forced in at a pressure of 20 lb/in^2.

(a) $W = (\text{work to raise water}) + (\text{work to push it in}) = mgh + pV$

$= (5 \text{ m}^3 \times 1000 \text{ kg/m}^3)(9.8 \text{ m/s}^2)(20 \text{ m}) + (1.5 \times 10^5 \text{ N/m}^2)(5 \text{ m}^3) = 1.73 \times 10^6 \text{ J}$

(b) $W = (100 \text{ ft}^3 \times 62.4 \text{ lb/ft}^3)(40 \text{ ft}) + (2880 \text{ lb/ft}^2)(100 \text{ ft}^3) = 5.38 \times 10^5 \text{ ft} \cdot \text{lb}$

14.11. A horizontal pipe has a constriction in it as shown in Fig. 14-4.   At point 1 the diameter is 6 cm, while at point 2 it is only 2 cm.   At point 1, $v_1 = 2 \text{ m/s}$ and $p_1 = 180 \text{ kPa}$.   Calculate v_2 and p_2.

Using Bernoulli's equation with $h_1 = h_2$,   we have

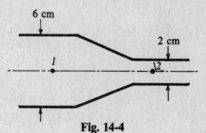

Fig. 14-4

$$p_1 + \tfrac{1}{2}\rho v_1^2 = p_2 + \tfrac{1}{2}\rho v_2^2 \qquad \text{or} \qquad p_1 + \tfrac{1}{2}\rho(v_1^2 - v_2^2) = p_2$$

However, $v_1 = 2 \text{ m/s}$ and the equation of continuity tells us that

$$v_2 = v_1 \frac{A_1}{A_2} = (2 \text{ m/s})\left(\frac{r_1}{r_2}\right)^2 = (2 \text{ m/s})(9) = 18 \text{ m/s}$$

Substituting,

$$1.8 \times 10^5 \text{ N/m}^2 + \tfrac{1}{2}(1000 \text{ kg/m}^3)\left[(2 \text{ m/s})^2 - (18 \text{ m/s})^2\right] = p_2$$

from which $p_2 = 0.20 \times 10^5 \text{ N/m}^2 = 20 \text{ kPa}$.

14.12. The pipe shown in Fig. 14-5 has a diameter of 16 cm at section 1 and 10 cm at section 2.   At section 1 the pressure is 200 kPa.   Point 2 is 6 m higher than point 1.   When oil of density 800 kg/m^3 flows at a rate of 0.03 m^3/s, find the pressure at point 2 if viscous effects are negligible.

From $Q = v_1 A_1 = v_2 A_2$ we have

$$v_1 = \frac{Q}{A_1} = \frac{0.03 \text{ m}^3/\text{s}}{\pi(8 \times 10^{-2} \text{ m})^2} = 1.49 \text{ m/s}$$

$$v_2 = \frac{Q}{A_2} = \frac{0.03 \text{ m}^3/\text{s}}{\pi(5 \times 10^{-2} \text{ m})^2} = 3.82 \text{ m/s}$$

We can now use Bernoulli's equation:

$$p_1 + \tfrac{1}{2}\rho v_1^2 + \rho g(h_1 - h_2) = p_2 + \tfrac{1}{2}\rho v_2^2$$

Setting $p_1 = 2 \times 10^5$ N/m², $h_2 - h_1 = 6$ m and $\rho = 800$ kg/m³ gives

$$p_2 = 2 \times 10^5 \text{ N/m}^3 + \tfrac{1}{2}(800 \text{ kg/m}^3)\left[(1.49 \text{ m/s})^2 - (3.82 \text{ m/s})^2\right] - (800 \text{ kg/m}^3)(9.8 \text{ m/s}^2)(6 \text{ m})$$

$$= 1.48 \times 10^5 \text{ N/m}^2 = 148 \text{ kPa}$$

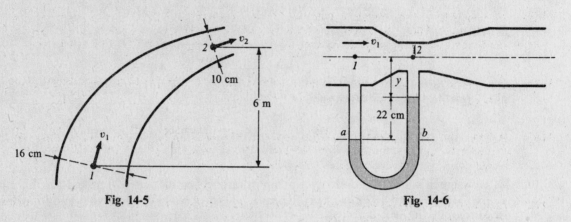

Fig. 14-5   Fig. 14-6

14.13. A venturi meter equipped with a differential mercury manometer is shown in Fig. 14-6. At the inlet, point 1, the diameter is 12 cm, while at the throat, point 2, the diameter is 6 cm. What is the flow Q of water through the meter if the manometer reading is 22 cm? The density of mercury is 13.6 g/cm³.

From the manometer reading we obtain

$$p_1 - p_2 = \rho g h = (13\,600 \text{ kg/m}^3)(9.8 \text{ m/s}^2)(0.22 \text{ m}) = 2.93 \times 10^4 \text{ N/m}^2$$

Since $Q = v_1 A_1 = v_2 A_2$, we have $v_1 = Q/A_1$ and $v_2 = Q/A_2$. Using Bernoulli's equation, with $h_1 - h_2 = 0$, gives

$$(p_1 - p_2) + \tfrac{1}{2}\rho(v_1^2 - v_2^2) = 0$$

$$2.93 \times 10^4 \text{ N/m}^2 + \tfrac{1}{2}(1000 \text{ kg/m}^3)\left(\frac{1}{A_1^2} - \frac{1}{A_2^2}\right)Q^2 = 0$$

Since

$$A_1 = \pi r_1^2 = \pi(0.06)^2 \text{ m}^2 = 0.01131 \text{ m}^2 \qquad A_2 = \pi r_2^2 = \pi(0.03)^2 \text{ m}^2 = 0.0028 \text{ m}^2$$

substitution gives $Q = 0.022$ m³/s.

Supplementary Problems

14.14. Oil flows through a 4 cm i.d. pipe at an average velocity of 2.5 m/s. Find the flow in m³/s and cm³/s. *Ans.* 3.14×10^{-3} m³/s = 3140 cm³/s

14.15. Compute the average velocity of water in a pipe having an i.d. of 2 inches and delivering 600 ft³ of water per hour. *Ans.* 7.64 ft/s

14.16. The velocity of glycerin in a 5 cm i.d. pipe is 0.54 m/s. Find the velocity in a 3 cm i.d. pipe that connects with it, both pipes flowing full. *Ans.* 1.50 m/s

14.17. How long will it take for 500 cm^3 of water to flow through a 15 cm long, 3 mm i.d. pipe, if the pressure differential across the pipe is 4 kPa? The viscosity of water is 0.80 cP.   *Ans.*   7.5 s

14.18. A molten plastic flows out of a tube 8 cm long at a rate of 13 cm^3/min when the pressure differential between the two ends of the tube is 18 cm of mercury. Find the viscosity of the plastic. The i.d. of the tube is 1.30 mm. The density of mercury is 13.6 g/cm^3.   *Ans.*   0.097 kg/m · s = 97 cP

14.19. In a horizontal pipe system, a pipe (i.d. 4 mm) 20 cm long connects in line to a pipe (i.d. 5 mm) 30 cm long. When a viscous fluid is being pushed through the pipes at a steady rate, what is the ratio of the pressure drop across the 20 cm pipe to that across the 30 cm pipe?   *Ans.*   1.63

14.20. How much work does the piston in a hydraulic system do during one 2 cm stroke if the end area of the piston is 0.75 cm^2 and the pressure in the hydraulic fluid is 50 kPa?   *Ans.*   0.075 J

14.21. A large open tank of nonviscous liquid springs a leak 4.5 m below the top of the liquid. What is the theoretical velocity of outflow from the hole? If the area of the hole is 0.25 cm^2, how much liquid would escape in 1 minute?   *Ans.*   9.4 m/s, 0.0141 m^3

14.22. Find the flow in liters/s of a nonviscous liquid through an opening 0.5 cm^2 in area and 2.5 m below the level of the liquid in an open tank.   *Ans.*   0.35 liter/s

14.23. Calculate the theoretical velocity of efflux of water from an aperture that is 8 m below the surface of water in a large tank, if an added pressure of 140 kPa is applied to the surface of the water.
Ans.   21 m/s

14.24. Repeat Problem 14.23 if the aperture is 25 ft below the surface and the added pressure is 20 lb/in^2.
Ans.   68 ft/s

14.25. What horsepower is required to force 8 m^3 of water per minute into a water main at a pressure of 220 kPa?   *Ans.*   39 hp

14.26. What horsepower is required to pump 200 ft^3 of water per minute into a main at a pressure of 25 lb/in^2?
Ans.   22 hp

14.27. A pump lifts water at the rate of 9000 cm^3/s from a lake through a 5 cm i.d. pipe and discharges it into the air at a point 16 m above the level of the water in the lake. What are the theoretical (*a*) velocity of the water at the point of discharge and (*b*) power delivered by the pump.
Ans.   (*a*) 4.6 m/s;   (*b*) 2.0 hp

14.28. Water flows steadily through a horizontal pipe of varying cross section. At one place the pressure is 130 kPa and the speed is 0.60 m/s. Determine the pressure at another place in the same pipe where the speed is 9.0 m/s.   *Ans.*   90 kPa

14.29. Repeat Problem 14.28 if, at the first point, $p = 20$ lb/in^2 and $v = 2$ ft/s, while $v = 30$ ft/s at the second point.   *Ans.*   14 lb/in^2

14.30. A pipe of varying inner diameter carries water. At point 1 the diameter is 20 cm and the pressure is 130 kPa. At point 2, which is 4 m higher than point 1, the diameter is 30 cm. If the flow is 0.080 m^3/s, what is the pressure at the second point?   *Ans.*   93 kPa

14.31. Fuel oil of density 820 kg/m^3 flows through a venturi meter having a throat diameter of 4 cm and an entrance diameter of 8 cm. The pressure drop between entrance and throat is 16 cm of mercury. Find the flow. The density of mercury is 13 600 kg/m^3.   *Ans.* 9.3×10^{-3} m^3/s

Chapter 15

Thermal Expansion

TEMPERATURE is commonly measured on the Celsius (or centigrade) scale and on the Fahrenheit scale. For scientific purposes, the absolute Kelvin scale and the Rankine scale are used.

The Kelvin and Celsius scales use the same intervals for degrees. The Fahrenheit and Rankine scales use an interval such that

$$1 \text{ Celsius degree} = \frac{180}{100} = \frac{9}{5} \text{ Fahrenheit degree}$$

A comparison of the four scales is given in Fig. 15-1.

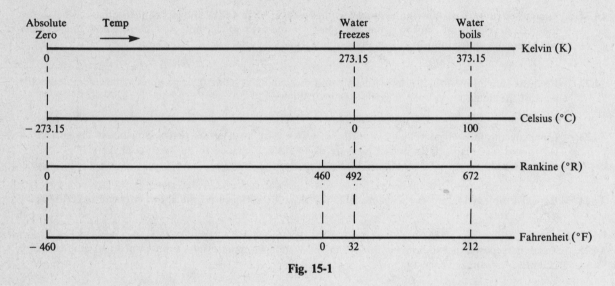

Fig. 15-1

Because the freezing point of water (at 1 atmosphere pressure) is 0 °C and 32 °F, and because 1 Celsius degree equals 9/5 Fahrenheit degree, the two scales can be converted by

$$\text{Fahrenheit temperature} = \frac{9}{5} \text{ (Celsius temperature)} + 32$$

$$\text{Celsius temperature} = \frac{5}{9} \text{ (Fahrenheit temperature} - 32)$$

LINEAR EXPANSION OF SOLIDS: When a solid is subjected to a rise in temperature (ΔT), its increase in length (ΔL) is very nearly proportional to its initial length (L_0) multiplied by ΔT. Therefore,

$$\Delta L = \alpha L_0 \Delta T$$

where the proportionality constant α is called the *coefficient of linear expansion*. The value of α depends on the nature of the substance.

From the above equation α is the change in length per unit initial length per degree change in temperature. For example, if a 1.000 000 cm length of brass becomes 1.000 019 cm long when the temperature is raised 1 °C, the linear expansion coefficient for brass is

$$\alpha = \frac{\Delta L}{L_0 \Delta T} = \frac{0.000\,019 \text{ cm}}{(1 \text{ cm})(1 \text{ °C})} = 1.9 \times 10^{-5} \text{ °C}^{-1}$$

118

VOLUME EXPANSION: If a volume V_0 expands to $V_0 + \Delta V$ when subjected to a temperature rise ΔT, then

$$\Delta V = \beta V_0 \Delta T$$

where β is the *coefficient of volume expansion* or *cubical expansion coefficient*. For many solids, $\beta = 3\alpha$, approximately.

Solved Problems

15.1.   Ethyl alcohol boils at 78.5 °C and freezes at -117 °C under a pressure of 1 atmosphere. Convert these temperatures to the (*a*) Kelvin scale and (*b*) Fahrenheit scale.

(*a*)   From Fig. 15-1, we see that

Kelvin temperature = Celsius temperature + 273.15 = 78.5 + 273.15 = 351.7 K

where the answer is read as "351.7 kelvins."   Similarly, for -117 °C,

Kelvin temperature = $-117 + 273.15 = 156$ K

(*b*)   We use   Fahrenheit temp = (9/5)(Celsius temp) + 32.

$$\text{boiling point} = \frac{9}{5}(78.5) + 32 = 141 + 32 = 173 \text{ °F}$$

$$\text{freezing point} = \frac{9}{5}(-117) + 32 = -211 + 32 = -179 \text{ °F}$$

15.2.   Mercury boils at 675 °F and solidifies at -38 °F under 1 atmosphere pressure.   Express these on the Rankine, Celsius, and Kelvin scales.

From Fig. 15-1 we see that

Rankine temperature = Fahrenheit temperature + 460

Therefore, on the Rankine scale,

freezing point = $-38 + 460 = 422$ °R
boiling point = $675 + 460 = 1135$ °R

From   Celsius temp = (5/9)(Fahrenheit temp $-$ 32),   the following Celsius temperatures apply:

$$\text{freezing point} = \frac{5}{9}(-38 - 32) = -38.9 \text{ °C}$$

$$\text{boiling point} = \frac{5}{9}(675 - 32) = 357 \text{ °C}$$

But   Kelvin temp = Celsius temp + 273.15   and so, on the Kelvin scale,

freezing point = $-38.9 + 273.15 = 234.3$ K
boiling point = $357 + 273.15 = 630$ K

15.3.   A copper bar is 80 cm long at 15 °C.   What is the increase in length when heated to 35 °C? The linear expansion coefficient for copper is 1.7×10^{-5} °C^{-1}.

$$\Delta L = \alpha L_0 \Delta T = (1.7 \times 10^{-5} \text{ °C}^{-1})(0.80 \text{ m})[(35 - 15) \text{ °C}] = 2.7 \times 10^{-4} \text{ m}$$

15.4.   A cylinder of diameter 1.000 00 cm at 30 °C is to be slid into a hole in a steel plate.   The hole has a diameter of 0.999 70 cm at 30 °C.   To what temperature must the plate be heated?   For steel, $\alpha = 1.1 \times 10^{-5}$ °C^{-1}.

The plate will expand in the same way whether or not there is a hole in it.   Hence the hole expands in the same way a circle of steel filling it would expand.   We want the diameter of the hole to change by

$$\Delta L = (1.000\,00 - 0.999\,70) \text{ cm} = 0.000\,30 \text{ cm}$$

Using $\Delta L = \alpha L_0 \Delta T$,

$$\Delta T = \frac{\Delta L}{\alpha L_0} = \frac{0.000\,30 \text{ cm}}{(1.1 \times 10^{-5}\,{}^{\circ}\text{C}^{-1})(0.999\,70 \text{ cm})} = 27.3\,{}^{\circ}\text{C}$$

The temperature of the plate must be $30 + 27.3 = 57.3\,{}^{\circ}\text{C}$.

15.5.   A steel tape is calibrated at 20 °C.   On a cold day when the temperature is $-15\,{}^{\circ}\text{C}$, what will be the percent error in the tape? $\alpha_{\text{steel}} = 1.1 \times 10^{-5}\,{}^{\circ}\text{C}^{-1}$.

For a temperature change from 20 °C to $-15\,{}^{\circ}\text{C}$, we have $\Delta T = -35\,{}^{\circ}\text{C}$.   Then,

$$\frac{\Delta L}{L_0} = \alpha\,\Delta T = (1.1 \times 10^{-5}\,{}^{\circ}\text{C}^{-1})(-35\,{}^{\circ}\text{C}) = -3.9 \times 10^{-4} = -0.039\,\%$$

15.6.   A steel tape measures the length of a copper rod as 90.00 cm when both are at 10 °C, the calibration temperature for the tape.   What would the tape read for the length of the rod when both are at 30 °C? $\alpha_{\text{steel}} = 1.1 \times 10^{-5}\,{}^{\circ}\text{C}^{-1}$, $\alpha_{\text{copper}} = 1.7 \times 10^{-5}\,{}^{\circ}\text{C}^{-1}$.

At 30 °C, the copper rod will be of length

$$L_0(1 + \alpha_c\,\Delta T)$$

while adjacent "centimeter" marks on the steel tape will be separated by a distance of

$$(1 \text{ cm})(1 + \alpha_s\,\Delta T)$$

Therefore, the number of "centimeters" read on the tape will be

$$\frac{L_0(1 + \alpha_c\,\Delta T)}{(1 \text{ cm})(1 + \alpha_s\,\Delta T)} = \frac{(90 \text{ cm})\left[1 + (1.7 \times 10^{-5}\,{}^{\circ}\text{C}^{-1})(20\,{}^{\circ}\text{C})\right]}{(1 \text{ cm})\left[1 + (1.1 \times 10^{-5}\,{}^{\circ}\text{C}^{-1})(20\,{}^{\circ}\text{C})\right]} = 90\,\frac{1 + 3.4 \times 10^{-4}}{1 + 2.2 \times 10^{-4}}$$

Using the approximation

$$\frac{1}{1 + x} \approx 1 - x$$

for x small compared to 1, we have

$$90\,\frac{1 + 3.4 \times 10^{-4}}{1 + 2.2 \times 10^{-4}} \approx 90(1 + 3.4 \times 10^{-4})(1 - 2.2 \times 10^{-4}) \approx 90(1 + 3.4 \times 10^{-4} - 2.2 \times 10^{-4})$$

$$= 90 + 0.0108$$

The tape will read 90.01 cm.

15.7.   A glass flask is filled "to the mark" with 50.00 cm^3 of mercury at 18 °C.   If the flask and its contents are heated to 38 °C, how much mercury will be above the mark? $\alpha_{\text{glass}} = 9 \times 10^{-6}\,{}^{\circ}\text{C}^{-1}$ and $\beta_{\text{mercury}} = 180 \times 10^{-6}\,{}^{\circ}\text{C}^{-1}$.

We shall take $\beta_{\text{glass}} = 3\alpha_{\text{glass}}$ as a good approximation.   The flask interior will expand just as though it were a solid piece of glass.   Thus

$$\text{volume above mark} = (\Delta V \text{ for mercury}) - (\Delta V \text{ for glass})$$

$$= \beta_m V_0 \Delta T - \beta_g V_0 \Delta T = (\beta_m - \beta_g)V_0\,\Delta T$$

$$= \left[(180 - 27) \times 10^{-6}\,{}^{\circ}\text{C}^{-1}\right](50 \text{ cm}^3)[(38 - 18)\,{}^{\circ}\text{C}]$$

$$= 0.15 \text{ cm}^3$$

15.8.   The density of mercury at 0 °C is 13 600 kg/m^3 and its cubical expansion coefficient is $1.82 \times 10^{-4}\,{}^{\circ}\text{C}^{-1}$.   Calculate the density of mercury at 50 °C.

Let $\rho_0 = $ density of mercury at 0 °C
        $\rho_1 = $ density of mercury at 50 °C
        $V_0 = $ volume of m kg of mercury at 0 °C
        $V_1 = $ volume of m kg of mercury at 50 °C

By conservation of mass, $\quad m = \rho_0 V_0 = \rho_1 V_1, \quad$ from which

$$\rho_1 = \rho_0 \frac{V_0}{V_1} = \rho_0 \frac{V_0}{V_0 + \Delta V} = \rho_0 \frac{1}{1 + (\Delta V / V_0)}$$

But

$$\frac{\Delta V}{V_0} = \beta \, \Delta T = (1.82 \times 10^{-4} \, °\text{C}^{-1})(50 \, °\text{C}) = 0.0091$$

Substitution gives

$$\rho_1 = (13\,600 \text{ kg/m}^3) \frac{1}{1 + 0.0091} = 13\,480 \text{ kg/m}^3$$

15.9. A steel wire of 2.0 mm² cross section is held straight (but under no tension) by attaching it firmly to two points a distance 1.50 m apart at 30 °C. If the temperature now decreases to -10 °C, and if the two tie points remain fixed, what will be the tension in the wire? For steel, $\alpha = 1.1 \times 10^{-5} \, °\text{C}^{-1}$ and $Y = 2.0 \times 10^{11} \text{ N/m}^2$.

If free to do so, the wire would contract a distance ΔL as it cooled, where

$$\Delta L = \alpha L_0 \, \Delta T = (1.1 \times 10^{-5} \, °\text{C}^{-1})(1.5 \text{ m})(40 \, °\text{C}) = 6.6 \times 10^{-4} \text{ m}$$

But the ends are fixed. As a result, forces at the ends must, in effect, stretch the wire this same length, ΔL. Therefore, from $Y = (F/A)/(\Delta L/L)$, we have

$$\text{tension} = F = \frac{YA \, \Delta L}{L}$$

$$= \frac{(2 \times 10^{11} \text{ N/m}^2)(2 \times 10^{-6} \text{ m}^2)(6.6 \times 10^{-4} \text{ m})}{1.5 \text{ m}} = 176 \text{ N}$$

Strictly, we should have substituted $(1.5 - 6.6 \times 10^{-4})$ m for L in the expression for the tension. However, the error incurred in not doing so is negligible.

15.10. When a building is constructed at -10 °C, a steel beam (cross-sectional area 45 cm²) is put in place with its ends cemented in pillars. If the sealed ends cannot move, what will be the compressional force in the beam when the temperature is 25 °C? For steel, $\alpha = 1.1 \times 10^{-5} \, °\text{C}^{-1}$ and $Y = 2 \times 10^{11} \text{ N/m}^2$.

We proceed much as in Problem 15.9.

$$\frac{\Delta L}{L_0} = \alpha \, \Delta T = (1.1 \times 10^{-5} \, °\text{C}^{-1})(35 \, °\text{C}) = 3.85 \times 10^{-4}$$

Then

$$F = YA \frac{\Delta L}{L_0} = (2 \times 10^{11} \text{ N/m}^2)(45 \times 10^{-4} \text{ m}^2)(3.85 \times 10^{-4}) = 3.5 \times 10^5 \text{ N}$$

Supplementary Problems

15.11. Change each of the following to the Celsius and Kelvin scales: 68 °F, 5 °F, 176 °F.
Ans. 20 °C, 293 K; -15 °C, 258 K; 80 °C, 353 K

15.12. Change each of the following to the Fahrenheit and Rankine scales: 30 °C, 5 °C, -20 °C.
Ans. 86 °F, 546 °R; 41 °F, 501 °R; -4 °F, 456 °R

15.13. At what temperature do the Celsius and Fahrenheit readings have the same numerical value?
Ans. -40 °C $= -40$ °F

15.14. Compute the increase in length of 50 m of copper wire when its temperature changes from 12 °C to 32 °C. For copper, $\alpha = 1.7 \times 10^{-5}$ °C^{-1}. *Ans.* 1.7 cm

15.15. A rod 3 m long is found to have expanded 0.091 cm in length for a temperature rise of 60 °C. What is α for the material of the rod? *Ans.* 5.1×10^{-6} °C^{-1}

15.16. At 15 °C, a bare wheel has a diameter 30.000 inches and the inside diameter of a steel rim is 29.930 inches. To what temperature must the rim be heated so as to slip over the wheel? For steel, $\alpha = 1.1 \times 10^{-5}$ °C^{-1}. *Ans.* 227 °C

15.17. An iron ball has a diameter of 6 cm and is 0.010 mm too large to pass through a hole in a brass plate when the ball and plate are at a temperature of 30 °C. At what temperature, the same for ball and plate, will the ball just pass through the hole? $\alpha = 1.2 \times 10^{-5}$ °C^{-1} and 1.9×10^{-5} °C^{-1}, for iron and brass respectively. *Ans.* 54 °C

15.18. (*a*) An aluminum measuring rod, which is correct at 5 °C, measures a certain distance as 88.42 cm at 35 °C. Determine the error in measuring the distance due to the expansion of the rod. (*b*) If this aluminum rod measures a length of steel as 88.42 cm at 35 °C, what is the correct length of the steel at 35 °C? The coefficient of linear expansion of aluminum is 22×10^{-6} °C^{-1}.
Ans. (*a*) 0.06 cm; (*b*) 88.48 cm

15.19. Calculate the increase in volume of 100 cm^3 of mercury when its temperature changes from 10 °C to 35 °C. The volume coefficient of expansion of mercury is 0.000 18 °C^{-1}. *Ans.* 0.45 cm^3

15.20. The coefficient of linear expansion of glass is 9×10^{-6} °C^{-1}. If a specific gravity bottle holds 50.000 cm^3 at 15 °C, find its capacity at 25 °C. *Ans.* 50.014 cm^3

15.21. Determine the change in volume of a block of cast iron 5 cm $\times$ 10 cm $\times$ 6 cm, when the temperature changes from 15 °C to 47 °C. The coefficient of linear expansion of cast iron is 0.000 010 °C^{-1}.
Ans. 0.29 cm^3

15.22. A glass vessel is filled with exactly 1 liter of turpentine at 20 °C. What volume of the liquid will overflow if the temperature is raised to 86 °C? The coefficient of linear expansion of glass is 9×10^{-6} °C^{-1}; the coefficient of cubical expansion of turpentine is 97×10^{-5} °C^{-1}. *Ans.* 62 cm^3

15.23. The density of gold is 19.30 g/cm^3 at 20 °C, and the coefficient of linear expansion is 14.3×10^{-6} °C^{-1}. Compute the density of gold at 90 °C. *Ans.* 19.24 g/cm^3

Chapter 16

Ideal Gases

AN IDEAL (OR PERFECT) GAS is one that obeys the ideal gas law, given below. At low to moderate pressures, and at temperatures not too low, the following common gases can be considered ideal: air, nitrogen, oxygen, helium, hydrogen, and neon. Almost any chemically stable gas behaves ideally if it is far removed from conditions under which it will liquefy or solidify.

THE IDEAL GAS LAW for a mass m of gas at pressure p confined to a volume V at absolute temperature T is:

$$\frac{pV}{T} = \text{constant for } m \text{ constant}$$

Notice that T must be measured on an absolute scale, a scale that has its zero at absolute zero. Usually T will be in kelvins, but degrees Rankine are also allowable (see Problem 16.7). Refer to Chapter 15 for a description of these temperature scales.

A MOLE OF SUBSTANCE is the amount of the substance that contains as many particles as there are atoms in exactly 12 grams of the isotope carbon-12. On a practical level, it is advantageous to use the *kilomole*, kmol. **A KILOMOLE OF SUBSTANCE** is the mass (in kg) that is numerically equal to the molecular (or atomic) weight of the substance. For example, the molecular weight of hydrogen gas, H_2, is 2; hence there are 2 kg in 1 kmol of H_2. Similarly, there are 32 kg in 1 kmol of O_2 and 28 kg in 1 kmol of N_2. We shall always use kilomoles and kilograms in our calculations.

THE UNIVERSAL GAS CONSTANT: If a mass m of ideal gas is confined to a volume V at pressure p and absolute temperature T, then the ideal gas law can be written as

$$pV = \left(\frac{m}{M}\right)RT$$

where M is the molecular weight of the gas and R is the *universal gas constant*. For m in kg and M in kg/kmol,

$$R = 8314 \frac{J}{\text{kmol} \cdot \text{K}}$$

Often the quantity m/M is replaced by n, the number of kilomoles of gas in the volume V.

STANDARD CONDITIONS (S.T.P.) are defined to be

$$T = 273.15 \text{ K} = 0 \text{ °C} \qquad p = 1.013 \times 10^5 \text{ Pa} = 1 \text{ atm}$$

Under standard conditions, 1 kmol of *ideal gas* occupies a volume of 22.4 m^3. Therefore, at S.T.P., 2 kg of H_2 occupies the same volume as 32 kg of O_2 or 28 kg of N_2, namely 22.4 m^3.

BOYLE'S LAW is obtained from the ideal gas law by holding T constant. Thus

$$pV = \text{const.} \qquad \text{(at constant } T\text{)}$$

for a fixed mass m of gas.

CHARLES' LAW is obtained from the ideal gas law by holding p constant. Thus

$$V = (\text{const.})T \qquad (\text{at constant } p)$$

for a fixed mass m of gas.

GAY-LUSSAC'S LAW is obtained from the ideal gas law by holding V constant. Thus

$$p = (\text{const.})T \qquad (\text{at constant } V)$$

for a fixed mass m of gas.

DALTON'S LAW OF PARTIAL PRESSURES: Define the *partial pressure* of one component of a gas mixture to be the pressure the component gas would exert if it alone occupied the entire volume. Then, the total pressure of a mixture of ideal, nonreactive gases is the sum of the partial pressures of the component gases.

GAS LAW PROBLEMS involving change of conditions from p_1, V_1, T_1 to p_2, V_2, T_2, are usually easily solved by writing the gas law as

$$\frac{p_1 V_1}{T_1} = \frac{p_2 V_2}{T_2} \qquad (\text{at constant } m)$$

Solved Problems

16.1. A mass of oxygen occupies 0.0200 m³ at atmospheric pressure, 101 kPa, and 5 °C. Determine its volume if its pressure is increased to 108 kPa while its temperature is changed to 30 °C.

$$\frac{p_1 V_1}{T_1} = \frac{p_2 V_2}{T_2} \qquad \text{or} \qquad V_2 = V_1 \left(\frac{p_1}{p_2} \right) \left(\frac{T_2}{T_1} \right)$$

But $T_1 = 5 + 273 = 278$ K, $T_2 = 30 + 273 = 303$ K, and so

$$V_2 = (0.020 \text{ m}^3) \left(\frac{101}{108} \right) \left(\frac{303}{278} \right) = 0.0204 \text{ m}^3$$

16.2. On a day when atmospheric pressure is 14.5 lb/in², the pressure gauge on a tank reads the pressure inside to be 130 lb/in². The gas in the tank has a temperature of 9 °C. If the tank is heated to 31 °C by the sun, and if no gas exits from it, what will the pressure gauge read?

$$\frac{p_1 V_1}{T_1} = \frac{p_2 V_2}{T_2} \qquad \text{or} \qquad p_2 = p_1 \left(\frac{T_2}{T_1} \right) \left(\frac{V_1}{V_2} \right)$$

But gauges on tanks usually read the difference in pressure between inside and outside; this is called the *gauge pressure*. Therefore

$$p_1 = 14.5 \text{ lb/in}^2 + 130 \text{ lb/in}^2 = 144.5 \text{ lb/in}^2$$

Also, $V_1 = V_2$. We then have

$$p_2 = (144.5 \text{ lb/in}^2) \left(\frac{273 + 31}{273 + 9} \right)(1) = 155.8 \text{ lb/in}^2$$

The gauge will read $155.8 - 14.5 = 141.3$ lb/in².

16.3. Gas at room temperature and pressure is confined to a cylinder by a piston. The piston is now pushed in so as to reduce the volume to one-eighth of its original value. After the gas temperature has returned to room temperature, what will be the gauge pressure of the gas? Local atmospheric pressure is 740 mm of mercury.

$$\frac{p_1 V_1}{T_1} = \frac{p_2 V_2}{T_2} \qquad \text{or} \qquad p_2 = p_1 \left(\frac{V_1}{V_2} \right) \left(\frac{T_2}{T_1} \right)$$

But $T_1 = T_2$, $p_1 = 740$ mmHg, $V_2 = V_1/8$. Substitution gives

$$p_2 = (740 \text{ mmHg})(8)(1) = 5920 \text{ mmHg}$$

Gauge pressure is the difference between actual and atmospheric pressure. Therefore,

$$\text{gauge pressure} = 5920 - 740 = 5180 \text{ mmHg}$$

Since 760 mmHg = 101 kPa, the gauge reading is

$$(5180 \text{ mmHg}) \left(\frac{101 \text{ kPa}}{760 \text{ mmHg}} \right) = 690 \text{ kPa}$$

16.4. To how many atmospheres pressure must a liter of gas measured at 1 atm and $-20\,°$C be subjected, to be compressed to $\frac{1}{2}$ liter when the temperature is 40 °C?

$$\frac{p_1 V_1}{T_1} = \frac{p_2 V_2}{T_2} \qquad \text{or} \qquad p_2 = p_1 \left(\frac{V_1}{V_2} \right) \left(\frac{T_2}{T_1} \right)$$

from which
$$p_2 = (1 \text{ atm}) \left(\frac{1}{1/2} \right) \left(\frac{273 + 40}{273 - 20} \right) = 2.47 \text{ atm}$$

16.5. The pressure in a car tire is 28 lb/in^2 at 0 °C. As the car runs on the road, the temperature of the tire increases. What is the temperature of the tire when the tire gauge reads 35 lb/in^2? Assume that the volume of the tire remains constant and that atmospheric pressure is 14.7 lb/in^2.

Since the true pressure is gauge pressure plus atmospheric pressure, we have

$$p_1 = (28 + 14.7) \text{ lb/in}^2 = 42.7 \text{ lb/in}^2 \qquad p_2 = (35 + 14.7) \text{ lb/in}^2 = 49.7 \text{ lb/in}^2$$

From $p_1 V_1/T_1 = p_2 V_2/T_2$,

$$T_2 = T_1 \left(\frac{p_2}{p_1} \right) \left(\frac{V_2}{V_1} \right) = [(273 + 0) \text{ K}] \left(\frac{49.7}{42.7} \right)(1) = 318 \text{ K}$$

The final temperature in the tire is $318 - 273 = 45\,°$C.

16.6. A certain mass of hydrogen gas occupies 370 cm^3 at 16 °C and 150 kPa. Find its volume at -21 °C and 420 kPa.

$$\frac{p_1 V_1}{T_1} = \frac{p_2 V_2}{T_2} \qquad \text{gives} \qquad V_2 = V_1 \left(\frac{p_1}{p_2} \right) \left(\frac{T_2}{T_1} \right)$$

$$V_2 = (370 \text{ cm}^3) \left(\frac{150}{420} \right) \left(\frac{273 - 21}{273 + 16} \right) = 115 \text{ cm}^3$$

16.7. A steel tank contains oxygen gas at 32 °F and a total pressure of 12 atm. Determine the gas pressure when the tank is heated to 212 °F.

$$\frac{p_1 V_1}{T_1} = \frac{p_2 V_2}{T_2} \qquad \text{or} \qquad p_2 = p_1 \left(\frac{V_1}{V_2} \right) \left(\frac{T_2}{T_1} \right)$$

We must have T on an absolute scale. In this case, the Rankine scale is most convenient since it allows us to work directly with Fahrenheit readings.

$$p_2 = (12 \text{ atm})(1) \frac{(212 + 460)\,°\text{R}}{(32 + 460)\,°\text{R}} = 16.4 \text{ atm}$$

16.8. The density of nitrogen is 1.25 kg/m^3 at S.T.P. Determine the density of nitrogen at $42 \,°C$ and 730 mm of mercury.

Since $\rho = m/V$, we have $V_1 = m/\rho_1$ and $V_2 = m/\rho_2$ for a given mass of gas under two sets of conditions. Then

$$\frac{p_1 V_1}{T_1} = \frac{p_2 V_2}{T_2} \qquad \text{gives} \qquad \frac{p_1}{\rho_1 T_1} = \frac{p_2}{\rho_2 T_2}$$

Since S.T.P. are 760 mmHg and 273 K,

$$\rho_2 = \rho_1 \left(\frac{p_2}{p_1} \right) \left(\frac{T_1}{T_2} \right) = (1.25 \text{ kg/m}^3) \left(\frac{730}{760} \right) \left(\frac{273}{273 + 42} \right) = 1.04 \text{ kg/m}^3$$

Notice that pressures in mmHg can be used here because the units cancel in the ratio p_2/p_1.

16.9. A 3000 cm^3 tank contains oxygen gas at $20 \,°C$ and a gauge pressure of 25×10^5 Pa. What mass of oxygen is in the tank? The molecular weight of oxygen gas is 32. Assume atmospheric pressure to be 1×10^5 Pa.

The absolute pressure of the gas is

$$p = (\text{gauge pressure}) + (\text{atmospheric pressure})$$
$$= (25 + 1) \times 10^5 \text{ N/m}^2 = 26 \times 10^5 \text{ N/m}^2$$

From the gas law, with $M = 32 \text{ kg/kmol}$,

$$pV = \left(\frac{m}{M} \right) RT$$

$$(26 \times 10^5 \text{ N/m}^2)(3 \times 10^{-3} \text{ m}^3) = \left(\frac{m}{32 \text{ kg/kmol}} \right) \left(8314 \frac{J}{\text{kmol} \cdot \text{K}} \right) (293 \text{ K})$$

Solving gives m, the mass of gas in the tank, as 0.102 kg.

16.10. Determine the volume occupied by 4.0 g of oxygen ($M = 32 \text{ kg/kmol}$) at S.T.P.

Method 1

Use the gas law directly.

$$pV = \left(\frac{m}{M} \right) RT$$

$$V = \left(\frac{1}{p} \right) \left(\frac{m}{M} \right) RT = \frac{(4 \times 10^{-3} \text{ kg})(8314 \text{ J/kmol} \cdot \text{K})(273 \text{ K})}{(1.01 \times 10^5 \text{ N/m}^2)(32 \text{ kg/kmol})} = 2.8 \times 10^{-3} \text{ m}^3 = 2800 \text{ cm}^3$$

Method 2

Under S.T.P., 1 kmol occupies 22.4 m^3. Therefore, 32 kg occupies 22.4 m^3, and so 4 g occupies

$$\left(\frac{4 \text{ g}}{32\,000 \text{ g}} \right) (22.4 \text{ m}^3) = 2.8 \times 10^{-3} \text{ m}^3$$

16.11. A 2 mg droplet of liquid nitrogen is present in a 30 cm^3 tube as it is sealed off at very low temperature. What will be the nitrogen pressure in the tube when it is warmed to $20 \,°C$? Express your answer in atmospheres. (M for nitrogen is 28 kg/kmol.)

Use $pV = (m/M)RT$ to find

$$p = \frac{mRT}{MV} = \frac{(2 \times 10^{-6} \text{ kg})(8314 \text{ J/kmol} \cdot \text{K})(293 \text{ K})}{(28 \text{ kg/kmol})(30 \times 10^{-6} \text{ m}^3)} = 5800 \text{ N/m}^2$$

$$= (5800 \text{ N/m}^2) \left(\frac{1 \text{ atm}}{1.01 \times 10^5 \text{ N/m}^2} \right) = 0.057 \text{ atm}$$

16.12. A tank of volume 590 liters contains oxygen at 20 °C and 5 atm pressure.   Calculate the mass of oxygen in the tank. $M = 32$ kg/kmol for oxygen.

Use $pV = (m/M)RT$ to give

$$m = \frac{pVM}{RT} = \frac{(5 \times 1.01 \times 10^5 \text{ N/m}^2)(0.59 \text{ m}^3)(32 \text{ kg/kmol})}{(8314 \text{ J/kmol} \cdot \text{K})(293 \text{ K})} = 3.9 \text{ kg}$$

16.13. At 18 °C and 765 mmHg, 1.29 liters of an ideal gas "weighs" 2.71 grams.   Compute the molecular weight of the gas.

Use $pV = (m/M)RT$ to give

$$M = \frac{mRT}{pV} = \frac{(0.00271 \text{ kg})(8314 \text{ J/kmol} \cdot \text{K})(291 \text{ K})}{\left[(765/760)(1.01 \times 10^5 \text{ N/m}^2)\right](0.00129 \text{ m}^3)} = 50 \text{ kg/kmol}$$

16.14. Compute the volume of 8.00 grams of helium ($M = 4$ kg/kmol) at 15 °C and 480 mmHg.

Use $pV = (m/M)RT$ to give

$$V = \frac{mRT}{Mp} = \frac{(0.0080 \text{ kg})(8314 \text{ J/kmol} \cdot \text{K})(288 \text{ K})}{(4 \text{ kg/kmol})\left[(480/760)(1.01 \times 10^5 \text{ N/m}^2)\right]} = 0.075 \text{ m}^3 = 75 \text{ liters}$$

16.15. Find the density of methane ($M = 16$ kg/kmol) at 20 °C and 5 atm.

Use $pV = (m/M)RT$ and $\rho = m/V$ to give

$$\rho = \frac{pM}{RT} = \frac{(5 \times 1.01 \times 10^5 \text{ N/m}^2)(16 \text{ kg/kmol})}{(8314 \text{ J/kmol} \cdot \text{K})(293 \text{ K})} = 3.32 \text{ kg/m}^3$$

16.16. In a gaseous mixture at 20 °C the partial pressures of the components are as follows: hydrogen, 200 mmHg; carbon dioxide, 150 mmHg; methane, 320 mmHg; ethylene, 105 mmHg.   What are (a) the total pressure of the mixture and (b) the mass fraction of hydrogen?   ($M_H = 2$, $M_{CO_2} = 44$, $M_{meth} = 16$, $M_{eth} = 30$ kg/kmol.)

(a) According to Dalton's law,

$$\text{total pressure} = \text{sum of partial pressures}$$
$$= 200 + 150 + 320 + 105 = 775 \text{ mmHg}$$

(b) From the gas law, $m = M(pV/RT)$.   The mass of hydrogen gas present is

$$m_H = M_H p_H (V/RT)$$

The total mass of gas present, m_t, is the sum of similar terms:

$$m_t = (M_H p_H + M_{CO_2} p_{CO_2} + M_{meth} p_{meth} + M_{eth} p_{eth})(V/RT)$$

The required fraction is

$$\frac{m_H}{m_t} = \frac{M_H p_H}{M_H p_H + M_{CO_2} p_{CO_2} + M_{meth} p_{meth} + M_{eth} p_{eth}}$$
$$= \frac{(2)(200)}{(2)(200) + (44)(150) + (16)(320) + (30)(105)} = 0.026$$

Supplementary Problems

16.17. A certain mass of an ideal gas occupied a volume of 4.00 m^3 at 758 mmHg. Compute its volume at 635 mmHg if the temperature remains unchanged. *Ans.* 4.77 m^3

16.18. A given mass of ideal gas occupies 38 cm^3 at 20 °C. If its pressure is held constant, what is the volume occupied at a temperature of 45 °C? *Ans.* 41.2 cm^3

16.19. On a day when atmospheric pressure is 75.83 cm of Hg, a pressure gauge on a tank of gas reads a pressure of 258.5 cm of Hg. What is the absolute pressure (in atmospheres and kPa) of the gas in the tank? *Ans.* 334.3 cm of Hg = 4.40 atm = 444 kPa

16.20. A tank of ideal gas is sealed off at 20 °C and 1 atm pressure. What will be the pressure (in kPa and mmHg) in the tank if the gas temperature is decreased to −35 °C? *Ans.* 82 kPa = 617 mmHg

16.21. Given 1000 ft^3 of helium at 15 °C and 763 mmHg, determine its volume at −6 °C and 420 mmHg. *Ans.* 1680 ft^3

16.22. If a mass of gas occupies 15.7 ft^3 at 60 °F and 14.7 lb/in^2, determine its volume at 100 °F and 25 lb/in^2. *Ans.* 9.94 ft^3

16.23. Calculate the final Celsius temperature required to change 10 liters of helium at 100 K and 0.1 atm to 20 liters at 0.2 atm. *Ans.* 127 °C

16.24. One kilomole of ideal gas occupies 22.4 m^3 at 0 °C and 1 atm. (*a*) What pressure is required to compress 1 kmol into a 5 m^3 container at 100 °C? (*b*) If 1 kmol was to be sealed in a 5 m^3 tank that could withstand a gauge pressure of only 3 atm, what would be the maximum temperature of the gas if the tank was not to burst? *Ans.* (*a*) 6.12 atm; (*b*) −30 °C

16.25. The U-tube shown in Fig. 16-1 confines an ideal gas in its sealed end by means of a mercury column. At the time shown, atmospheric pressure is 750 mmHg and the gas has a volume of 50 cm^3 and is at 30 °C. Find its volume at S.T.P. *Ans.* 50.38 cm^3

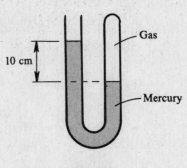

16.26. On a day when the barometer reads 75.23 cm, a reaction vessel holds 250 cm^3 of ideal gas at 20 °C. An oil manometer ($\rho = 810$ kg/m^3) reads the pressure in the vessel to be 41 cm of oil and below atmospheric pressure. What volume will the gas occupy under S.T.P.? *Ans.* 233.1 cm^3

Fig. 16-1

16.27. A 5000 cm^3 tank contains an ideal gas ($M = 40$ kg/kmol) at a gauge pressure of 530 kPa and a temperature of 25 °C. Assuming atmospheric pressure to be 100 kPa, what mass of gas is in the tank? *Ans.* 0.0509 kg

16.28. The pressure of air in a reasonably good vacuum might be 2×10^{-5} mmHg. What mass of air exists in a 250 cm^3 volume at this pressure and 25 °C? You can take $M = 28$ kg/kmol for air. *Ans.* 7.5×10^{-12} kg

16.29. What volume will 1.216 grams of SO_2 gas ($M = 64.1$ kg/kmol) occupy at 18 °C and 755 mmHg if it acts like an ideal gas?   *Ans.* 457 cm^3

16.30. Compute the density of H_2S gas ($M = 34.1$ kg/kmol) at 27 °C and 2.00 atm, assuming it to be ideal. *Ans.* 2.76 kg/m^3

16.31. A 30 cm^3 tube contains 0.25 gram of water vapor ($M = 18$ kg/kmol) at a temperature of 340 °C. Assuming the gas to be ideal, what is its pressure?   *Ans.* 2.36 MPa

16.32. One method for estimating the temperature at the center of the sun is based on the ideal gas law. If the center is assumed to consist of gases whose average M is 0.7 kg/kmol, and if the density and pressure are 90×10^3 kg/m^3 and 1.4×10^{11} atm respectively, calculate the temperature.   *Ans.* 1.3×10^7 K

16.33. A 500 cm^3 sealed flask contains nitrogen at a pressure of 76.00 cm of Hg. A tiny glass tube lies at the bottom of the flask. Its volume is 0.50 cm^3 and it contains hydrogen gas at a pressure of 4.5 atm. Suppose the glass tube is now broken so that the hydrogen fills the flask. What is the new pressure in the flask?   *Ans.* 76.34 cm of Hg

16.34. As shown in Fig. 16-2, two flasks are connected by an initially closed stopcock. One flask contains krypton gas at 500 mmHg, while the other contains helium at 950 mmHg. The stopcock is now opened so that the gases mix. What is the final pressure in the system? Assume constant temperature.
Ans. 789 mmHg

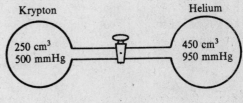

Fig. 16-2

16.35. An air bubble of volume V_0 is released near the bottom of a lake at a depth of 11.0 m. What will be its new volume at the surface? Assume its temperature to be 4 °C at the release point and 12 °C at the surface. The water has a density of 1000 kg/m^3 and atmospheric pressure is 75 cm of Hg.
Ans. 2.14 V_0

16.36. A cylindrical diving bell (a vertical cylinder with open bottom end and closed top end) 12 m high is lowered in a lake until water within the bell rises 8 m from the bottom end. Determine the distance from the top of the bell to the surface of the lake. (Atmospheric pressure = 1 atm.)
Ans. 20.6 m − 4.0 m = 16.6 m

Chapter 17

Kinetic Theory

THE KINETIC THEORY considers matter to be composed of discrete particles or molecules in continual motion. In a gas, the molecules are in random motion with a wide distribution of speeds ranging from zero to very large values.

THE AVERAGE TRANSLATIONAL KINETIC ENERGY of a gas molecule is $3kT/2$, where T is the absolute temperature of the gas and $k = 1.381 \times 10^{-23}$ J/K is *Boltzmann's constant*. Then, for a molecule of mass m_0,

$$\left(\text{average of } \frac{1}{2} m_0 v^2\right) = \frac{3}{2} kT$$

THE ROOT MEAN SQUARE speed of a gas molecule is the square root of the average of v^2 for a molecule over a prolonged time. Equivalently, the average may be taken over all molecules of the gas at a given instant. From the expression for the average kinetic energy, one has

$$\text{rms speed} = v_{\text{rms}} = \sqrt{\frac{3kT}{m_0}}$$

AVOGADRO'S NUMBER (N_A) is the number of particles (or molecules or atoms) in 1 kmol of substance. For all substances,

$$N_A = 6.0222 \times 10^{26} \text{ particles per kmol}$$

As examples, $M = 2$ kg/kmol for H_2 and $M = 32$ kg/kmol for O_2. Therefore, 2 kg of H_2 and 32 kg of O_2 each contain 6.02×10^{26} molecules.

THE MASS OF A MOLECULE (or atom) can be found from the molecular (or atomic) weight, M, of the substance and Avogadro's number, N_A. Since M kg of substance contains N_A particles, the mass, m_0, of a particle is given by

$$m_0 = \frac{M}{N_A}$$

THE PRESSURE of an ideal gas is related to the properties of its molecules by

$$pV = \frac{1}{3} N m_0 v_{\text{rms}}^2$$

where N is the number of molecules in the volume V. As before, m_0 is the mass of a molecule. Since Nm_0 is the mass m of the substance, and since $\rho = m/V$, we have

$$p = \frac{1}{3} \rho v_{\text{rms}}^2$$

THE ABSOLUTE TEMPERATURE of an ideal gas has a meaning found from $\frac{1}{2} m_0 v_{\text{rms}}^2 = \frac{3}{2} kT$. Solving gives

$$T = \left(\frac{2}{3k}\right)\left(\frac{1}{2} m_0 v_{\text{rms}}^2\right)$$

The absolute temperature of an ideal gas is a measure of its average translational KE per molecule.

A RELATION BETWEEN N_A, k, and R can be obtained by combining the previous equations with $pV = (m/M)RT$.   It is

$$k = \frac{R}{N_A}$$

THE MEAN FREE PATH of a gas molecule is the average distance such a molecule goes between collisions.   For an ideal gas of spherical molecules with radius b,

$$\text{mean free path} = \frac{1}{4\pi\sqrt{2}\, b^2 (N/V)}$$

where N/V is the number of molecules per unit volume.

Solved Problems

17.1.   Find the mass of a nitrogen molecule, N_2.   The molecular weight is 28 kg/kmol.

$$m_0 = \frac{M}{N_A} = \frac{28 \text{ kg/kmol}}{6.02 \times 10^{26} \text{ kmol}^{-1}} = 4.65 \times 10^{-26} \text{ kg}$$

17.2.   How many helium atoms, He, are there in 2.0 g of helium? $M = 4$ kg/kmol for He.

Method 1

One kilomole of He is 4 kg and it contains N_A atoms.   But 2 g is equivalent to

$$\frac{0.002 \text{ kg}}{4 \text{ kg/kmol}} = 0.0005 \text{ kmol}$$

of helium.   Therefore

$$\text{no. of atoms in 2 g} = (0.0005 \text{ kmol})N_A$$
$$= (0.0005 \text{ kmol})(6.02 \times 10^{26} \text{ kmol}^{-1}) = 3.01 \times 10^{23}$$

Method 2

The mass of a helium atom is

$$m_0 = \frac{M}{N_A} = \frac{4 \text{ kg/kmol}}{6.02 \times 10^{26} \text{ kmol}^{-1}} = 6.64 \times 10^{-27} \text{ kg}$$

so

$$\text{no. in 2 g} = \frac{0.002 \text{ kg}}{6.64 \times 10^{-27} \text{ kg}} = 3.01 \times 10^{23}$$

17.3.   A droplet of mercury has a radius of 0.5 mm.   How many mercury atoms are in the droplet? For Hg, $M = 202$ kg/kmol   and $\rho = 13\,600$ kg/m^3.

The volume of the droplet is

$$V = \frac{4\pi r^3}{3} = \left(\frac{4\pi}{3}\right)(5 \times 10^{-4} \text{ m})^3 = 5.24 \times 10^{-10} \text{ m}^3$$

The mass of the droplet is

$$m = \rho V = (13\,600 \text{ kg/m}^3)(5.24 \times 10^{-10} \text{ m}^3) = 7.1 \times 10^{-6} \text{ kg}$$

The mass of a mercury atom is

$$m_0 = \frac{M}{N_A} = \frac{202 \text{ kg/kmol}}{6.02 \times 10^{26} \text{ kmol}^{-1}} = 3.36 \times 10^{-25} \text{ kg}$$

The number of atoms in the droplet is

$$\text{number} = \frac{m}{m_0} = \frac{7.1 \times 10^{-6} \text{ kg}}{3.36 \times 10^{-25} \text{ kg}} = 2.1 \times 10^{19}$$

17.4. How many molecules are there in 70 cm^3 of benzene? For benzene, $\rho = 0.88$ g/cm^3 and $M = 78$ kg/kmol.

$$\text{mass of 70 cm}^3 = m = \rho V = (880 \text{ kg/m}^3)(70 \times 10^{-6} \text{ m}^3) = 0.0616 \text{ kg}$$

$$m_0 = \frac{M}{N_A} = \frac{78 \text{ kg/kmol}}{6.02 \times 10^{26} \text{ kmol}^{-1}} = 1.30 \times 10^{-25} \text{ kg}$$

$$\text{number in 70 cm}^3 = \frac{m}{m_0} = \frac{0.0616 \text{ kg}}{1.30 \times 10^{-25} \text{ kg}} = 4.8 \times 10^{23}$$

17.5. Find the rms speed of a nitrogen molecule ($M = 28$ kg/kmol) in air at 0 °C.

We know that $\frac{1}{2} m_0 v_{\text{rms}}^2 = \frac{3}{2} kT$ and so

$$v_{\text{rms}} = \sqrt{\frac{3kT}{m_0}}$$

But

$$m_0 = \frac{M}{N_A} = \frac{28 \text{ kg/kmol}}{6.02 \times 10^{26} \text{ kmol}^{-1}} = 4.65 \times 10^{-26} \text{ kg}$$

Therefore

$$v_{\text{rms}} = \sqrt{\frac{3(1.38 \times 10^{-23} \text{ J/K})(273 \text{ K})}{4.65 \times 10^{-26} \text{ kg}}} = 490 \text{ m/s}$$

17.6. A gas molecule at the surface of the earth happens to have the rms speed for that gas at 0 °C. If it were to go straight up without colliding with other molecules, how high would it rise?

The molecule's KE is initially

$$\text{KE} = \frac{1}{2} m_0 v_{\text{rms}}^2 = \frac{3}{2} kT$$

The molecule will rise until its KE has been changed to GPE. Therefore, calling the height h,

$$\frac{3}{2} kT = m_0 gh$$

Solving gives

$$h = \left(\frac{1}{m_0}\right)\left(\frac{3kT}{2g}\right) = \left(\frac{1}{m_0}\right)\frac{(3)(1.38 \times 10^{-23} \text{ J/K})(273 \text{ K})}{2(9.8 \text{ m/s}^2)} = \frac{5.8 \times 10^{-22} \text{ kg} \cdot \text{m}}{m_0}$$

where m_0 is in kg. The height varies inversely with the mass of the molecule. For an N_2 molecule, $m_0 = 4.65 \times 10^{-26}$ kg (Problem 17.5), and in this case h turns out to be 12.4 km.

17.7. Find the following ratios for hydrogen ($M = 2$ kg/kmol) and nitrogen ($M = 28$ kg/kmol) gases at the same temperature:

$\quad$ (a) $\quad$ (KE)$_{\text{H}}$/(KE)$_{\text{N}}$ $\qquad$ (b) $\quad$ (rms speed)$_{\text{H}}$/(rms speed)$_{\text{N}}$

(a) The average translational KE of a molecule, $\frac{3}{2} kT$, depends only on temperature. Therefore the ratio is unity.

(b)

$$\frac{(v_{\text{rms}})_{\text{H}}}{(v_{\text{rms}})_{\text{N}}} = \sqrt{\frac{3kT/m_{0\text{H}}}{3kT/m_{0\text{N}}}} = \sqrt{\frac{m_{0\text{N}}}{m_{0\text{H}}}}$$

But $m_0 = M/N_A$, and so

$$\frac{(v_{\text{rms}})_{\text{H}}}{(v_{\text{rms}})_{\text{N}}} = \sqrt{\frac{M_{\text{N}}}{M_{\text{H}}}} = \sqrt{\frac{28}{2}} = 3.74$$

17.8. Certain ideal gas molecules behave like spheres of radius 3×10^{-10} m. Find the mean free path of these molecules under S.T.P.

Method 1

We know that at S.T.P. one kilomole of substance occupies 22.4 m^3. Therefore, in 22.4 m^3 there are $N_A = 6.02 \times 10^{26}$ molecules. The mean free path is given by

$$\text{m.f.p.} = \frac{1}{4\pi\sqrt{2}\, b^2(N/V)} = \frac{1}{4\pi\sqrt{2}\,(3 \times 10^{-10}\text{ m})^2}\left(\frac{22.4\text{ m}^3}{6.02 \times 10^{26}}\right) = 2.4 \times 10^{-8}\text{ m}$$

Method 2

Using the facts that $M = m_0 N_A = m_0(R/k)$ and $m = N m_0$,

$$pV = \left(\frac{m}{M}\right)RT \qquad \text{becomes} \qquad pV = NkT$$

and so

$$\frac{N}{V} = \frac{p}{kT} = \frac{1 \times 10^5 \text{ N/m}^2}{(1.38 \times 10^{-23}\text{ J/K})(273\text{ K})} = 2.65 \times 10^{25}\text{ m}^{-3}$$

We then use the mean free path equation as in the previous method.

17.9. At what pressure will the mean free path be 50 cm for spherical molecules of radius 3×10^{-10} m? Assume an ideal gas at 20 °C.

From the expression for the mean free path, we obtain

$$\frac{N}{V} = \frac{1}{4\pi\sqrt{2}\, b^2(\text{m.f.p.})}$$

Combining this with the ideal gas law, in the form $pV = NkT$ (see Problem 17.8), gives

$$p = \frac{kT}{4\pi\sqrt{2}\, b^2(\text{m.f.p.})} = \frac{(1.38 \times 10^{-23}\text{ J/K})(293\text{ K})}{4\pi\sqrt{2}\,(3 \times 10^{-10}\text{ m})^2(0.50\text{ m})} = 5.1 \times 10^{-3}\text{ N/m}^2$$

Supplementary Problems

17.10. Find the mass of a neon atom. The atomic weight of neon is 20.2 kg/kmol. *Ans.* 3.36×10^{-26} kg

17.11. A typical polymer molecule in polyethylene might have a molecular weight of 15 000. (*a*) What is the mass of such a molecule? (*b*) How many such molecules would make up 2 g of polymer? *Ans.* (*a*) 2.5×10^{-23} kg; (*b*) 8×10^{19}

17.12. A certain strain of tobacco mosaic virus has $M = 4.0 \times 10^7$ kg/kmol. How many molecules of the virus are present in 1 cm^3 of a solution that contains 0.10 mg of virus per cm^3? *Ans.* 1.5×10^{12}

17.13. An electronic vacuum tube was sealed off during manufacture at a pressure of 1.2×10^{-7} mmHg at 27 °C. Its volume is 100 cm^3. (*a*) What is the pressure in the tube (in Pa)? (*b*) How many gas molecules remain in the tube? *Ans.* (*a*) 1.6×10^{-5} Pa; (*b*) 3.8×10^{11}

17.14. In a certain region of outer space there are an average of only 5 molecules per cm^3. The temperature there is about 3 K. What is the average pressure of this very dilute gas? *Ans.* 2×10^{-16} Pa

17.15. A cube of aluminum has a volume of 1 cm^3 and a mass of 2.70 grams. (*a*) How many aluminum atoms are there in the cube? (*b*) How large a volume is associated with each atom? (*c*) If each atom were a cube, what would be its edge length? $M = 108$ kg/kmol for aluminum. *Ans.* (*a*) 1.5×10^{22}; (*b*) 6.6×10^{-29} m^3; (*c*) 4.0×10^{-10} m

17.16. The rms velocity of nitrogen molecules in the air at S.T.P. is about 490 m/s. Find their mean free path and the average time between collisions. The radius of a nitrogen molecule can be taken to be 2×10^{-10} m. *Ans.* 5.2×10^{-8} m, 1.1×10^{-10} s

17.17. What is the mean free path of a gas molecule (radius 2.5×10^{-10} m) in an ideal gas at 500 °C when the pressure is 7×10^{-6} mmHg? *Ans.* 10 m

Chapter 18

Calorimetry, Fusion, Vaporization

HEAT is a form of energy. The units most commonly used in measuring heat are the joule, calorie, kilocalorie, and Btu.

1 *calorie* (cal) = 4.184 J, exactly. The calorie is very nearly the quantity of heat energy needed to increase the temperature of 1 g of water by 1 °C.

1 *kilocalorie* (kcal) = 1000 cal. This unit is also called the *large calorie*. When nutritionists give the energy content of foods in "calories," the unit they are actually using is the kilocalorie.

1 *British thermal unit* (Btu) = 252 cal. It is very nearly the quantity of heat energy required to raise the temperature of 1 pound of water 1 °F. Whenever the Btu is based on a weight unit, it is understood that g, the acceleration due to gravity, is to be given its standard value, 32.17 ft/s^2. One Btu is then equivalent to 778 ft·lb of energy.

SPECIFIC HEAT (c) (or *specific heat capacity*) of a substance is the quantity of heat required to heat unit mass (or weight) of a substance through a temperature rise of one degree.

If a quantity of heat ΔQ is required to produce a temperature increase ΔT in a mass m (or weight w) of a substance, then the specific heat of that substance is

$$c = \frac{\Delta Q}{m \, \Delta T} \qquad \text{or} \qquad c = \frac{\Delta Q}{w \, \Delta T}$$

In calorimetry, the traditional units for specific heat are cal/g·°C and Btu/lb·°F. The SI units would be J/kg·K or kJ/kg·K.

Because of the way the calorie and the Btu are defined, the numerical values for c in cal/g·°C and in Btu/lb·°F are the same (provided that $g = 32.17$ ft/s^2). The specific heat of water is 1.00 cal/g·°C or 1.00 Btu/lb·°F.

Each substance has its own specific heat, which varies slightly with temperature. Most substances have c-values that are less than that for water.

HEAT CAPACITY (or *water equivalent*) of a body is the quantity of heat required to raise the temperature of the whole body by one degree. From its definition, the heat capacity of an object of mass m (or weight w) and specific heat c is mc (or wc).

THE HEAT GAINED (OR LOST) by a body of mass m and specific heat c whose state does not change is, for a temperature change ΔT,

$$\Delta Q = mc \, \Delta T$$

when calories or joules are used and is

$$\Delta Q = wc \, \Delta T$$

when Btu's are used.

THE HEAT OF FUSION (H_f) of a crystalline solid is the quantity of heat required to melt unit mass (or weight) of the solid at constant temperature. It is also equal to the quantity of heat given off by unit mass of the molten solid as it crystallizes at this same temperature. The heat of fusion of water at 0 °C is about 80 cal/g or 144 Btu/lb.

THE HEAT OF VAPORIZATION (H_v) of a liquid is the quantity of heat required to vaporize unit mass (or weight) of the liquid at constant temperature. For water at 100 °C, H_v is about 540 cal/g or 970 Btu/lb.

THE HEAT OF SUBLIMATION of a solid substance is the quantity of heat required to convert unit mass (or weight) of it from the solid to the gaseous state at constant temperature.

CALORIMETRY PROBLEMS involve the sharing of heat energy among initially hot objects and cold objects. Since energy must be conserved, one can always write the following equation:

heat lost by hot objects = heat gained by cold objects

This assumes that no heat energy is lost from the system.

ABSOLUTE HUMIDITY is the mass of water vapor present per unit volume of gas (usually the atmosphere). Typical units are kg/m^3 and g/cm^3.

RELATIVE HUMIDITY (R.H.) is the ratio obtained by dividing the *mass of water vapor per unit volume present in the air* by the *mass of water vapor per unit volume present in saturated air at the same temperature*. When expressed in percent, the above ratio is multiplied by 100.

DEW POINT: Cooler air at saturation contains less water than warmer air does at saturation. When air is cooled, it eventually reaches a temperature at which it is saturated. This temperature is called the *dew point*. At temperatures lower than this, water condenses out of the air.

Solved Problems

18.1. (a) How much heat is required to heat 250 cm^3 of water from 20 °C to 35 °C? (b) How much heat is lost by the water as it cools back down to 20 °C?

Since 250 cm^3 of water has a mass of 250 g, and since $c = 1.00$ cal/g·°C for water, we have

(a) $$\Delta Q = mc\,\Delta T = (250 \text{ g})(1 \text{ cal/g} \cdot °C)(15 \text{ °C}) = 3750 \text{ cal}$$

(b) $$\Delta Q = mc\,\Delta T = (250 \text{ g})(1 \text{ cal/g} \cdot °C)(-15 \text{ °C}) = -3750 \text{ cal}$$

18.2. How much heat does 25 g of aluminum give off as it cools from 100 °C to 20 °C? For aluminum, $c = 0.21$ cal/g·°C.

$$\Delta Q = mc\,\Delta T = (25 \text{ g})(0.21 \text{ cal/g} \cdot °C)(-80 \text{ °C}) = -420 \text{ cal}$$

18.3. How much heat is needed to heat 5 lb of aluminum from 40 °F to 212 °F?

In Problem 18.2 we were told that for aluminum, $c = 0.21$ cal/g·°C. Because of the numerical equivalence of c in the two sets of units, $c = 0.21$ Btu/lb·°F. Therefore,

$$\Delta Q = wc\,\Delta T = (5 \text{ lb})(0.21 \text{ Btu/lb} \cdot °F)(172 \text{ °F}) = 181 \text{ Btu}$$

18.4. A certain amount of heat is added to a mass of aluminum ($c = 0.21$ cal/g·°C) and its temperature is raised 57 °C. Suppose that the same amount of heat is added to the same mass of copper ($c = 0.093$ cal/g·°C). How much does the temperature of the copper rise?

From $\Delta Q = mc\,\Delta T$, the temperature rise is inversely proportional to the specific heat. Thus

$$\Delta T_{Cu} = \left(\frac{0.21}{0.093}\right)(57 \text{ °C}) = 129 \text{ °C}$$

18.5. A thermos bottle contains 250 g of coffee at 90 °C. To this is added 20 g of milk at 5 °C. After equilibrium is established, what is the temperature of the liquid? Assume no heat loss to the thermos bottle.

Water, coffee, and milk all have the same value of c, 1.00 cal/g·°C. The law of energy conservation allows us to write

$$\text{heat lost by coffee} = \text{heat gained by milk}$$

$$(cm\,\Delta T)_{\text{coffee}} = (cm\,\Delta T)_{\text{milk}}$$

If the final temperature of the liquid is t, then

$$\Delta T_{\text{coffee}} = 90\ °C - t \qquad \Delta T_{\text{milk}} = t - 5\ °C$$

Substituting in the equation and canceling c gives

$$(250\ \text{g})(90\ °C - t) = (20\ \text{g})(t - 5\ °C)$$

Solving gives $t = 83.7\ °C$.

18.6. A thermos bottle contains 150 g of water at 4 °C. Into this is placed 90 g of metal at 100 °C. After equilibrium is established, the temperature of the water and metal is 21 °C. What is the specific heat of the metal? Assume no heat loss to the thermos bottle.

$$\text{heat lost by metal} = \text{heat gained by water}$$

$$(cm\,\Delta T)_{\text{metal}} = (cm\,\Delta T)_{\text{water}}$$

$$c_{\text{metal}}(90\ \text{g})(79\ °C) = (1\ \text{cal/g·}°C)(150\ \text{g})(17\ °C)$$

Solving gives $c_{\text{metal}} = 0.36\ \text{cal/g·}°C$.

18.7. A 200 g copper calorimeter can contains 150 g of oil at a temperature of 20 °C. To the oil is added 80 g of aluminum at 300 °C. What will be the temperature of the system after equilibrium is established? $c_{\text{Cu}} = 0.093\ \text{cal/g·}°C$, $c_{\text{Al}} = 0.21\ \text{cal/g·}°C$, $c_{\text{oil}} = 0.37\ \text{cal/g·}°C$.

$$\text{heat lost by aluminum} = \text{heat gained by can and oil}$$

$$(cm\,\Delta T)_{\text{Al}} = (cm\,\Delta T)_{\text{Cu}} + (cm\,\Delta T)_{\text{oil}}$$

$$\left(0.21\ \frac{\text{cal}}{\text{g·}°C}\right)(80\ \text{g})(300\ °C - t) = \left(0.093\ \frac{\text{cal}}{\text{g·}°C}\right)(200\ \text{g})(t - 20\ °C) + \left(0.37\ \frac{\text{cal}}{\text{g·}°C}\right)(150\ \text{g})(t - 20\ °C)$$

Solving gives t as 72 °C.

18.8. Exactly 3 g of carbon was burned to CO_2 in a copper calorimeter. The mass of the calorimeter is 1500 g, and there is 2000 g of water in the calorimeter. The initial temperature was 20 °C and the final temperature is 31 °C. Calculate the heat given off per gram of carbon. $c_{\text{Cu}} = 0.093\ \text{cal/g·}°C$. Neglect the small heat capacity of the carbon and carbon dioxide.

The law of energy conservation tells us that

$$\text{heat lost by carbon} = (\text{heat gained by calorimeter}) + (\text{heat gained by water})$$

$$= (0.093\ \text{cal/g·}°C)(1500\ \text{g})(11\ °C) + (1\ \text{cal/g·}°C)(2000\ \text{g})(11\ °C)$$

$$= 23\,500\ \text{cal}$$

Therefore, the heat given off by one gram of carbon as it burns is

$$\frac{23\,500\ \text{cal}}{3\ \text{g}} = 7800\ \text{cal/g}$$

18.9. Determine the resulting temperature, t, when 150 g of ice at 0 °C is mixed with 300 g of water at 50 °C.

From energy conservation

$$\text{heat gained by ice} = \text{heat lost by water}$$

$$(\text{heat to melt ice}) + (\text{heat to warm ice water}) = \text{heat lost by water}$$

$$(mH_f)_{\text{ice}} + (cm\,\Delta T)_{\text{ice water}} = (cm\,\Delta T)_{\text{water}}$$

$$(150\ \text{g})(80\ \text{cal/g}) + (1\ \text{cal/g·}°C)(150\ \text{g})(t - 0°\ C) = (1\ \text{cal/g·}°C)(300\ \text{g})(50\ °C - t)$$

from which $t = 6.7\ °C$.

18.10. How much heat is given up when 20 g of steam at 100 °C is condensed and cooled to 20 °C?

heat lost = (condensation heat loss) + (heat lost by water during cooling)

$$= mH_v + cm \Delta T$$

$$= (20 \text{ g})(540 \text{ cal/g}) + (1 \text{ cal/g} \cdot {}^{\circ}\text{C})(20 \text{ g})(100 \text{ °C} - 20 \text{ °C})$$

$$= 12\,400 \text{ cal}$$

18.11. A 20 g piece of aluminum ($c = 0.21$ cal/g · °C) at 90 °C is dropped into a cavity in a large block of ice at 0 °C. How much ice does the aluminum melt?

heat lost by Al as it cools to 0 °C = heat gained by mass m of ice melted

$$(mc \Delta T)_{\text{Al}} = (H_f m)_{\text{ice}}$$

$$(20 \text{ g})(0.21 \text{ cal/g} \cdot {}^{\circ}\text{C})(90 \text{ °C} - 0 \text{ °C}) = (80 \text{ cal/g})m$$

from which $m = 4.7$ g is the quantity of ice melted.

18.12. (a) Convert the heat of fusion of ice, 80 cal/g, to its equivalent in Btu/lb. (b) How many Btu are absorbed by a refrigerator in changing 5.0 lb of water at 60 °F to ice at 32 °F?

(a)   Under standard gravity, a mass of 454 grams weighs 1 pound. Hence

$$H_f = 80 \frac{\text{cal}}{\text{g}} \times \frac{1 \text{ Btu}}{252 \text{ cal}} \times \frac{454 \text{ g}}{1 \text{ lb}} = 144 \text{ Btu/lb}$$

(b)   heat absorbed = (heat to cool water) + (heat to change to ice)

$$= wc \Delta T + wH_f$$

$$= (5 \text{ lb})(1 \text{ Btu/lb} \cdot {}^{\circ}\text{F})(28 \text{ °F}) + (5 \text{ lb})(144 \text{ Btu/lb}) = 860 \text{ Btu}$$

18.13. In a calorimeter can (water equivalent = 40 g) are 200 g of water and 50 g of ice, all at 0 °C. Into this is poured 30 g of water at 90 °C. What will be the final condition of the system?

Let us start by assuming (perhaps incorrectly) that the final temperature is $t > 0$ °C. Then

$$\begin{pmatrix} \text{heat lost by} \\ \text{hot water} \end{pmatrix} = \begin{pmatrix} \text{heat to} \\ \text{melt ice} \end{pmatrix} + \begin{pmatrix} \text{heat to warm} \\ \text{250 g of water} \end{pmatrix} + \begin{pmatrix} \text{heat to warm} \\ \text{calorimeter} \end{pmatrix}$$

$$(30 \text{ g})(1 \text{ cal/g} \cdot {}^{\circ}\text{C})(90 \text{ °C} - t) = (50 \text{ g})(80 \text{ cal/g}) + (250 \text{ g})(1 \text{ cal/g} \cdot {}^{\circ}\text{C})(t - 0 \text{ °C})$$

$$+ (40 \text{ g})(1 \text{ cal/g} \cdot {}^{\circ}\text{C})(t - 0 \text{ °C})$$

Solving gives $t = -4.1$ °C, contrary to our assumption that the final temperature is above 0 °C. Apparently, not all the ice melts. Therefore, $t = 0$ °C.

In order to find how much ice melts, we have

heat lost by hot water = heat gained by melting ice

$$(30 \text{ g})(1 \text{ cal/g} \cdot {}^{\circ}\text{C})(90 \text{ °C}) = (80 \text{ cal/g})m$$

where m is the mass of ice that melts. Solving gives $m = 34$ g. The final system has $50 - 34 = 16$ g of ice not melted.

18.14. A thermometer in a 10 m × 8 m × 4 m room reads 22 °C and a humidistat reads the R.H. to be 35%. What mass of water vapor is in the room? Saturated air at 22 °C contains 19.33 g H_2O/m^3.

$$\%\text{R.H.} = \frac{\text{mass of water/m}^3}{\text{mass of water/m}^3 \text{ of saturated air}} \times 100$$

$$35 = \frac{\text{mass/m}^3}{0.01933 \text{ kg/m}^3} \times 100$$

from which   mass/m$^3 = 6.77 \times 10^{-3}$ kg/m^3.

But the room in question has a volume of $10 \times 8 \times 4 = 320$ m^3. Therefore, the water in it is

$$(320 \text{ m}^3)(6.77 \times 10^{-3} \text{ kg/m}^3) = 2.17 \text{ kg}$$

18.15. On a certain day when the temperature is 28 °C, moisture forms on the outside of a glass of cold drink if the glass is at a temperature of 16 °C or lower.   What is the R.H. on that day?   Saturated air at 28 °C contains 26.93 g/m^3 of water, while, at 16 °C, it contains 13.50 g/m^3.

Dew forms at a temperature of 16 °C or lower, so the dew point is 16 °C.   The air is saturated at that temperature and therefore contains 13.50 g/m^3.   Then

$$\text{R.H.} = \frac{\text{mass present/m}^3}{\text{mass/m}^3 \text{ in saturated air}} = \frac{13.50}{26.93} = 0.50 = 50\%$$

18.16. Outside air at 5 °C and 20% relative humidity is introduced into a heating and air conditioning plant where it is heated to 20 °C and the relative humidity is increased to a comfortable 50%.   How many grams of water must be evaporated into a cubic meter of outside air to accomplish this?   Saturated air at 5 °C contains 6.8 g/m^3 of water, and at 20 °C it contains 17.3 g/m^3.

mass/m^3 of water vapor in air at 5 °C = 0.20 × 6.8 g/m^3 = 1.36 g/m^3

comfortable mass/m^3 at 20 °C = 0.50 × 17.3 g/m^3 = 8.65 g/m^3

one m^3 of air at 5 °C expands to   (293/278) m^3 = 1.054 m^3 at 20 °C

mass of water vapor in 1.054 m^3 at 20 °C = 1.054 m^3 × 8.65 g/m^3 = 9.12 g

mass of water vapor to be added to each m^3 of air at 5 °C = (9.12 − 1.36) g = 7.76 g

Supplementary Problems

18.17. How many calories are required to heat each of the following from 15 °C to 65 °C?   (*a*) 3 g aluminum, (*b*) 5 g pyrex glass, (*c*) 20 g platinum.   The specific heats, in cal/g · °C, for aluminum, pyrex, and platinum are 0.21, 0.20, and 0.032 respectively.   *Ans.*   (*a*) 31.5 cal;   (*b*) 50 cal;   (*c*) 32 cal

18.18. How many Btu are removed in cooling each of the following from 212 °F to 68 °F?   (*a*) 1 lb water, (*b*) 2 lb leather (*c* = 0.36 cal/g · °C), (*c*) 3 lb asbestos (*c* = 0.20 cal/g · °C).   *Ans.*   (*a*) 144 Btu;   (*b*) 104 Btu;   (*c*) 86 Btu

18.19. A normal diet might furnish 2000 nutritionist's calories (2000 kcal) to a 60 kg person in a day.   If this energy was used to heat the person with no losses to the surroundings, how much would the person's temperature increase?   *c* = 0.83 cal/g · °C   for the average person.   *Ans.*   40 °C

18.20. When 5 g of a certain type of coal is burned, it raises the temperature of 1000 cm^3 of water from 10 °C to 47 °C.   Calculate the heat energy produced per gram of coal.   Neglect the small heat capacity of the coal.   *Ans.*   7400 cal/g

18.21. Furnace oil has a heat of combustion of 19 000 Btu/lb.   Assuming that 70% of the heat is useful, how many pounds of oil are required to heat 1000 lb of water from 50 °F to 190 °F?   *Ans.*   10.5 lb

18.22. What will be the final temperature if 50 g of water at 0 °C is added to 250 g of water at 90 °C?   *Ans.*   75 °C

18.23. A 50 g piece of metal at 95 °C is dropped into 250 g of water at 17.0 °C and warms it to 19.4 °C.   What is the specific heat of the metal?   *Ans.*   0.16 cal/g · °C

18.24. A 55 g copper calorimeter ($c = 0.093$ cal/g·°C) contains 250 g of water at 18.0 °C. When 75 g of an alloy at 100 °C is dropped into the calorimeter, the resulting temperature is 20.4 °C. What is the specific heat of the alloy?   *Ans.*   0.103 cal/g·°C

18.25. Determine the resulting temperature when 1 kg of ice at 0 °C is mixed with 9 kg of water at 50 °C. *Ans.*   37 °C

18.26. How much heat is required to change 10 g of ice at 0 °C to steam at 100 °C?   *Ans.*   7.2 kcal

18.27. Ten pounds of steam at 212 °F is condensed in 500 lb of water at 40 °F. What is the resulting temperature?   *Ans.*   62.4 °F

18.28. The heat of combustion of ethane gas is 373 kcal per mole. Assuming that 60% of the heat is useful, how many liters of ethane, measured at standard temperature and pressure, must be burned to convert 50 kg of water at 10 °C to steam at 100 °C? One mole of a gas occupies 22.4 liters at 0 °C and 1 atm. *Ans.*   3150 liters

18.29. Calculate the heat of fusion of ice from the following data for ice at 0 °C added to water:

Mass of calorimeter	60 g
Mass of calorimeter plus water	460 g
Mass of calorimeter plus water and ice	618 g
Initial temperature of water	38 °C
Final temperature of mixture	5 °C
Specific heat of calorimeter	0.10 cal/g·°C

Ans.   79.8 cal/g

18.30. A calorimeter whose water equivalent is 5 lb contains 45 lb of water and 10 lb of ice at 32 °F. What will be the final temperature if 5 lb of steam at 212 °F is admitted to the calorimeter and contents? *Ans.*   98 °F

18.31. Determine the final result when 200 g of water and 20 g of ice at 0 °C are in a calorimeter whose water equivalent is 30 g and into which is passed 100 g of steam at 100 °C.
Ans.   49 g of steam condensed, final temperature 100 °C

18.32. Determine the final result when 400 g of water and 100 g of ice at 0 °C are in a calorimeter whose water equivalent is 50 g and into which is passed 10 g of steam at 100 °C.
Ans.   80 g of ice melted, final temperature 0 °C

18.33. On a certain day the temperature is 20 °C and the dew point is 5 °C. What is the relative humidity? Saturated air at 20° and 5° contains 17.12 and 6.80 g/m³ water respectively.   *Ans.*   40%

18.34. How much water vapor exists in a 105 m³ room on a day when the relative humidity in the room is 32% and the room temperature is 20 °C? Saturated air at 20 °C contains 17.12 g/m³ water.   *Ans.*   575 g

18.35. Air at 30 °C and 90% relative humidity is drawn into an air conditioning unit and cooled to 20 °C. The relative humidity is simultaneously reduced to 50%. How many grams of water are removed from a cubic meter of 30 degree air by the air conditioner? Saturated air contains 30.4 g/m³ and 17.1 g/m³, at 30 °C and 20 °C respectively.   *Ans.*   19.1 g/m³

Chapter 19

Transfer of Heat Energy

HEAT ENERGY IS TRANSMITTED by *conduction*, *convection*, and *radiation*.

CONDUCTION occurs when heat energy moves through a material as a result of the collisions between the molecules of the material. The hotter a substance, the higher the average KE of its molecules. When a temperature difference exists between materials in contact, the higher-energy molecules in the warmer substance transfer energy to the lower-energy molecules in the cooler substance when molecular collisions between the two kinds occur. A flow of energy, heat energy, thus occurs from hot to cold.

Consider the slab of material shown in Fig. 19-1. Its thickness is L and its cross-sectional area is A. The temperatures of its two faces are T_1 and T_2, with $T_1 > T_2$. We call the quantity $(T_1 - T_2)/L$ the *temperature gradient*. It is the rate of change of temperature with distance.

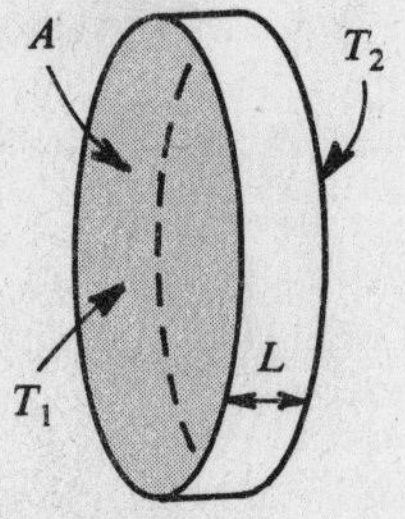

The quantity of heat, ΔQ, transmitted from face 1 through face 2 in time Δt is proportional to Δt, the face area A, and the temperature gradient. Therefore

$$\Delta Q = k(\Delta t)A\,\frac{T_1 - T_2}{L}$$

or

$$\frac{\Delta Q}{\Delta t} = kA\,\frac{T_1 - T_2}{L}$$

Fig. 19-1

where k is a proportionality constant which depends only on the nature of the substance. It is called the *coefficient of thermal conductivity*, or simply the *thermal conductivity*, of the substance. For a given substance, k is the amount of heat transmitted per unit time per unit perpendicular area per unit temperature gradient.

Typical units for k are cal/s·cm·°C and W/m·K, where 1 watt (W) = 1 J/s. Or, in the British system, the common units are Btu/hr·ft·°F. Great care must be taken to use the units of ΔQ, t, L, T, and A appropriate to the value of k used.

CONVECTION of heat energy occurs when a warm material is transported so as to displace a cooler material. Typical examples are the flow of warm air from a register in a heating system and the flow of warm water in the Gulf Stream.

RADIATION is the mode of transport of heat energy through vacuum and the empty space between molecules. It is an electromagnetic wave phenomenon and its nature is discussed in Chapters 36 through 44.

A *blackbody* is one that absorbs all the radiant energy falling on it. At thermal equilibrium, a body emits as much energy as it absorbs. Hence a good absorber of radiation is also a good emitter of radiation.

The *Stefan-Boltzmann law* states that the total energy radiated per second by unit area of a surface is proportional to the absolute temperature of the surface raised to the fourth power. For a blackbody, the relation is

$$\text{energy radiated per second per square meter} = \sigma T^4$$

where $\sigma = 5.67 \times 10^{-8}$ W/m²·K⁴ is the *Stefan-Boltzmann constant* and T is the Kelvin temperature.

Solved Problems

19.1. An iron plate 2 cm thick has a cross section of 5000 cm^2. One side is at 150 °C and the other side is at 140 °C. How much heat passes through the plate each second? For iron, k is 0.115 cal/s·cm·°C.

$$\frac{\Delta Q}{\Delta t} = k\frac{A(T_1 - T_2)}{L} = \left(0.115\frac{\text{cal}}{\text{s·cm·°C}}\right)\frac{(5000\text{ cm}^2)(10\text{ °C})}{2\text{ cm}} = 2880\text{ cal/s}$$

19.2. A metal plate 4 mm thick has a temperature difference of 32 °C between its faces. It transmits 200 kcal/h through an area of 5 cm^2. Calculate the thermal conductivity of this metal in W/m·K.

$$k = \frac{\Delta Q}{\Delta t}\frac{L}{A(T_1 - T_2)} = \frac{(2\times10^5\text{ cal})(4.184\text{ J/cal})}{(1\text{ h})(3600\text{ s/h})}\frac{4\times10^{-3}\text{ m}}{(5\times10^{-4}\text{ m}^2)(32\text{ K})}$$

$$= 58.5\text{ W/m·K}$$

19.3. Two metal plates are soldered together as shown in Fig. 19-2. It is known that $A = 80$ cm^2, $L_1 = L_2 = 3.0$ mm, $T_1 = 100$ °C, $T_2 = 0$ °C. For the plate on the left, $k_1 = 0.115$ cal/s·cm·°C; for the plate on the right, $k_2 = 0.163$ cal/s·cm·°C. Find the heat flow rate through the plates and the temperature t of the soldered junction.

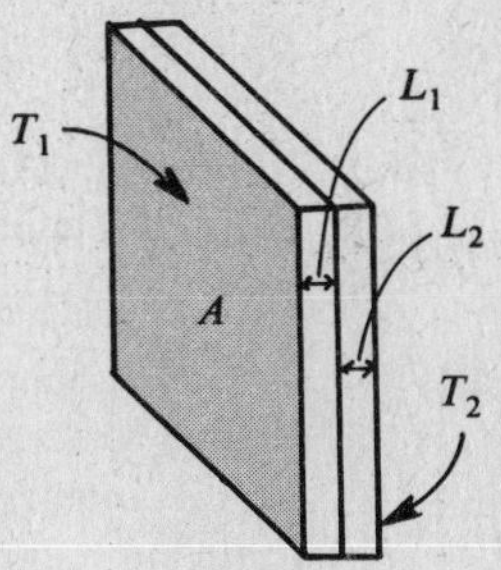

We assume equilibrium conditions so that the heat flowing through plate 1 equals that through plate 2. Then

$$k_1A\frac{T_1 - t}{L_1} = k_2A\frac{t - T_2}{L_2}$$

Fig. 19-2

But $L_1 = L_2$, so this becomes

$$k_1(100\text{ °C} - t) = k_2(t - 0\text{ °C})$$

from which

$$t = (100\text{ °C})\frac{k_1}{k_1 + k_2} = (100\text{ °C})\frac{0.115}{0.115 + 0.163} = 41.4\text{ °C}$$

The rate is then

$$\frac{\Delta Q}{\Delta t} = k_1A\frac{T_1 - t}{L_1} = \left(0.115\frac{\text{cal}}{\text{s·cm·°C}}\right)(80\text{ cm}^2)\frac{(100 - 41.4)\text{ °C}}{0.30\text{ cm}} = 1800\text{ cal/s}$$

19.4. A copper tube (length, 3 m; inner diameter, 1.500 cm; outer diameter, 1.700 cm) passes through a vat of rapidly circulating water maintained at 20 °C. Live steam at 100 °C passes through the tube. (a) What is the heat flow rate from the steam into the vat? (b) How much steam is condensed each minute? For copper, $k = 1.0$ cal/s·cm·°C.

Because the thickness of the tube is much smaller than its radius, the inner surface area of the tube,

$$2\pi r_iL = 2\pi(0.750\text{ cm})(300\text{ cm}) = 1410\text{ cm}^2$$

nearly equals its outer surface area,

$$2\pi r_oL = 2\pi(0.850\text{ cm})(300\text{ cm}) = 1600\text{ cm}^2$$

As an approximation, we can consider the tube to be a plate of thickness 0.100 cm and area given by

$$A = \frac{1}{2}(1410\text{ cm}^2 + 1600\text{ cm}^2) = 1500\text{ cm}^2$$

(a) $$\frac{\Delta Q}{\Delta t} = k\frac{A(T_1 - T_2)}{L} = \left(1.0\frac{\text{cal}}{\text{s·cm·°C}}\right)\frac{(1500\text{ cm}^2)(80\text{ °C})}{(0.100\text{ cm})} = 1.2\times10^6\text{ cal/s}$$

(b) In one minute, the heat conducted from the tube is

$$\Delta Q = (1.2 \times 10^6 \text{ cal/s})(60 \text{ s}) = 72 \times 10^6 \text{ cal}$$

It takes a 540 cal heat loss to condense one gram of steam at 100 °C.   Therefore

$$\text{steam condensed per min} = \frac{72 \times 10^6 \text{ cal}}{540 \text{ cal/g}} = 13.3 \times 10^4 \text{ g} = 133 \text{ kg}$$

In practice, various factors would greatly reduce this theoretical value.

19.5. A spherical body of 2 cm diameter is maintained at 600 °C.   Assuming that it radiates as if it were a blackbody, at what rate (in watts) is energy radiated from the sphere?

$$A = \text{surface area} = 4\pi r^2 = 4\pi(0.01 \text{ m})^2 = 1.26 \times 10^{-3} \text{ m}^2$$

$$\text{rate} = A\sigma T^4 = (1.26 \times 10^{-3} \text{ m}^2)(5.67 \times 10^{-8} \text{ W/m}^2 \cdot \text{K}^4)(873 \text{ K})^4 = 41.4 \text{ W}$$

Supplementary Problems

19.6. What temperature gradient must exist in an aluminum rod in order to transmit 8 cal per second per cm^2 of cross section down the rod? k for aluminum is 0.50 cal/s $\cdot$ cm $\cdot$ °C.   *Ans.*   16 °C/cm

19.7. A single-thickness glass window 80 cm $\times$ 40 cm $\times$ 3 mm on a house actually has layers of stagnant air on its two surfaces.   But if it did not, how much heat would flow out of it each hour on a day when the outside temperature was 0 °C and the inside temperature was 18 °C?   For glass, k is 0.20 cal/s $\cdot$ m $\cdot$ °C (notice the units of k).   *Ans.*   1400 kcal/h

19.8. How many grams of water at 100 °C could be evaporated per hour per cm^2 by the heat transmitted through a steel plate 0.2 cm thick, if the temperature difference between the plate faces is 100 °C?   For steel, k is 0.11 cal/s $\cdot$ cm $\cdot$ °C.   *Ans.*   367 g/h $\cdot$ cm^2

19.9. Masonite sheeting transmits 0.33 Btu per hour through 1 ft^2 section when the temperature gradient is 1 °F per inch of thickness.   How many Btu will be transmitted per day through a sheet having dimensions 3 ft $\times$ 6 ft $\times$ 3/4 inch if one face is at 38 °F and the opposite face is at 63 °F?   *Ans.*   4800 Btu/d

19.10. A certain double-pane window consists of two glass sheets, each 80 cm $\times$ 80 cm $\times$ 0.30 cm, separated by a 0.30 cm stagnant air space.   The indoor surface temperature is 20 °C, while the outdoor surface temperature is 0 °C.   How much heat passes through the window each second? $k = 2 \times 10^{-3}$ cal/s $\cdot$ cm $\cdot$ °C   for glass and about 2×10^{-4} cal/s $\cdot$ cm $\cdot$ °C   for air.   *Ans.*   71 cal/s

19.11. A small hole in a furnace acts like a blackbody.   (Why?)   Its area is 1 cm^2 and its temperature is the same as that for the interior of the furnace, 1727 °C.   How many calories are radiated out the hole each second?   *Ans.*   21.7 cal/s

19.12. An incandescent lamp filament has an area 50 mm^2 and operates at a temperature of 2127 °C.   Assume that all the energy furnished to the bulb is radiated from it.   If the filament acts like a blackbody, how much power must be furnished to the bulb when operating?   *Ans.*   94 W

19.13. A sphere of 3 cm radius acts like a blackbody.   It is in equilibrium with its surroundings and absorbs 30 kW of power radiated to it from the surroundings.   What is the temperature of the sphere?   *Ans.*   2600 K

Chapter 20

First Law of Thermodynamics

HEAT ENERGY (ΔQ) is the energy that flows from one body to another body because of their temperature difference. Heat always flows from hot to cold. For two objects in contact to be in thermal equilibrium with each other (i.e. for no net heat transfer from one to the other), their temperatures must be the same. If each of two objects is in thermal equilibrium with a third body, then the two are in thermal equilibrium with each other. (This fact is often referred to as the *zeroth law* of thermodynamics.)

THE INTERNAL ENERGY (U) of a system is the total energy content of the system. It is the sum of the kinetic, potential, chemical, electrical, nuclear, and all other forms of energy possessed by the atoms and molecules of the system. An important fact about the internal energy of an ideal gas is that it is purely kinetic and *depends only on the temperature* ($\frac{1}{2} m_0 v_{rms}^2 = \frac{3}{2} kT$).

THE WORK DONE BY A SYSTEM (ΔW) is positive if the system thereby loses energy to its surroundings. When the surroundings do work on the system so as to give it energy, then ΔW is a negative quantity. In a small expansion ΔV, a fluid with constant pressure p does work given by

$$\Delta W = p \, \Delta V$$

THE FIRST LAW OF THERMODYNAMICS is a statement of the law of conservation of energy. It states that if an amount of heat energy ΔQ flows into a system, then this energy must appear as increased internal energy ΔU for the system and/or work ΔW done by the system on its surroundings. As an equation, the first law is

$$\Delta Q = \Delta U + \Delta W$$

Here ΔQ, ΔU, and ΔW must all be in the same units: joules, ft·lb, calories, or Btu.

AN ISOBARIC PROCESS is a process carried out at constant pressure.

AN ISOVOLUMIC PROCESS is a process carried out at constant volume. When a gas undergoes such a process,

$$\Delta W = p \, \Delta V = 0$$

and so the first law of thermodynamics becomes

$$\Delta Q = \Delta U$$

Any heat that flows into the system appears as increased internal energy of the system.

AN ISOTHERMAL PROCESS is a constant-temperature process. In the case of an ideal gas, $\Delta U = 0$ in an isothermal process. However, this is not true for many other systems. For example, $\Delta U \neq 0$ as ice melts to water at 0° C, even though the process is isothermal.

For an ideal gas, $\Delta U = 0$ in an isothermal change and so the first law is

$$\Delta Q = \Delta W \qquad \text{(ideal gas)}$$

It is also true that, for an ideal gas changing from (p_1, V_1) to (p_2, V_2), where $p_1 V_1 = p_2 V_2$,

$$\Delta Q = \Delta W = p_1 V_1 \ln\left(\frac{V_2}{V_1}\right) = 2.30 \, p_1 V_1 \log\left(\frac{V_2}{V_1}\right)$$

Here, ln and log are logarithms to the base e and base 10 respectively.

AN ADIABATIC PROCESS is one in which no heat is transferred to or from the system. For it, $\Delta Q = 0$. Hence, in an adiabatic process, the first law becomes

$$0 = \Delta U + \Delta W$$

Any work done by the system is done at the expense of the internal energy. Any work done on the system serves to increase the internal energy.

For an ideal gas changing from conditions (p_1, V_1, T_1) to (p_2, V_2, T_2) in an adiabatic process,

$$p_1 V_1^\gamma = p_2 V_2^\gamma \qquad \text{and} \qquad T_1 V_1^{\gamma-1} = T_2 V_2^{\gamma-1}$$

where $\gamma = c_p/c_v$ is discussed below.

SPECIFIC HEATS OF GASES: When a gas is heated at *constant volume*, the entire heat energy supplied goes to increase the internal energy of the gas molecules. But when a gas is heated at *constant pressure*, the heat supplied not only increases the internal energy of the molecules but also does mechanical work in expanding the gas against the opposing constant pressure. Hence the specific heat of a gas at constant pressure, c_p, is greater than its specific heat at constant volume, c_v. It can be shown that for an ideal gas of molecular weight M,

$$c_p - c_v = \frac{R}{M} \qquad \text{(ideal gas)}$$

where R is the universal gas constant. If $R = 8314 \text{ J/kmol} \cdot \text{K}$ and M is in kg/kmol, then c_p and c_v must be in $\text{J/kg} \cdot \text{K} = \text{J/kg} \cdot {}^\circ\text{C}$. Some people use $R = 1.98 \text{ cal/mol} \cdot {}^\circ\text{C}$ and M in g/mol, in which case c_p and c_v will be in cal/g $\cdot {}^\circ\text{C}$.

SPECIFIC HEAT RATIO ($\gamma = c_p/c_v$): As discussed above, this ratio is greater than unity for a gas. The kinetic theory of gases indicates that for monatomic gases (such as He, Ne, Ar), $\gamma = 1.67$. For diatomic gases (such as O_2, N_2), $\gamma = 1.40$ at ordinary temperatures.

WORK IS RELATED TO AREA in a p-V diagram. The work done by a fluid in an expansion is equal to the area beneath the expansion curve on a p-V diagram.

In a cyclic process, the output work per cycle done by a fluid is equal to the area enclosed by the p-V diagram representing the cycle.

THE EFFICIENCY OF A HEAT ENGINE is defined as

$$\text{eff} = \frac{\text{work output}}{\text{heat input}}$$

The *Carnot cycle* is the most efficient cycle possible for a heat engine. An engine that operates in accordance to this cycle between a hot reservoir (T_h) and a cold reservoir (T_c) has efficiency

$$\text{eff}_{\max} = 1 - \frac{T_c}{T_h}$$

Here, Kelvin temperatures must be used.

Solved Problems

20.1. In a certain process, 8000 cal of heat is furnished to the system while the system does 6000 J of work. By how much did the internal energy of the system change during the process?

We have

$$\Delta Q = (8000 \text{ cal})\left(4.184 \frac{\text{J}}{\text{cal}}\right) = 33.5 \text{ kJ} \qquad \text{and} \qquad \Delta W = 6 \text{ kJ}$$

Therefore, from the first law, $\Delta Q = \Delta U + \Delta W$,

$$\Delta U = \Delta Q - \Delta W = 33.5 \text{ kJ} - 6 \text{ kJ} = 27.5 \text{ kJ}$$

20.2. The specific heat of water is 1 cal/g·°C. By how many joules does the internal energy of 50 g of water change as it is heated from 21 °C to 37 °C?

To heat the water, the heat added is

$$\Delta Q = cm\, \Delta T = (1 \text{ cal/g} \cdot {}^{\circ}\text{C})(50 \text{ g})(16 \text{ }^{\circ}\text{C}) = 800 \text{ cal}$$

$$= (800 \text{ cal})(4.184 \text{ J/cal}) = 3350 \text{ J}$$

If we ignore the slight expansion of the water, no work was done on the surroundings and so $\Delta W = 0$. Then, the first law, $\Delta Q = \Delta U + \Delta W$, tells us that

$$\Delta U = \Delta Q = 3350 \text{ J}$$

20.3. An 8 g metal bullet ($c = 0.16$ cal/g·°C) is moving at 70 m/s when it strikes a block of wood and lodges in it. If 60% of the work done by the stopping forces goes into heating the metal, how much will the bullet's temperature rise in the process?

In situations such as this, it is often easiest to bypass the first-law equation and work directly in terms of energy conservation.

The work done by the stopping forces acting on the bullet is

work done = loss in KE

$$= \tfrac{1}{2}mv^2 = \tfrac{1}{2}(0.008 \text{ kg})(70 \text{ m/s})^2 = 19.6 \text{ J}$$

Of this, it is given that 60% goes into heating the bullet. (This is what we have referred to previously as *frictional heating*.) In effect, heat has been added in the amount

$$\Delta Q = (0.60)(19.6 \text{ J}) = 11.8 \text{ J}$$

$$= (11.8 \text{ J})\left(\frac{1 \text{ cal}}{4.184 \text{ J}}\right) = 2.81 \text{ cal}$$

Now we can use $\Delta Q = cm\, \Delta T$ to find

$$\Delta T = \frac{\Delta Q}{cm} = \frac{2.81 \text{ cal}}{(0.16 \text{ cal/g} \cdot {}^{\circ}\text{C})(8 \text{ g})} = 2.2 \text{ }^{\circ}\text{C}$$

20.4. How much does the internal energy of 5 g of ice at 0 °C increase as it is changed to water at 0 °C? Neglect the small change in volume.

In order to melt the ice, the heat needed is

$$\Delta Q = mH_f = (5 \text{ g})(80 \text{ cal/g}) = 400 \text{ cal}$$

No external work was done by the ice as it melted and so $\Delta W = 0$. Therefore, the first law, $\Delta Q = \Delta U + \Delta W$, tells us that

$$\Delta U = \Delta Q = (400 \text{ cal})(4.184 \text{ J/cal}) = 1670 \text{ J}$$

20.5. Find ΔW and ΔU for a 6 cm cube of iron as it is heated from 20 °C to 300 °C. For iron, $c = 0.11$ cal/g·°C and the volume coefficient of thermal expansion is 3.6×10^{-5} °C^{-1}. The mass of the cube is 1700 g.

$$\Delta Q = cm\, \Delta T = (0.11 \text{ cal/g} \cdot {}^{\circ}\text{C})(1700 \text{ g})(280 \text{ }^{\circ}\text{C}) = 52\,000 \text{ cal}$$

The volume of the cube is $V = (6 \text{ cm})^3 = 216 \text{ cm}^3$. Using $(\Delta V)/V = \beta \Delta T$, we have

$$\Delta V = V\beta \Delta T = (216 \times 10^{-6} \text{ m}^3)(3.6 \times 10^{-5} \text{ }^\circ\text{C}^{-1})(280 \text{ }^\circ\text{C}) = 2.18 \times 10^{-6} \text{ m}^3$$

Then, assuming atmospheric pressure to be 1×10^5 Pa,

$$\Delta W = p \Delta V = (1 \times 10^5 \text{ N/m}^2)(2.18 \times 10^{-6} \text{ m}^3) = 0.22 \text{ J}$$

But the first law tells us that

$$\Delta U = \Delta Q - \Delta W$$
$$= (52\,000 \text{ cal})(4.184 \text{ J/cal}) - 0.22 \text{ J}$$
$$= 218\,000 \text{ J} - 0.22 \text{ J} \approx 218\,000 \text{ J}$$

Notice how very small the work of expansion against the atmosphere is in comparison to ΔU and ΔQ. Often ΔW can be neglected when dealing with liquids and solids.

20.6. One pound of fuel, having a heat of combustion of 10 000 Btu/lb, was burned in an engine that raised 6000 lb of water 110 ft. What percentage of the heat was transformed into useful work?

$$\text{efficiency} = \frac{\text{work done by engine}}{\text{work equivalent of heat supplied}}$$

$$= \frac{(6000 \text{ lb})(110 \text{ ft})}{(10\,000 \text{ Btu})(778 \text{ ft} \cdot \text{lb/Btu})} = 0.085 = 8.5\%$$

20.7. A motor supplies 0.4 hp to stir 5 kg of water. Assuming that all of the work goes into heating the water by friction losses, how long will it take to increase the temperature of the water 6 °C?

$$\text{heat required to heat water} = \Delta Q = mc \Delta T$$
$$= (5000 \text{ g})(1 \text{ cal/g} \cdot {}^\circ\text{C})(6 \text{ }^\circ\text{C}) = 30\,000 \text{ cal}$$

This was actually supplied by friction work.

$$\text{friction work done} = \Delta Q = (30\,000 \text{ cal})(4.184 \text{ J/cal}) = 1.26 \times 10^5 \text{ J}$$

and this equals the work done by the motor. But

$$\text{work done by motor in time } t = (\text{power})(t) = (0.4 \text{ hp} \times 746 \text{ W/hp})(t)$$

Equating this to our previous value for the work done gives

$$t = \frac{1.26 \times 10^5 \text{ J}}{(0.4 \times 746) \text{ W}} = 420 \text{ s} = 7 \text{ min}$$

20.8. In each of the following situations, find the change in internal energy of the system. (a) A system absorbs 500 cal of heat and at the same time does 400 J of work. (b) A system absorbs 300 cal and at the same time 420 J of work is done on it. (c) Twelve hundred calories is removed from a gas held at constant volume.

(a) $\Delta U = \Delta Q - \Delta W = (500 \text{ cal})(4.184 \text{ J/cal}) - 400 \text{ J} = 1700 \text{ J}$

(b) $\Delta U = \Delta Q - \Delta W = (300 \text{ cal})(4.184 \text{ J/cal}) - (-420 \text{ J}) = 1680 \text{ J}$

(c) $\Delta U = \Delta Q - \Delta W = (-1200 \text{ cal})(4.184 \text{ J/cal}) - 0 = -5000 \text{ J}$

Notice that ΔQ is positive when heat is added to the system and ΔW is positive when the system does work. In the reverse cases, ΔQ and ΔW must be taken negative.

20.9. For each of the following adiabatic processes, find the change in internal energy. (a) A gas does 5 J of work while expanding adiabatically. (b) During an adiabatic compression, 80 J of work is done on a gas.

During an adiabatic process, no heat is transferred to or from the system.

(a) $\Delta U = \Delta Q - \Delta W = 0 - 5 \text{ J} = -5 \text{ J}$

(b) $\Delta U = \Delta Q - \Delta W = 0 - (-80 \text{ J}) = +80 \text{ J}$

20.10. One pound of steam at 212 °F and 14.7 lb/in^2 occupies 26.8 ft^3. (a) What fraction of the observed heat of vaporization of water (970 Btu/lb at 212 °F and 14.7 lb/in^2) is accounted for by the external work done in expanding water into steam at 212 °F against atmospheric pressure of 14.7 lb/in^2? The density of water at 212 °F is 60 lb/ft^3. (b) Determine the increase in internal energy when 1 lb of steam is formed at 212 °F.

(a) One pound of water occupies $1 \text{ lb}/(60 \text{ lb/ft}^3) \approx 0.02 \text{ ft}^3$ and expands to 26.8 ft^3 of steam under the conditions stated. The work done in expanding 1 lb water to steam at constant pressure $p = 1$ atm is then

$$p \, \Delta V = (14.7 \text{ lb/in}^2 \times 144 \text{ in}^2/\text{ft}^2)(26.8 \text{ ft}^3 - 0.02 \text{ ft}^3) = 56\,700 \text{ ft} \cdot \text{lb}$$

$$\text{heat equivalent of } 56\,700 \text{ ft} \cdot \text{lb} = (56\,700 \text{ ft} \cdot \text{lb})\left(\frac{1 \text{ Btu}}{778 \text{ ft} \cdot \text{lb}} \right) = 73 \text{ Btu}$$

$$\text{required fraction} = \frac{73 \text{ Btu}}{970 \text{ Btu}} = 0.075 = 7.5\%$$

(b)
$$\Delta U = \Delta Q - \Delta W = 970 \text{ Btu} - 73 \text{ Btu} = 897 \text{ Btu}$$

20.11. The temperature of 5 kg of N$_2$ gas is raised from 10 °C to 130 °C. If this is done at constant pressure, find the increase in internal energy, ΔU, and the external work, ΔW, done by the gas. For N$_2$ gas, $c_v = 0.177$ cal/g $\cdot$ °C and $c_p = 0.248$ cal/g $\cdot$ °C.

At constant pressure, the first law gives

$$(\Delta Q)_p = (\Delta U)_p + (\Delta W)_p \qquad \text{or} \qquad mc_p \, \Delta T = (\Delta U)_p + (\Delta W)_p$$

For the same temperature increase carried out at constant volume,

$$(\Delta Q)_v = (\Delta U)_v + (\Delta W)_v \qquad \text{or} \qquad mc_v \, \Delta T = (\Delta U)_p$$

since $(\Delta W)_v = 0$ and since $(\Delta U)_v = (\Delta U)_p$ for the same change in temperature. Subtracting the two equations gives

$$\Delta W_p = m(c_p - c_v) \, \Delta T = (5000 \text{ g})\left[(0.248 - 0.177) \frac{\text{cal}}{\text{g} \cdot \text{°C}} \right](120 \text{ °C})$$

$$= 43 \text{ kcal} = 180 \text{ kJ}$$

Furthermore
$$(\Delta U)_p = mc_v \, \Delta T = (5000 \text{ g})\left(0.177 \frac{\text{cal}}{\text{g} \cdot \text{°C}} \right)(120 \text{ °C}) = 106 \text{ kcal}$$

20.12. For nitrogen gas, $c_v = 0.177$ cal/g $\cdot$ °C. Find its specific heat at constant pressure, c_p. The molecular weight of nitrogen is 28.0 kg/kmol.

Method 1

$$c_p = c_v + \frac{R}{M} = 0.177 \frac{\text{cal}}{\text{g} \cdot \text{°C}} + \frac{8314 \text{ J/kmol} \cdot \text{K}}{28 \text{ kg/kmol}} \left(\frac{1 \text{ cal}}{4.184 \text{ J}} \right)\left(\frac{1 \text{ kg}}{1000 \text{ g}} \right)$$

$$= 0.177 \frac{\text{cal}}{\text{g} \cdot \text{°C}} + 0.071 \frac{\text{cal}}{\text{g} \cdot \text{°C}} = 0.248 \frac{\text{cal}}{\text{g} \cdot \text{°C}}$$

Method 2
Since N$_2$ is a diatomic gas and since $c_p/c_v \approx 1.40$ for such a gas,

$$c_p = (1.40)c_v = (1.40)\left(0.177 \frac{\text{cal}}{\text{g} \cdot \text{°C}} \right) = 0.248 \frac{\text{cal}}{\text{g} \cdot \text{°C}}$$

20.13. How much work is done by an ideal gas in expanding isothermally from an initial volume of 3 liters at 20 atm to a final volume of 24 liters?

For an isothermal expansion by an ideal gas,

$$W = p_1 V_1 \ln\left(\frac{V_2}{V_1} \right) = 2.30 \, p_1 V_1 \log\left(\frac{V_2}{V_1} \right)$$

$$= (2.30)(20 \times 1.01 \times 10^5 \text{ N/m}^2)(3 \times 10^{-3} \text{ m}^3) \log\left(\frac{24}{3} \right) = 1.26 \times 10^4 \text{ J}$$

20.14. The p-V diagram shown in Fig. 20-1 applies to a gas undergoing a cyclic change in a piston-cylinder arrangement. What is the work done by the gas in (a) portion AB of the cycle? (b) portion BC? (c) portion CD? (d) portion DA?

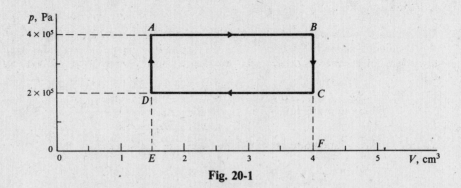

Fig. 20-1

In expansion, the work done is equal to the area under the pertinent portion of the p-V curve. In contraction, the work is numerically equal to the area but is negative.

(a)   work = area $ABFEA = [(4 - 1.5) \times 10^{-6} \text{ m}^3](4 \times 10^5 \text{ N/m}^2) = 1.00$ J

(b)   work = area under $BC = 0$

In portion BC, the volume does not change; therefore $p \, \Delta V = 0$.

(c) This is a contraction and so the work is negative.

work $= -(\text{area } CDEFC) = -(2.5 \times 10^{-6} \text{ m}^3)(2 \times 10^5 \text{ N/m}^2) = -0.50$ J

(d)   work = 0

20.15. For the thermodynamic cycle shown in Fig. 20-1, find (a) net output work of the gas during the cycle, (b) net heat flow into the gas per cycle.

(a) **Method 1**
From Problem 20.14, the net work done is   1.00 J − 0.50 J = 0.50 J.

Method 2
The net work done is equal to the area enclosed by the p-V diagram.

work = area $ABCDA = (2 \times 10^5 \text{ N/m}^2)(2.5 \times 10^{-6} \text{ m}^3) = 0.50$ J

(b) Suppose the cycle starts at point A. The gas returns to this point at the end of the cycle, so there is no difference in the gas at its start and end point. For one complete cycle, ΔU is therefore zero. We have then, if the first law is applied to a complete cycle,

$$\Delta Q = \Delta U + \Delta W = 0 + 0.50 \text{ J} = 0.50 \text{ J} = 0.120 \text{ cal}$$

20.16. What is the net output work per cycle for the thermodynamic cycle shown in Fig. 20-2?

We know that the net output work per cycle is the area enclosed by the p-V diagram. We estimate that in area $ABCA$ there are 22 squares, each of area

$$(0.5 \times 10^5 \text{ N/m}^2)(0.1 \text{ m}^3) = 5 \times 10^3 \text{ J}$$

Therefore,

area enclosed by cycle $\approx (22)(5 \times 10^3 \text{ J})$

$$= 110 \times 10^3 \text{ J}$$

The net output work per cycle is 110 kJ.

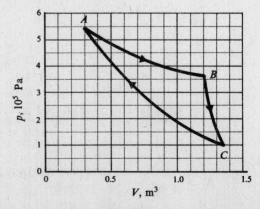

Fig. 20-2

20.17. Twenty cubic centimeters of monatomic gas at 12 °C and 100 kPa is suddenly (and adiabatically) compressed to 0.5 cm^3.   What are its new pressure and temperature?

For an adiabatic change involving an ideal gas, $p_1 V_1^\gamma = p_2 V_2^\gamma$,   where $\gamma = 1.67$ for a monatomic gas.   Hence,

$$p_2 = p_1 \left(\frac{V_1}{V_2} \right)^\gamma = (1 \times 10^5 \text{ N/m}^2) \left(\frac{20}{0.5} \right)^{1.67} = 4.74 \times 10^7 \text{ N/m}^2$$

If you do not have a calculator that will perform the operation x^y, then you can evaluate $(40)^{1.67}$ as follows:

$$\log \left[(40)^{1.67} \right] = 1.67 \log 40 = (1.67)(1.602) = 2.675$$

and $$(40)^{1.67} = \text{antilog } 2.675 = 474$$

To find its final temperature, we could use $p_1 V_1 / T_1 = p_2 V_2 / T_2$.   Instead, let us use

$$T_1 V_1^{\gamma-1} = T_2 V_2^{\gamma-1} \qquad \text{or} \qquad T_2 = T_1 \left(\frac{V_1}{V_2} \right)^{\gamma-1}$$

$$T_2 = (285 \text{ K}) \left(\frac{20}{0.5} \right)^{0.67} = (285 \text{ K})(11.8) = 3370 \text{ K}$$

As a check,

$$\frac{p_1 V_1}{T_1} = \frac{p_2 V_2}{T_2}$$

$$\frac{(1 \times 10^5 \text{ N/m}^2)(20 \text{ cm}^3)}{285 \text{ K}} = \frac{(4.74 \times 10^7 \text{ N/m}^2)(0.5 \text{ cm}^3)}{3370 \text{ K}}$$

$$7000 = 7000 \quad \surd$$

20.18. Compute the maximum possible efficiency of a heat engine operating between the temperature limits of 100 °C and 400 °C.

The most efficient engine is the Carnot engine, for which

$$\text{efficiency} = 1 - \frac{T_c}{T_h} = 1 - \frac{373}{673} = 0.45 = 45\%$$

20.19. A steam engine operating between a boiler temperature of 220 °C and a condenser temperature of 35 °C delivers 8 hp.   If its efficiency is 30% of that for a Carnot engine operating between these temperature limits, how many calories are absorbed each second by the boiler? How many calories are exhausted to the condenser each second?

$$\text{actual efficiency} = (0.30)(\text{Carnot efficiency}) = (0.30)\left(1 - \frac{308}{493} \right) = 0.113$$

But from the relation

$$\text{efficiency} = \frac{\text{output work}}{\text{input heat}}$$

$$\text{input heat/s} = \frac{\text{output work/s}}{\text{efficiency}} = \frac{(8 \text{ hp})(746 \text{ W/hp})\left(\frac{1 \text{ cal/s}}{4.184 \text{ W}} \right)}{0.113} = 12\,700 \text{ cal/s}$$

To find the energy rejected to the condenser, we use the law of conservation of energy:

$$\text{input energy} = (\text{output work}) + (\text{rejected energy})$$

Thus,

$$\text{rejected energy/s} = (\text{input energy/s}) - (\text{output work/s})$$

$$= (\text{input energy/s})[1 - (\text{efficiency})]$$

$$= (12\,700 \text{ cal/s})(1 - 0.113) = 11\,300 \text{ cal/s}$$

20.20. A cylinder of ideal gas is closed by an 8 kg movable piston (area 60 cm^2) as shown in Fig. 20-3. Atmospheric pressure is 100 kPa. When the gas is heated from 30 °C to 100 °C, the piston rises 20 cm. The piston is then fastened in place and the gas is cooled back to 30 °C. Calling ΔQ_1 the heat added to the gas in the heating process and ΔQ_2 the heat lost during cooling, find the difference between ΔQ_1 and ΔQ_2.

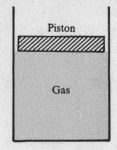

Fig. 20-3

During the heating process, the internal energy changed by ΔU_1 and work ΔW_1 was done. The gas pressure was

$$p = \frac{8(9.8)\ \text{N}}{60 \times 10^{-4}\ \text{m}^2} + 1.00 \times 10^5\ \text{N/m}^2 = 1.13 \times 10^5\ \text{N/m}^2$$

Therefore

$$\Delta Q_1 = \Delta U_1 + \Delta W_1 = \Delta U_1 + p\,\Delta V$$

$$= \Delta U_1 + (1.13 \times 10^5\ \text{N/m}^2)(0.20 \times 60 \times 10^{-4}\ \text{m}^3) = \Delta U_1 + 136\ \text{J}$$

During the cooling process, $\Delta W = 0$ and so (ΔQ_2 is heat *lost*)

$$-\Delta Q_2 = \Delta U_2$$

But the ideal gas returns to its original temperature and so its internal energy is the same as at the start. Therefore, $\Delta U_2 = -\Delta U_1$, or $\Delta Q_2 = \Delta U_1$. It follows that ΔQ_1 exceeds ΔQ_2 by 136 J = 32.5 cal.

Supplementary Problems

20.21. A 2.0 kg metal block ($c = 0.137$ cal/g·°C) is heated from 15 °C to 90 °C. By how much did its internal energy change? *Ans.* 86 kJ

20.22. By how much does the internal energy of 50 g of oil ($c = 0.32$ cal/g·°C) change as the oil is cooled from 100 °C to 25 °C? *Ans.* −1200 cal

20.23. A 70 g metal block moving at 200 cm/s slides across a tabletop a distance of 83 cm before it comes to rest. Assuming 75% of the friction heat goes into the block, how much does the temperature of the block rise? For the metal, $c = 0.106$ cal/g·°C. *Ans.* 3.4×10^{-3} °C

20.24. If a certain mass of water falls a distance of 854 m and all the energy is effective in heating the water, what will be the temperature rise of the water? *Ans.* 2.00 °C

20.25. How many Btu of heat per hour are produced in a motor that is 75% efficient and requires 0.25 hp to run it? *Ans.* 160 Btu/h

20.26. A 100 g bullet ($c = 0.030$ cal/g·°C) is initially at 20 °C. It is fired straight upward with a speed of 420 m/s, and on returning to the starting point strikes a cake of ice at 0 °C. How much ice is melted? Neglect air friction. *Ans.* 26 g

20.27. In order to determine the specific heat of an oil, an electrical heating coil is placed in a calorimeter with 380 g of the oil at 10 °C. The coil consumes energy (and gives off heat) at the rate of 84 W. After 3 min, the oil temperature is 40 °C. If the water equivalent of the calorimeter and coil is 20 g, what is the specific heat of the oil? *Ans.* 0.26 cal/g·°C

20.28. How much external work is done by an ideal gas in expanding from a volume of 3 liters to a volume of 30 liters against a constant pressure of 2 atm? *Ans.* 5450 J

20.29. As 3 liters of ideal gas at 27 °C is heated, it expands at a constant pressure of 2 atm. How much work is done by the gas as its temperature is changed from 27 °C to 227 °C? *Ans.* 404 J

20.30. As an ideal gas is compressed isothermally, the compressing agent does 36 J of work. How much heat flows from the gas during the compression process?   *Ans.* 8.6 cal

20.31. The specific heat of air at constant volume is 0.175 cal/g·°C. (*a*) By how much does the internal energy of 5 g of air change as it is heated from 20 °C to 400 °C? (*b*) Suppose that 5 g of air is adiabatically compressed so as to raise its temperature from 20 °C to 400 °C. How much work must be done on the air to compress it?   *Ans.* (*a*) 333 cal;   (*b*) 1390 J

20.32. Water is boiled at 100 °C and 1 atm. Under these conditions, 1 g of water occupies 1 cm³, 1 g of steam occupies 1670 cm³, and $H_v = 540$ cal/g. Find (*a*) the external work done when 1 g of steam is formed at 100 °C and (*b*) the increase in internal energy.   *Ans.* (*a*) 169 J;   (*b*) 500 cal

20.33. The temperature of 3 kg of krypton gas is raised from −20 °C to 80 °C. (*a*) If this is done at constant volume, compute the heat added, the work done, and the change in internal energy. (*b*) Repeat if the heating process is at constant pressure. For the monatomic gas Kr, $c_v = 0.0357$ cal/g·°C and $c_p = 0.0595$ cal/g·°C.   *Ans.* (*a*) 10.7 kcal, 0, 44.8 kJ;   (*b*) 17.85 kcal, 30 kJ, 44.8 kJ

20.34. (*a*) Compute c_v for the monatomic gas argon, given $c_p = 0.125$ cal/g·°C and $\gamma = 1.67$. (*b*) Compute c_p for the diatomic gas nitric oxide (NO), given $c_v = 0.166$ cal/g·°C and $\gamma = 1.40$.   *Ans.* (*a*) 0.0749 cal/g·°C;   (*b*) 0.232 cal/g·°C

20.35. Compute the work done in an isothermal compression of 30 liters of ideal gas at 1 atm to a volume of 3 liters.   *Ans.* 6990 J

20.36. Five moles of neon gas at 2 atm and 27 °C is adiabatically compressed to one-third of its initial volume. Find the final pressure, final temperature, and the external work done on the gas. For neon, $\gamma = 1.67$, $c_v = 0.148$ cal/g·°C, and $M = 20.18$ kg/kmol.   *Ans.* 1.27 MPa, 626 K, 20.4 kJ

20.37. Determine the work done by the gas in portion AB of the thermodynamic cycle shown in Fig. 20-2. Repeat for portion CA.   *Ans.* 4×10^5 J, -3.2×10^5 J

20.38. Find the net output work per cycle for the thermodynamic cycle shown in Fig. 20-4.   *Ans.* 2120 J

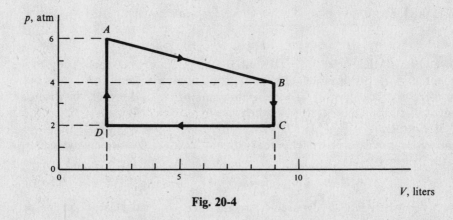

Fig. 20-4

20.39. Figure 20-4 is the p-V diagram for 25 g of an enclosed ideal gas. At A the gas temperature is 200 °C. The value of c_v for the gas is 0.150 cal/g·°C. (*a*) What is the temperature of the gas at point B? (*b*) Find ΔU for the portion of the cycle from A to B. (*c*) Find ΔW for this same portion. (*d*) Find ΔQ for this same portion.
Ans. (*a*) 1420 K;   (*b*) 3550 cal = 14 900 J;   (*c*) 3540 J;   (*d*) 18 440 J = 4400 cal

Chapter 21

Entropy and the Second Law

THE SECOND LAW OF THERMODYNAMICS can be stated in three equivalent ways:

(1) Heat energy flows spontaneously from a hotter to a colder object, but not vice versa.

(2) If a system undergoes spontaneous change, it will change in such a way that its disorder will increase or, at best, stay the same.

(3) If a system undergoes spontaneous change, it will change in such a way that its entropy will increase or, at best, remain constant.

The second law tells us the direction a spontaneous change will follow, while the first law tells us whether or not the change is possible according to energy conservation.

ENTROPY (S) is a *state variable* for a system at equilibrium. By this is meant that S (the entropy) is always the same for the system when it is in a given equilibrium state. Like p, V, and U, the entropy is a characteristic of the system at equilibrium.

When heat ΔQ enters a system at absolute temperature T, then the change in entropy of the system is

$$\Delta S = \frac{\Delta Q}{T}$$

provided the system changes in a reversible way.

A *reversible change* (or process) is one in which the values of p, V, T, and U are well-defined during the change. If the process is reversed, then p, V, T, and U will follow the same values in reverse order throughout the change. To be reversible, a process must usually be slow, and always the system must be close to equilibrium during the entire change.

Another, fully equivalent, definition of entropy can be given from a detailed molecular analysis of the system. If a system can achieve the same state (i.e. the same values of p, V, T, and U) in Ω (omega) different ways (different arrangements of the molecules, for example), then the entropy of the state is

$$S = k \ln \Omega$$

where ln is the logarithm to base e and k is Boltzmann's constant, 1.38×10^{-23} J/K.

ENTROPY IS A MEASURE OF DISORDER: A state that can occur in only one way (one arrangement of its molecules, for example) is a state of high order. But a state that can occur in many ways is a more disordered state. To associate a number with disorder, the disorder of a state is taken proportional to Ω, the number of ways the state can occur. Because $S = k \ln \Omega$, the entropy is a measure of disorder.

Spontaneous processes in systems that contain many molecules always occur in a direction from

$$\begin{pmatrix} \text{state that can exist} \\ \text{in only a few ways} \end{pmatrix} \rightarrow \begin{pmatrix} \text{state that can exist} \\ \text{in many ways} \end{pmatrix}$$

Hence systems, when left to themselves, retain their original state of order or else increase their disorder.

THE MOST PROBABLE STATE of a system is the state with the largest entropy. It is also the state with the most disorder and the state that can occur in the largest number of ways.

Solved Problems

21.1. Twenty grams of ice at 0 °C melts to water at 0 °C. How much does the entropy of the 20 g change in this process?

By slowly adding heat to the ice, we can melt it in a reversible way. The heat needed is

$$\Delta Q = mH_f = (20 \text{ g})(80 \text{ cal/g}) = 1600 \text{ cal}$$

Therefore,

$$\Delta S = \frac{\Delta Q}{T} = \frac{1600 \text{ cal}}{273 \text{ K}} = 5.86 \text{ cal/K} = 24.5 \text{ J/K}$$

Notice that, upon melting, the entropy (and disorder) increased.

21.2. As shown in Fig. 21-1, an ideal gas is confined to a cylinder by a piston. The piston is slowly pushed down so that the gas temperature remains at 20 °C. During the compression, 730 J of work was done on the gas. Find the entropy change of the gas.

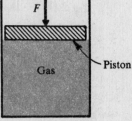

Fig. 21-1

The first law tells us that

$$\Delta Q = \Delta U + \Delta W$$

Because the process was isothermal, the internal energy of the ideal gas did not change. Therefore, $\Delta U = 0$ and

$$\Delta Q = \Delta W = -730 \text{ J}$$

(Because the gas was compressed, the gas did negative work, hence the minus sign.) Now we can write

$$\Delta S = \frac{\Delta Q}{T} = \frac{-730 \text{ J}}{293 \text{ K}} = -2.49 \text{ J/K}$$

Notice that the entropy change is negative. Disorder of the gas decreased as it was pushed into a smaller volume.

21.3. As shown in Fig. 21-2, a container is separated into two, equal-volume compartments. The two compartments contain equal masses of the same gas, 0.74 g in each, and c_v for the gas is 0.178 cal/g·°C. At the start, the hot gas is at 67 °C, while the cold gas is at 20 °C. No heat can leave or enter the compartments except slowly through the partition AB. Find the entropy change of each compartment as the hot gas cools from 67 °C to 65 °C.

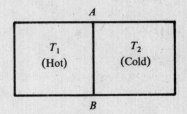

Fig. 21-2

The heat lost by the hot gas in the process is

$$\Delta Q = mc_v \, \Delta T = (0.74 \text{ g})(0.178 \text{ cal/g·°C})(2 \text{ °C}) = 0.263 \text{ cal} = 1.10 \text{ J} \quad \text{lost}$$

This is equal also to the heat gained by the cold gas. For the hot gas, approximately,

$$\Delta S_h = \frac{\Delta Q}{T_h} \approx \frac{-1.10 \text{ J}}{(273 + 66) \text{ K}} = -3.24 \times 10^{-3} \text{ J/K}$$

For the cold gas, approximately,

$$\Delta S_c = \frac{\Delta Q}{T_c} \approx \frac{1.10 \text{ J}}{(273 + 21) \text{ K}} = 3.75 \times 10^{-3} \text{ J/K}$$

As we see, the entropy changes were different for the two compartments. The total entropy of the universe increased as a result of this process.

21.4. The ideal gas in the cylinder of Fig. 21-1 is initially at conditions p_1, V_1, T_1. It is slowly expanded at constant temperature by allowing the piston to rise. Its final conditions are p_2, V_2, T_1, where $V_2 = 3V_1$. Find the change in entropy of the gas during this expansion. The mass of gas is 1.5 g and its molecular weight is 28.

Recall from Chapter 20 that, for an isothermal expansion of an ideal gas,

$$\Delta W = \Delta Q = p_1 V_1 \ln \frac{V_2}{V_1}$$

Consequently,

$$\Delta S = \frac{\Delta Q}{T} = \frac{p_1 V_1}{T_1} \ln\left(\frac{V_2}{V_1}\right) = \left(\frac{m}{M}\right) R \ln\left(\frac{V_2}{V_1}\right)$$

where we have used the ideal gas law. Substituting the data,

$$\Delta S = \left(\frac{1.5 \times 10^{-3} \text{ kg}}{28 \text{ kg/kmol}}\right)\left(8314 \frac{\text{J}}{\text{kmol} \cdot \text{K}}\right)(\ln 3) = 0.49 \text{ J/K}$$

21.5. A system consists of 3 coins that can come up either heads or tails. In how many different ways can the system have (*a*) all heads up? (*b*) all tails up? (*c*) one tail and two heads up? (*d*) two tails and one head up?

(*a*) There is only one way all coins can be heads-up: each coin must be heads-up.

(*b*) Here, too, there is only one way.

(*c*) There are three ways, corresponding to the three choices for the coin showing the tail.

(*d*) By symmetry with (*c*), there are three ways.

21.6. Find the entropy of the three-coin system described in Problem 21.5 if (*a*) all coins are heads-up, (*b*) two coins are heads-up.

We use the Boltzmann relation $S = k \ln \Omega$, where Ω is the number of ways the state can occur and $k = 1.38 \times 10^{-23}$ J/K.

(*a*) Since this state can occur in only one way,

$$S = k \ln 1 = (1.38 \times 10^{-23} \text{ J/K})(0) = 0$$

(*b*) Since this state can occur in three ways,

$$S = (1.38 \times 10^{-23} \text{ J/K}) \ln 3 = (1.38 \times 10^{-23} \text{ J/K})(2.30 \log 3) = 1.52 \times 10^{-23} \text{ J/K}$$

Supplementary Problems

21.7. Compute the entropy change of 5 g of water at 100 °C as it changes to steam at 100 °C under standard pressure. *Ans.* 7.24 cal/K = 30.3 J/K

21.8. By how much does the entropy of 300 g of metal ($c = 0.093$ cal/g · °C) change as it is cooled from 90 °C to 70 °C? You may approximate $T = \frac{1}{2}(T_1 + T_2)$. *Ans.* −1.58 cal/K = −6.6 J/K

21.9. An ideal gas is slowly expanded from 2.00 m³ to 3.00 m³ at a constant temperature of 30 °C. The entropy change of the gas was +47 J/K during the process. (*a*) How much heat was added to the gas during the process? (*b*) How much work did the gas do during the process?
Ans. (*a*) 3.4 kcal; (*b*) 14.2 kJ

21.10. Starting at standard conditions, 3 kg of ideal gas ($M = 28$ kg/kmol) is isothermally compressed to 1/5 its original volume. Find the change in entropy of the gas. *Ans.* −1430 J/K

21.11. Four poker chips are red on one side and white on the other. How many different ways can (*a*) 3 reds come up? (*b*) two reds come up? *Ans.* (*a*) 4; (*b*) 6

21.12. When 100 coins are tossed, there is 1 way that all can come up heads. There are 100 ways that only one tail is up. There are about 1×10^{29} ways that 50 heads can come up. One hundred coins are placed in a box with only one head up. They are shaken and then there are 50 heads up. What was the change in entropy of the coins caused by the shaking? *Ans.* 8.6×10^{-22} J/K

Chapter 22

Wave Motion

WAVES CARRY ENERGY in the direction of propagation of the wave. As the wave passes through a region, periodic vibrations occur there.

In *transverse waves*, the vibration direction is perpendicular to the propagation direction. Typical transverse waves are waves on a string and radio waves.

In *longitudinal* (or *compressional*) *waves*, the vibration direction is parallel to the direction of propagation. Sound waves are of this type.

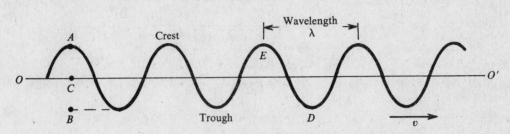

Fig. 22-1

TERMINOLOGY for waves can be understood by referring to the wave on a string shown in Fig. 22-1. The undisturbed string lies along line OO'. As a wave passes down the string, energy is carried to the right along OO', the *line* (or *direction*) *of propagation* of the wave. A tiny particle of the string, such as the one at A, vibrates up and down from A to B and back to A. Since the particle vibrates perpendicular to the direction of propagation, this is a transverse wave.

The *period* of vibration, T, is the time taken for a particle such as the one at A to move from A to B and back to A. The *frequency* of vibration (f or v) is the number of such vibrations executed by the particle each second. One has

$$f = 1/T$$

If T is in seconds, then f is in *hertz* (Hz), where

$$1 \text{ Hz} = 1 \text{ vibration/s} = 1 \text{ cycle/s} = 1 \text{ s}^{-1}$$

The period and frequency of the wave are the same as the period and frequency of the vibration.

At any instant, the string may appear as shown in Fig. 22-1. Points such as A, the top points, are called *crests* of the wave. *Troughs* of the wave are points such as D, the bottom points of the wave. As time goes on, a crest (or trough) of the wave moves in the direction of propagation with speed v, the speed of the wave.

The *wavelength* λ is the distance between two adjacent crests of the wave (or between any two adjacent corresponding points). In a time T a crest, moving with speed v, will move a distance λ. Therefore, $s = \bar{v}t$ gives

$$\lambda = vT$$

The *amplitude* of a wave is the maximum disturbance undergone during a vibration cycle. In Fig. 22-1, the amplitude is distance AC (not AB!).

IN-PHASE VIBRATIONS exist at two vibration points of a wave if the points undergo vibration in the same direction together. For example, the particles of the string at points A and E in Fig. 22-1 vibrate in phase, since they move up and down together. Vibrations are in phase if the points are a whole number of wavelengths apart. The pieces of the string at A and D vibrate opposite to each other. The vibrations there are said to be 180°, or half a cycle, *out of phase*.

SPEED OF A TRANSVERSE WAVE on a stretched string or wire is

$$v = \sqrt{\frac{\text{tension in string}}{\text{mass per unit length of string}}}$$

SPEED OF A LONGITUDINAL WAVE (compressional wave) depends on the elasticity and density of the medium.

in liquids: $v = \sqrt{\dfrac{\text{bulk modulus}}{\text{density of liquid}}} = \sqrt{\dfrac{B}{\rho}}$

in solid rods: $v = \sqrt{\dfrac{\text{Young's modulus}}{\text{density of solid}}} = \sqrt{\dfrac{Y}{\rho}}$

in gases: $v = \sqrt{\dfrac{\gamma \times (\text{pressure of gas})}{\text{density of gas}}} = \sqrt{\dfrac{\gamma p}{\rho}}$

where γ, the ratio of specific heats c_p/c_v, is about 1.67 for monatomic gases such as helium and neon. It is about 1.40 for diatomic gases such as N_2, O_2, and H_2.

STANDING (OR STATIONARY) WAVES: A standing wave may be set up in a region when two waves of equal frequencies and amplitudes move in opposite directions through the region. When a wave is repeatedly reflected back and forth on a string, or in other similar situations, a standing wave of large amplitude is often produced. In such a case, a standing wave results only if the wavelength has certain special values compared to the path length for the wave.

A *standing wave on a string* causes the string to sweep out a pattern such as the one shown in Fig. 22-2. The string vibrates between the limits indicated. Points A, B, C, D are called *displacement nodes*. A displacement node is a point on a standing wave where the amplitude of vibration is zero. Points F, G, H, I, J are *displacement antinodes*. A displacement antinode is a point on a standing wave where the amplitude of vibration is greatest. The distance between two adjacent nodes (or antinodes) is a half wavelength. The region between two nodes is called a *segment*.

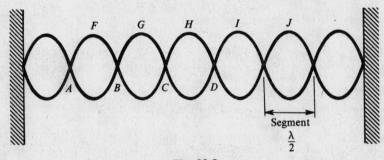

Fig. 22-2

CONDITION FOR RESONANCE: When a standing wave exists on a system, the system is said to be undergoing *resonance*. The wavelength must be such that the wave segments fit properly on the resonating system. A proper fit occurs when nodes and antinodes exist at positions demanded by the constraints on the system.

RESONANCE OF STRINGS: A stretched string of length L is held firmly at its two ends. The ends are therefore at nodes. Consequently, if a standing wave is to exist on the string, it must be 1 segment, or 2 segments, or 3 segments, etc., in length. Since a segment is $\lambda/2$ long, a string will resonate to a wavelength λ only if $L = n(\lambda/2)$,   where n is an integer.

The standing wave patterns for $n = 1, 2,$ and 3 are shown in Fig. 22-3. We call the simplest resonance form the *fundamental vibration* and its frequency is f_0. Since $\lambda = vT = v/f$, the shorter the segments, the higher the resonance frequency. In this case, the resonance frequencies are $f_0, 2f_0, 3f_0$, and so on.

RESONANCE OF AIR IN PIPES: At resonance, a displacement node exists at a closed end, while an antinode is close to an open end. Sound waves (compressional waves) with wavelengths that satisfy these requirements will resonate in a pipe. Since the distance from node to antinode is $\lambda/4$, the fundamental vibration in a pipe of length L will have a wavelength given by:

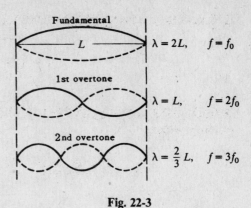

Fig. 22-3

pipe open at both ends:	$L = 2(\lambda/4) = \lambda/2$
pipe open at only one end:	$L = \lambda/4$
pipe closed at both ends:	$L = \lambda/2$

Solved Problems

22.1. Suppose that Fig. 22-1 represents a 50 Hz wave on a string. Take distance AB to be 3 mm and distance AE to be 40 cm. Find the following for the wave: (*a*) amplitude, (*b*) wavelength, (*c*) speed.

(*a*) By definition, the amplitude is distance AC in this case and is 1.5 mm.

(*b*) The distance between adjacent crests is the wavelength and so $\lambda = 20$ cm.

(*c*) From $\lambda = vT$ and $T = 1/f$, we have

$$v = \frac{\lambda}{T} = \lambda f = (0.2 \text{ m})(50 \text{ s}^{-1}) = 10 \text{ m/s}$$

22.2. Measurements show that the wavelength of a sound wave in a certain material is 18 cm. The frequency of the wave is 1900 Hz. What is the speed of the sound wave?

From $\lambda = vT = v/f$,

$$v = \lambda f = (0.18 \text{ m})(1900 \text{ s}^{-1}) = 342 \text{ m/s}$$

22.3. A horizontal cord 5 m long "weighs" 1.45 g. What must be the tension in the cord if the wavelength of a 120 Hz wave on it is to be 60 cm? How large a mass must be hung from its end to give it this tension?

We know that

$$v = \sqrt{(\text{tension})/(\text{mass per unit length})}$$

from which, since $v = \lambda/T = \lambda f$,

$$\text{tension} = (\text{mass/u.l.})v^2 = \left(\frac{1.45 \times 10^{-3} \text{ kg}}{5 \text{ m}} \right)(0.6 \text{ m} \times 120 \text{ s}^{-1})^2 = 1.50 \text{ N}$$

The tension in the cord balances the weight of the mass hung at its end. Therefore

$$\text{tension} = mg \qquad \text{or} \qquad m = \frac{\text{tension}}{g} = \frac{1.50 \text{ N}}{9.8 \text{ m/s}^2} = 0.153 \text{ kg}$$

22.4. A wire 8 ft long weighs 0.020 lb. What is the speed of transverse waves along the wire when a 5 lb weight is hung from its end?

The tension in the wire is 5 lb. Its mass per unit length is

$$(0.020/32.2) \div 8 = 7.8 \times 10^{-5} \text{ slug/ft}$$

Therefore,

$$v = \sqrt{\frac{\text{tension}}{\text{mass/u.l.}}} = \sqrt{\frac{5 \text{ lb}}{7.8 \times 10^{-5} \text{ slug/ft}}} = 254 \text{ ft/s}$$

22.5. What is the speed of compressional waves (sound waves) in water? The bulk modulus for water is 2.2×10^9 N/m².

$$v = \sqrt{\frac{\text{bulk modulus}}{\text{density}}} = \sqrt{\frac{2.2 \times 10^9 \text{ N/m}^2}{1000 \text{ kg/m}^3}} = 1480 \text{ m/s}$$

22.6. The speed of compressional waves in a metal rod is 6000 m/s. What is Young's modulus for the material of the rod if the density of the material is 8.2 g/cm³?

For a rod, $v = \sqrt{Y/\rho}$ and so

$$Y = \rho v^2 = (8200 \text{ kg/m}^3)(6000 \text{ m/s})^2 = 3 \times 10^{11} \text{ N/m}^2$$

22.7. Calculate the speed of sound waves in air at 0 °C and a pressure of 76 cm of mercury. The density of air under these conditions is 1.293 kg/m³, and ρ for mercury is 13.6 g/cm³. γ for air is 1.40.

$$p = \rho g h = (13\,600 \text{ kg/m}^3)(9.8 \text{ m/s}^2)(0.76 \text{ m}) = 1.01 \times 10^5 \text{ N/m}^2$$

$$v = \sqrt{\frac{\gamma p}{\rho}} = \sqrt{\frac{(1.4)(1.01 \times 10^5 \text{ N/m}^2)}{1.293 \text{ kg/m}^3}} = 331 \text{ m/s}$$

22.8. The metal string shown in Fig. 22-3 is under a tension of 88.2 N. Its length is 50 cm and its mass is 0.50 g. (a) Compute v for transverse waves on the string. (b) Determine the frequencies of its fundamental, first overtone, and second overtone.

(a)

$$v = \sqrt{\frac{\text{tension}}{\text{mass/u.l.}}} = \sqrt{\frac{88.2 \text{ N}}{5 \times 10^{-4} \text{ kg/0.50 m}}} = 297 \text{ m/s}$$

(b) Recall that the length of a segment is $\lambda/2$ and use $\lambda = v/f$. For the fundamental:

$$\lambda = 1.00 \text{ m} \qquad f = \frac{297 \text{ m/s}}{1.00 \text{ m}} = 297 \text{ Hz}$$

For first overtone:

$$\lambda = 0.50 \text{ m} \qquad f = \frac{297 \text{ m/s}}{0.50 \text{ m}} = 594 \text{ Hz}$$

For the second overtone:

$$\lambda = 0.33 \text{ m} \qquad f = \frac{297 \text{ m/s}}{0.33 \text{ m}} = 891 \text{ Hz}$$

22.9. A string 2 m long is driven by a 240 Hz vibrator at its end. The string resonates in four segments. What is the speed of transverse waves on the string?

Since each segment is $\lambda/2$ long, we have

$$4(\lambda/2) = L \qquad \text{or} \qquad \lambda = \frac{L}{2} = \frac{2 \text{ m}}{2} = 1 \text{ m}$$

Using $\lambda = vT = v/f$, we have

$$v = f\lambda = (240 \text{ s}^{-1})(1 \text{ m}) = 240 \text{ m/s}$$

22.10. A banjo string 30 cm long resonates in its fundamental to a frequency of 256 Hz. What is the tension in the string if 80 cm of the string "weighs" 0.75 g?

First we shall find v and then find the tension. We know that the string vibrates in one segment when $f = 256$ Hz. Therefore,

$$\frac{\lambda}{2} = L \qquad \text{or} \qquad \lambda = (0.30 \text{ m})(2) = 0.60 \text{ m}$$

and

$$v = f\lambda = (256 \text{ s}^{-1})(0.6 \text{ m}) = 154 \text{ m/s}$$

The mass per unit length of the string is

$$\frac{0.75 \times 10^{-3} \text{ kg}}{0.8 \text{ m}} = 9.4 \times 10^{-4} \text{ kg/m}$$

Then, from $v = \sqrt{(\text{tension})/(\text{mass/u.l.})}$,

$$\text{tension} = v^2(\text{mass/u.l.}) = (154 \text{ m/s})^2(9.4 \times 10^{-4} \text{ kg/m}) = 22 \text{ N}$$

22.11. A rod 200 cm long is clamped 50 cm from one end, as shown in Fig. 22-4. It is set into longitudinal vibration by an electrical driving mechanism at one end. As the frequency of the driver is slowly increased from a very low value, the rod is first found to resonate at 3000 Hz. What is the speed of sound (compressional waves) in the rod?

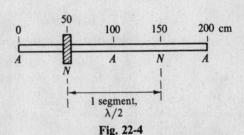

Fig. 22-4

The clamped point remains stationary and so a node exists there. Since the ends of the rod are free, antinodes exist there. The lowest-frequency resonance occurs when the rod is vibrating in its longest possible segments. In Fig. 22-4 we show the mode of vibration that corresponds to this condition. If we recall that a segment is the length from one node to the next, then the length from A to N in the figure is one half segment. Therefore, the rod is two segments long. This resonance form satisfies our restrictions about positions of nodes and antinodes, as well as the condition that the bar vibrate in the longest segments possible. Since one segment is $\lambda/2$ long,

$$L = 2(\lambda/2) \qquad \text{or} \qquad \lambda = L = 200 \text{ cm}$$

Then, using $\lambda = vT = v/f$,

$$v = \lambda f = (2 \text{ m})(3000 \text{ s}^{-1}) = 6000 \text{ m/s}$$

22.12. A string vibrates in 5 segments to a frequency of 460 Hz. What frequency will cause it to vibrate in 2 segments?

The frequency is proportional to the number of segments (see Fig. 22-3) and so

$$f = \frac{2}{5}(460 \text{ Hz}) = 184 \text{ Hz}$$

22.13. A metal rod 40 cm long is dropped, end first, onto a wooden floor and rebounds into the air. Compressional waves of many frequencies are thereby set up in the bar. If the speed of compressional waves in the bar is 5500 m/s, to what lowest-frequency compressional wave will the bar resonate as it rebounds?

Both ends of the bar will be free and so antinodes will exist there. In its lowest resonance form (i.e. the one with longest segments), only one node will exist on the bar, at its center, as shown in Fig. 22-5. We then have

$$L = 2(\lambda/4) \qquad \text{or} \qquad \lambda = 2L = 2(0.40 \text{ m}) = 0.80 \text{ m}$$

Then, from $\lambda = vT = v/f$,

$$f = \frac{v}{\lambda} = \frac{5500 \text{ m/s}}{0.80 \text{ m}} = 6875 \text{ Hz}$$

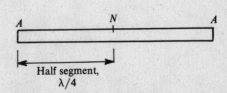

Fig. 22-5

22.14. Compressional waves (sound waves) are sent down an air-filled tube 90 cm long and closed at one end. The tube resonates to several frequencies, the lowest of which is 95 Hz. Find the speed of sound waves in air.

The tube and several of its various resonance forms are shown in Fig. 22-6. Recall that the distance between a node and an adjacent antinode is $\lambda/4$. In our case, the top resonance form applies, since the segments are longest for it and its frequency is therefore lowest. For it,

$$L = \frac{\lambda}{4}$$

or

$$\lambda = 4L = 4(0.9 \text{ m}) = 3.6 \text{ m}$$

Using $\lambda = vT = v/f$ gives

$$v = \lambda f = (3.6 \text{ m})(95 \text{ s}^{-1}) = 342 \text{ m/s}$$

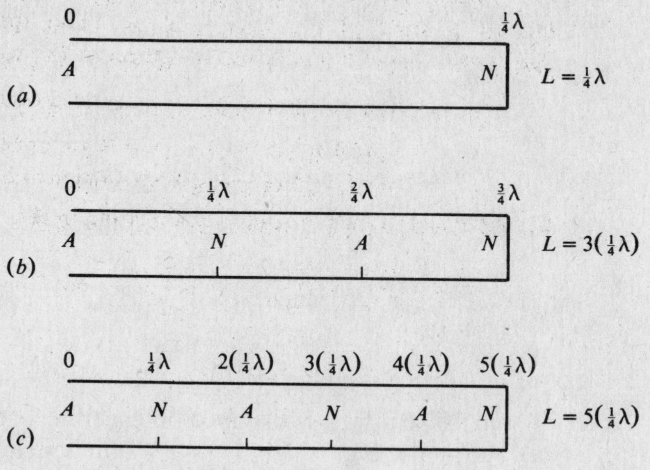

Fig. 22-6

22.15. To what other frequencies will the tube described in Problem 22.14 resonate?

The first few resonances are shown in Fig. 22-6. We see that, at resonance,

$$L = n\left(\frac{1}{4}\lambda_n\right)$$

where $n = 1, 3, 5, 7, \ldots$, an odd integer, and λ_n are the resonance wavelengths. But $\lambda_n = v/f_n$ and so

$$L = n\frac{v}{4f_n} \qquad \text{or} \qquad f_n = n\frac{v}{4L} = nf_1$$

where, from Problem 22.14, $f_1 = 95$ Hz. The first few resonance frequencies are thus 95, 285, 475, ..., Hz.

22.16. (a) Determine the shortest length of pipe closed at one end that will resonate in air to a sound source of frequency 160 Hz. Take the speed of sound in air to be 340 m/s. (b) Repeat for a pipe open at both ends.

(a) Figure 22-6(a) applies in this case. The shortest pipe will be $\lambda/4$ long. Therefore

$$L = \frac{1}{4}\lambda = \frac{1}{4}\left(\frac{v}{f}\right) = \frac{340 \text{ m/s}}{4(160 \text{ s}^{-1})} = 0.53 \text{ m}$$

(b) In this case the pipe will have antinodes at both ends and a node at its center.

$$L = 2\left(\frac{1}{4}\lambda\right) = \frac{1}{2}\left(\frac{v}{f}\right) = \frac{340 \text{ m/s}}{2(160 \text{ s}^{-1})} = 1.06 \text{ m}$$

22.17. A pipe 90 cm long is open at both ends. How long must a second pipe, closed at one end, be if it is to have the same fundamental resonance frequency as the open pipe?

The two pipes in their fundamental resonances are shown in Fig. 22-7. As we see,

$$L_o = 2\left(\frac{1}{4}\lambda\right) \qquad L_c = \frac{1}{4}\lambda$$

from which $L_c = \frac{1}{2}L_o = 45$ cm.

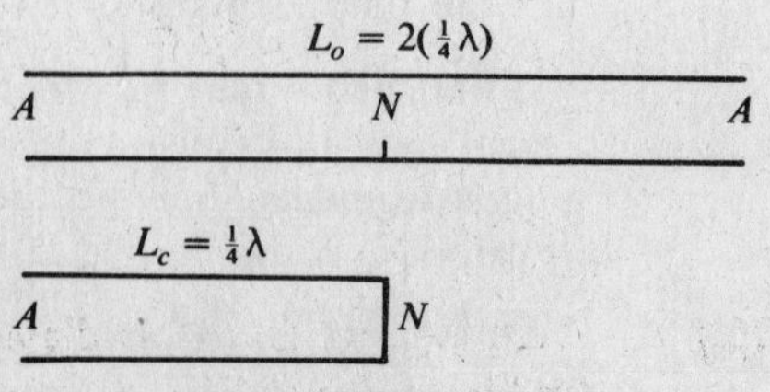

Fig. 22-7

22.18. A *Kundt's tube* is employed to measure the speed of sound in a steel rod. The rod, clamped at its center, is stroked to vibrate longitudinally and emits its fundamental tone. The length of the rod is 90 cm and the distance between the nodes (powder heaps) of the standing wave in the air column of the tube is 6 cm. The speed of sound in air at room temperature is 340 m/s. What is the speed of sound in the steel rod?

The rod vibrates with a node at its clamped center and antinodes at both free ends.

wavelength of sound in rod = 2 × (length of rod) = 180 cm

wavelength of sound in air column = 2 × (distance between nodes) = 12 cm

But the frequency is the same in the rod and in the tube. Thus

$$f = \frac{v_{rod}}{\lambda_{rod}} = \frac{v_{air}}{\lambda_{air}} \qquad \text{and} \qquad v_{rod} = v_{air}\frac{\lambda_{rod}}{\lambda_{air}} = (340 \text{ m/s})\frac{180 \text{ cm}}{12 \text{ cm}} = 5100 \text{ m/s}$$

Supplementary Problems

22.19. The average person can hear sound waves ranging in frequency from about 20 to 20 000 Hz. Determine the wavelengths at these limits, taking the speed of sound to be 340 m/s.   *Ans.* 17 m, 1.7 cm

22.20. Radio station WJR in Detroit broadcasts at 760 kHz. The speed of radio waves is 3×10^8 m/s. What is the wavelength of WJR's waves?   *Ans.* 395 m

22.21. Radar waves with 3.4 cm wavelength are sent out from a transmitter. Their speed is 3×10^8 m/s. What is their frequency?   *Ans.* 8.8×10^9 Hz = 8.8 GHz

22.22. When driven by a 120 Hz vibrator, a string has transverse waves of 31 cm wavelength traveling along it. (*a*) What is the speed of the waves on the string?   (*b*) If the tension in the string is 1.20 N, what is the mass of 50 cm of the string?   *Ans.* (*a*) 37 m/s;   (*b*) 0.43 g

22.23. The wave shown in Fig. 22-8 is being sent out by a 60 cycle/s vibrator. Find the following for the wave: (*a*) amplitude, (*b*) frequency, (*c*) wavelength, (*d*) speed, (*e*) period.
Ans. (*a*) 3 mm;   (*b*) 60 Hz;   (*c*) 2 cm;   (*d*) 1.20 m/s;   (*e*) 0.0167 s

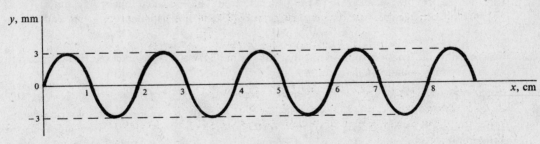

Fig. 22-8

22.24. A copper wire 2.4 mm in diameter is 3 m long and is used to suspend a 2 kg mass from a beam. If a transverse disturbance is sent along the wire by striking it lightly with a pencil, how fast will the disturbance travel? The density of copper is 8920 kg/m³.   *Ans.* 22 m/s

22.25. An increase in pressure of 100 kPa causes a certain volume of water to decrease by 5×10^{-3} percent of its original volume. (*a*) What is the bulk modulus of water? (*b*) What is the speed of sound (compressional waves) in water?   *Ans.* (*a*) 2×10^9 N/m²; (*b*) 1400 m/s

22.26. Helium is a monatomic gas that has a density of 0.179 kg/m³ at a pressure of 76 cm of mercury and a temperature of 0 °C. Find the speed of compressional waves (sound) in helium at this temperature and pressure.   *Ans.* 970 m/s

22.27. Find the speed of compressional waves in a metal rod if the material of the rod has a Young's modulus of 1.2×10^{10} N/m² and a density of 8920 kg/m³.   *Ans.* 1160 m/s

22.28. A string 180 cm long resonates in three segments to transverse waves sent down it by a 270 Hz vibrator. What is the speed of the waves on the string?   *Ans.* 324 m/s

22.29. A string resonates in 3 segments to a frequency of 165 Hz. What frequency must be used if it is to resonate in 4 segments?   *Ans.* 220 Hz

22.30. A flexible cable, 100 ft long and weighing 16 lb, is stretched between two poles with a force of 450 lb. If the cable is struck sideways at one end, how long will it take the transverse wave to travel to the other end and return?   *Ans.* $\frac{2}{3}$ s

22.31. A wire under tension vibrates with a fundamental frequency of 256 Hz. What would be the fundamental frequency if the wire were half as long, twice as thick, and under one-fourth the tension?
Ans. 128 Hz

22.32. Steel and silver wires of the same diameter and same length are stretched with equal tensions. The densities are 7.8 and 10.6 g/cm³ respectively. What is the fundamental frequency of the silver wire, that of the steel being 200 Hz?   *Ans.* 172 Hz

22.33. A string has mass 3 grams and length 60 cm. What must be the tension so that when vibrating transversely its first overtone has frequency 200 Hz?   *Ans.* 72 N

22.34. (*a*) At what point should a stretched string be plucked to make its fundamental tone most prominent? At what point should it be plucked and then at what point touched (*b*) to make its first overtone most prominent?   (*c*) to make its second overtone most prominent?
Ans. (*a*) center; (*b*) plucked at 1/4 of its length from one end, then touched at center; (*c*) plucked at 1/6 of its length from one end, then touched at 1/3 of its length from that end

22.35. What must be the length of an iron rod that has the fundamental frequency 320 Hz when clamped at its center? Assume longitudinal vibration at speed 5000 m/s.   *Ans.* 7.81 m

22.36. A bar, 1 cm² × 200 cm and mass 2 kg, is clamped at its center. When vibrating longitudinally it emits its fundamental tone in unison with a tuning fork making 1000 vibrations/s. How much will the bar be elongated if, when clamped at one end, a stretching force of 980 N is applied at the other end?
Ans. 0.1225 m

22.37. A rod 120 cm long is clamped at the center and is stroked in such a way as to give its first overtone. Make a drawing showing the location of the nodes and antinodes, and determine at what other points the rod might be clamped and still emit the same tone.   *Ans.* 20 cm from either end

22.38.   A metal bar 6 m long, clamped at its center and vibrating longitudinally in such a manner that it gives its first overtone, vibrates in unison with a tuning fork marked 1200 vibrations/s.   Compute the speed of sound in the metal.   *Ans.*   4800 m/s

22.39.   In a Kundt's tube experiment, the metal rod, 120 cm long, is clamped at its center and emits its fundamental tone.   The distance between the first and seventh nodes of the standing wave in the air column of the tube is 60 cm.   The speed of sound in air at the room temperature is 345 m/s.   Compute the speed of sound in the rod.   *Ans.*   4140 m/s

22.40.   A Kundt's tube is filled with air and the distance between the powder heaps is 10.0 cm.   It is then filled with hydrogen and the powder heaps are 38.5 cm apart.   (*a*) Determine the speed of sound in hydrogen if the speed of sound in air is 340 m/s.   (*b*) How many vibrations per second are produced in the rod?   *Ans.*   (*a*) 1310 m/s;   (*b*) 1700 Hz

22.41.   Determine the length of the shortest air column in a cylindrical jar that will strongly reinforce the sound of a tuning fork having a vibration rate of 512 Hz.   Use $v = 340$ m/s   for the speed of sound in air.   *Ans.*   16.6 cm

22.42.   A long, narrow pipe closed at one end does not resonate to a tuning fork having a frequency of 300 Hz until the length of the air column reaches 28 cm.   (*a*) What is the speed of sound in air at the room temperature?   (*b*) What is the next length of column that will resonate to the fork?   *Ans.*   (*a*) 336 m/s;   (*b*) 84 cm

22.43.   An organ pipe closed at one end is 61 cm long.   What are the frequencies of the first three overtones if v for sound is 342 m/s?   *Ans.*   420, 700, 980 Hz

Chapter 23

Sound

SOUND WAVES are compressional waves in a material medium such as air, water, or steel. When the compressions and rarefactions of the waves strike the eardrum, they result in the sensation of sound, provided the frequency of the waves is between about 20 Hz and 20 000 Hz. Waves with frequencies above 20 kHz are called *ultrasonic* waves. Those with frequencies below 20 Hz are called *infrasonic* waves.

SPEED OF SOUND IN AIR at 0 °C is 331 m/s or 1087 ft/s. The speed increases with temperature, about 0.6 m/s or 2 ft/s for each degree Celsius rise. More precisely, the speeds v_1 and v_2 at absolute temperatures T_1 and T_2 are related by

$$\frac{v_1}{v_2} = \sqrt{\frac{T_1}{T_2}}$$

The speed of sound is essentially independent of pressure, frequency, and wavelength.

INTENSITY (I) of a sound wave is the power carried by the wave through unit area erected perpendicular to the direction of propagation of the wave. Suppose that in a time Δt an amount of energy ΔE is carried through an area ΔA erected perpendicular to the propagation direction of the wave. Then

$$I = \frac{\Delta E}{\Delta A\, \Delta t}$$

It may be shown that for a sound wave with amplitude a_0 and frequency f, traveling with speed v in a material of density ρ,

$$I = 2\pi^2 f^2 \rho v a_0^2$$

If f is in Hz, ρ in kg/m^3, v in m/s, and a_0 in m, then I is in W/m^2.

LOUDNESS is a term used to measure the human perception of sound. Although a sound wave of high intensity is perceived as louder than a wave of lower intensity, the relation is far from linear. The sensation of sound is roughly proportional to the logarithm of the sound intensity. But the exact relation between loudness and intensity is complicated and not the same for all individuals.

INTENSITY (OR LOUDNESS) LEVEL is defined by an arbitrary scale that corresponds roughly to the sensation of loudness. The zero on this scale is taken at the sound wave intensity $I_0 = 1 \times 10^{-12}$ W/m^2, which corresponds roughly to the weakest audible sound. The scale is then defined by

$$\text{intensity level of } I \text{ in decibels} = 10 \log\left(\frac{I}{I_0}\right)$$

The *decibel* (dB) is a dimensionless unit. The normal ear can distinguish between intensities that differ by an amount down to about 1 dB.

164

BEATS: The alternations of maximum and minimum sound intensity produced by the superposition of two sound waves of slightly different frequencies are called *beats*. The number of beats per second is equal to the difference between the frequencies of the two sound waves that are combined.

DOPPLER EFFECT: Suppose that a moving sound source emits a sound of frequency f_0. Let V be the speed of sound and let the source approach the listener or observer at speed v_s, measured relative to the medium conducting the sound. Suppose further that the observer is moving toward the source at speed v_o, also measured relative to the medium. Then the observer will hear a sound of frequency f given by

$$\text{observed frequency } f = f_0 \frac{V + v_o}{V - v_s}$$

In case the source or observer is moving away from the other, the sign on its v in the equation must be changed.

When the source and observer are approaching each other, more wave crests strike the ear each second than when both are at rest. This causes the ear to perceive a higher frequency than that emitted by the source. When the two are receding, the opposite effect occurs; the frequency appears to be lowered.

INTERFERENCE EFFECTS: Two sound waves of the same frequency and amplitude may give rise to easily observed interference effects at a point through which they both pass. If the crests of one wave fall on the crests of the other, the two waves are said to be *in phase*. In that case, they reinforce each other and give rise to high intensity at the point.

However, if the crests of one wave fall on the troughs of the other, the two waves will exactly cancel each other. No sound will then be heard at the point. We say that the two waves are then 180° (or a half wavelength) *out of phase*.

Intermediate effects are observed if the two waves are neither in phase nor 180° out of phase, but have a fixed phase relationship somewhere in between.

Solved Problems

23.1. An explosion occurs at a distance of 6.0 km from a person. How long after the explosion does the person hear it? Assume the temperature is 14 °C.

Because the speed of sound increases by about 0.6 m/s for each °C, we have

$$v = 331 \text{ m/s} + (0.6)(14) \text{ m/s} = 339 \text{ m/s}$$

Using $s = vt$, the time taken is

$$t = \frac{s}{v} = \frac{6000 \text{ m}}{339 \text{ m/s}} = 17.7 \text{ s}$$

23.2. To find the distance away of a lightning flash, a rough rule is the following: "Divide the time in seconds between the flash and the sound by three. The result equals the distance away in km of the flash." Justify this.

The speed of sound is $v \approx 333 \text{ m/s} \approx \frac{1}{3} \text{ km/s}$, so the distance of the flash is

$$s = vt \approx \frac{t}{3}$$

where t, the travel time of the sound, is in seconds and s is in kilometers. The light from the flash travels so fast, 3×10^8 m/s, that it reaches the observer almost instantaneously. Hence t is essentially equal to the time between *seeing* the flash and hearing the thunder. Thus the rule.

23.3. Compute the speed of sound in helium gas ($M = 4$ kg/kmol), at 800 °C and a pressure of 2.3 atm. For helium, $\gamma = 1.66$.

Recall from Chapter 22 that the speed of compressional waves in gases is given by

$$v = \sqrt{\frac{\gamma p}{\rho}}$$

For an ideal gas, $pV = (m/M)RT$, from which

$$\rho = \frac{m}{V} = \frac{pM}{RT}$$

Substitution gives

$$v = \sqrt{\frac{\gamma RT}{M}} = \sqrt{\frac{1.66(8314 \text{ J/kmol} \cdot \text{K})(1073 \text{ K})}{4 \text{ kg/kmol}}} = 1920 \text{ m/s}$$

23.4. A tuning fork makes 284 vib/s in air. Compute the wavelength of the tone emitted at 25 °C.

$$v \text{ at } 25 \text{ °C} = 331 \text{ m/s} + (0.6)(25) \text{ m/s} = 346 \text{ m/s}$$

Using $\lambda = vT = v/f$,

$$\lambda = \frac{v}{f} = \frac{346 \text{ m/s}}{284 \text{ s}^{-1}} = 1.22 \text{ m}$$

23.5. An uncomfortably loud sound might have an intensity of 0.54 W/m^2. Find the amplitude of such a sound wave if its frequency is 800 Hz. Take the density of air to be 1.29 kg/m^3 and the speed of sound to be 340 m/s.

From $I = 2\pi^2 f^2 \rho v a_0^2$,

$$a_0 = \frac{1}{\pi f} \sqrt{\frac{I}{2\rho v}} = \frac{1}{800\pi} \sqrt{\frac{0.54}{(2)(1.29)(340)}} \text{ m} = 9.9 \times 10^{-6} \text{ m}$$

23.6. A sound has an intensity of 3×10^{-8} W/m^2. What is the sound level in dB?

$$\text{intensity level in dB} = 10 \log\left(\frac{I}{1 \times 10^{-12} \text{ W/m}^2}\right)$$

$$= 10 \log\left(\frac{3 \times 10^{-8}}{1 \times 10^{-12}}\right) = 10 \log(3 \times 10^4) = 10(4 + \log 3)$$

$$= 10(4 + 0.477) = 44.8 \text{ dB}$$

23.7. A noise-level meter reads the sound level in a room to be 85 dB. What is the sound intensity in the room?

$$\text{level in dB} = 10 \log\left(\frac{I}{1 \times 10^{-12} \text{ W/m}^2}\right)$$

$$\log\left(\frac{I}{1 \times 10^{-12} \text{ W/m}^2}\right) = \frac{\text{dB}}{10} = \frac{85}{10} = 8.5$$

$$\frac{I}{1 \times 10^{-12} \text{ W/m}^2} = \text{antilog } 8.5 = 3.16 \times 10^8$$

$$I = (1 \times 10^{-12} \text{ W/m}^2)(3.16 \times 10^8) = 3.16 \times 10^{-4} \text{ W/m}^2$$

23.8. Two sound waves have intensities of 10 and 500 microwatts/cm². How many decibels is the louder sound above the other?

Call the 10 μW/cm² sound A, and the other B. Then

$$dB_A = 10 \log \left(\frac{I_A}{I_0} \right) = 10(\log I_A - \log I_0)$$

$$dB_B = 10 \log \left(\frac{I_B}{I_0} \right) = 10(\log I_B - \log I_0)$$

Subtracting,

$$dB_B - dB_A = 10 \left(\log I_B - \log I_A \right) = 10 \log \left(\frac{I_B}{I_A} \right)$$

$$= 10 \log \left(\frac{500}{10} \right) = 10 \log 50 = (10)(1.70)$$

$$= 17 \text{ decibels difference}$$

23.9. Find the ratio of the intensities of two sounds if one is 8 dB louder than the other.

We saw in Problem 23.8 that

$$dB_B - dB_A = 10 \log \left(\frac{I_B}{I_A} \right)$$

In the present case this becomes

$$8 = 10 \log \left(\frac{I_B}{I_A} \right) \qquad \text{or} \qquad \frac{I_B}{I_A} = \text{antilog } 0.8 = 6.3$$

23.10. An automobile moving at 30 m/s is approaching a factory whistle that has a frequency of 500 Hz. (a) If the speed of sound in air is 340 m/s, what is the apparent frequency of the whistle as heard by the driver? (b) Repeat for the case of the car leaving the factory at the same speed.

(a)

$$f = f_0 \frac{V + v_o}{V - v_s} = (500 \text{ Hz}) \frac{340 + 30}{340 - 0} = 544 \text{ Hz}$$

(b)

$$f = f_0 \frac{V + v_o}{V - v_s} = (500 \text{ Hz}) \frac{340 + (-30)}{340 - 0} = 456 \text{ Hz}$$

23.11. A car moving at 20 m/s with its horn blowing ($f = 1200$ Hz) is chasing another car going at 15 m/s. What is the apparent frequency of the horn as heard by the driver being chased? Take the speed of sound to be 340 m/s.

$$f = f_0 \frac{V + v_o}{V - v_s} = (1200 \text{ Hz}) \frac{340 + (-15)}{340 - 20} = 1219 \text{ Hz}$$

23.12. Two tuning forks when sounded simultaneously give one beat each 0.3 s. (a) By how much do their frequencies differ? (b) A tiny piece of chewing gum is placed on a prong of one fork. Now there is one beat each 0.4 s. Was this tuning fork the lower- or the higher-frequency fork?

The number of beats per second equals the frequency difference.

(a)

$$\text{frequency difference} = \frac{1}{0.3 \text{ s}} = 3.3 \text{ Hz}$$

(b)

$$\text{frequency difference} = \frac{1}{0.4 \text{ s}} = 2.5 \text{ Hz}$$

Adding gum to the prong increases its mass and thereby decreases its vibration frequency. This lowering of frequency caused it to come closer to the frequency of the other fork. Hence the fork in question had the higher frequency.

23.13. A tuning fork of frequency 400 Hz is moved away from an observer and toward a flat wall with a speed of 6 ft/s. What is the apparent frequency (*a*) of the unreflected sound waves coming directly to the observer, and (*b*) of the sound waves coming to the observer after reflection? (*c*) How many beats per second are heard? Assume the speed of sound in air to be 1100 ft/s.

(*a*) The fork is receding from the observer.

$$f = f_0 \frac{V + v_o}{V - v_s} = (400 \text{ Hz}) \frac{1100 + 0}{1100 - (-6)} = 398 \text{ Hz}$$

(*b*) The wave crests reaching the wall are closer together than normally because the fork is moving toward the wall. Therefore, the reflected wave appears to come from an approaching source.

$$f = f_0 \frac{V + v_o}{V - v_s} = (400 \text{ Hz}) \frac{1100 + 0}{1100 - 6} = 402 \text{ Hz}$$

(*c*)   beats/s = difference between frequencies = (402 − 398) Hz = 4 per s

23.14. In Fig. 23-1, S_1 and S_2 are identical sound sources. They send out their wave crests simultaneously (the sources are in phase). For what values of $L_1 - L_2$ will a loud sound be heard at point P?

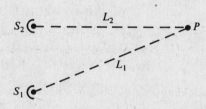

Fig. 23-1

If $L_1 = L_2$, then the waves from the two sources will take equal times to reach P. Crests from one will arrive there at the same times as crests from the other. The waves will therefore be in phase at P and loudness will result.

If $L_1 = L_2 + \lambda$, then the wave from S_1 will be one wavelength behind the one from S_2 when they reach P. But because the wave repeats each wavelength, a crest from S_1 will still reach P at the same time a crest from S_2 does. Once again the waves are in phase at P and loudness will exist there.

In general, loudness will be heard at P when $L_1 - L_2 = \pm n\lambda$, where n is an integer.

Supplementary Problems

23.15. Three seconds after a gun is fired, the person who fired the gun hears an echo. How far away was the surface that reflected the sound of the shot? Use 340 m/s for the speed of sound. *Ans.* 510 m

23.16. What is the speed of sound in air when the air temperature is 31 °C? *Ans.* 350 m/s

23.17. A shell fired at a target half a mile distant (2640 ft) was heard to strike it 5.0 s after leaving the gun. Compute the average horizontal velocity of the shell. The air temperature is 20 °C. *Ans.* 990 ft/s

23.18. In an experiment to determine the speed of sound, two observers, A and B, were stationed 5 km apart. Each was equipped with gun and stopwatch. Observer A heard the report of B's gun 15.5 s after seeing its flash. Later, A fired his gun and B heard the report 14.5 s after seeing the flash. Determine the speed of sound and the component of the speed of the wind along the line joining A to B. *Ans.* 334 m/s, 11.1 m/s

23.19. Determine the frequency and wavelength of the tone produced by a siren disk when a jet of air is blown against it. The disk has 40 holes around its circumference and is rotating at 1200 rpm. The temperature is 15 °C. *Ans.* 800 Hz, 0.425 m

23.20. Determine the speed of sound in carbon dioxide ($M = 44$ kg/kmol, $\gamma = 1.30$) at a pressure of 0.5 atm and a temperature of 400 °C. *Ans.* 407 m/s

23.21. Compute the molecular weight of a gas for which $\gamma = 1.40$ and in which the speed of sound is 1260 m/s at 0 °C. *Ans.* 2.00 kg/kmol (hydrogen)

23.22. At S.T.P., the speed of sound in air is 331 m/s. Determine the speed of sound in hydrogen at S.T.P. if the specific gravity of hydrogen relative to air is 0.069 and if $\gamma = 1.40$ for both gases.
Ans. 1260 m/s

23.23. A sound has an intensity of 5×10^{-7} W/m². What is its decibel sound level? *Ans.* 57 dB

23.24. A person riding a power mower may be subjected to a sound of intensity 2×10^{-2} W/m². To what dB level is the person subjected? *Ans.* 103 dB

23.25. A loud band might easily produce a sound level of 107 dB in a room. What is the sound intensity at 107 dB? *Ans.* 0.050 W/m²

23.26. A whisper has an intensity level of about 15 dB. What sound intensity corresponds to this dB level?
Ans. 3.2×10^{-11} W/m²

23.27. What sound intensity is 3 dB louder than a sound of intensity 10 μW/cm²? *Ans.* 20 μW/cm²

23.28. Calculate the intensity of a sound wave in air at 0 °C and 1 atm if its amplitude is 0.002 mm and its wavelength is 66.2 cm. The density of air at S.T.P. is 1.293 kg/m³. *Ans.* 8.4×10^{-3} W/m²

23.29. What is the amplitude of vibration in a 8000 Hz sound beam if its dB level is 62 dB? Assume that the air is at 15 °C and its density is 1.29 kg/m³. *Ans.* 1.7×10^{-9} m

23.30. A certain organ pipe is tuned to emit a frequency of 196.00 Hz. When it and the G string of a violin are sounded together, ten beats are heard in a time of 8 s. The beats become slower as the violin string is slowly tightened. What was the original frequency of the violin string? *Ans.* 194.75 Hz

23.31. A locomotive moving at 30 m/s approaches and passes a person standing beside the track. Its whistle is emitting a note of frequency 2000 Hz. What frequency will the person hear (*a*) as the train approaches and (*b*) as it recedes? The speed of sound is 340 m/s.
Ans. (*a*) 2190 Hz; (*b*) 1840 Hz

23.32. Two cars are heading straight at each other with the same speed v. The horn of one ($f = 3000$ Hz) is blowing, and is heard to have a frequency of 3400 Hz by the people in the other car. Find v if the speed of sound is 340 m/s. *Ans.* 21 m/s

23.33. To determine the speed of a harmonic oscillator, a sound beam is sent along the line of the oscillator's motion. The sound is then reflected straight back by the oscillator to a detector system. The sound has an emitted frequency of 8000.0 Hz. The detector observes that the reflected beam oscillates in frequency between the limits 8003.1 Hz and 7996.9 Hz. What is the maximum speed of the oscillator? Take the speed of sound to be 340 m/s. *Ans.* 0.132 m/s

23.34. In Fig. 23-1 are shown two identical sound sources sending waves to point P. They send out wave crests simultaneously (they are in phase) and the wavelength of the wave is 60 cm. If $L_2 = 200$ cm, give the values of L_1 for which (a) maximum sound is heard at P and (b) minimum sound is heard at P. *Ans.* (a) $(200 \pm 60n)$ cm, where $n = 0, 1, 2, \ldots$; (b) $(230 \pm 60n)$ cm, where $n = 0, 1, 2, \ldots$

23.35. The two sound sources shown in Fig. 23-2 send out identical sound beams ($\lambda = 80$ cm) toward each other. Each sends out a crest at the same time as the other (the sources are in phase). Point P is a position of loud sound. As one moves from P toward Q, the sound decreases in intensity. (a) How far from P will a sound minimum first be heard? (b) How far from P will a loud sound be heard once again? *Ans.* (a) 20 cm; (b) 40 cm

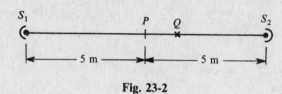

Fig. 23-2

Chapter 24

Coulomb's Law and Electric Fields

COULOMB'S LAW: Suppose that two point charges, q and q', are a distance r apart in vacuum. If q and q' have the same sign, the two charges repel each other; if they have opposite signs, then they attract each other. The force experienced by one charge due to the other is given by *Coulomb's law*,

$$F = k\,\frac{qq'}{r^2}$$

where k is a positive constant. Note that a positive F tends to increase r.

We use only one set of units in electricity, the SI (or mks) system. In this system, distances are in meters, forces are in newtons, and charges are in a unit called the *coulomb* (C). To avoid very small numbers in practical work, the *microcoulomb* ($1\ \mu C = 10^{-6}$ C) and the *nanocoulomb* ($1\ nC = 10^{-9}$ C) are frequently employed.

The fundamental, smallest charge found in nature is denoted e. Its value is $1.602\,19 \times 10^{-19}$ C. All other charges are integer multiples of e. The electron has a charge $-e$, while the proton charge is $+e$.

In SI units, the Coulomb constant k has the following value for charges in vacuum:

$$k = 8.988 \times 10^9\ \text{N} \cdot \text{m}^2/\text{C}^2 \approx 9 \times 10^9\ \text{N} \cdot \text{m}^2/\text{C}^2$$

Often k is replaced by $1/4\pi\epsilon_0$, where $\epsilon_0 = 8.85 \times 10^{-12}$ C^2/N$\cdot$m^2 is called the *permittivity of free space*. In terms of it, Coulomb's law becomes, for vacuum,

$$F = \frac{1}{4\pi\epsilon_0}\ \frac{qq'}{r^2}$$

When the surrounding medium is not a vacuum, forces caused by induced charges in the material reduce the force between point charges. If the material has a *dielectric constant* K, then ϵ_0 in Coulomb's law must be replaced by $K\epsilon_0 = \epsilon$, where ϵ is called the *permittivity* of the material. Then

$$F = \frac{1}{4\pi\epsilon}\ \frac{qq'}{r^2}$$

with $\epsilon = K\epsilon_0$. For vacuum, $K = 1$; for air, $K = 1.0006$.

Coulomb's law also applies to uniform spherical shells or spheres of charge. In that case, r, the distance between the centers of the spheres, must be larger than the sum of the sphere radii; that is to say, the charges must be separate.

THE TEST CHARGE CONCEPT: It is often convenient to make use of a fictitious charge called a *test charge*. This charge is similar to a real charge except in one respect: the test charge is defined to exert no force on other charges. It therefore does not disturb the charges in the vicinity. In practical situations, a test charge can be approximated by a charge of nearly negligible magnitude.

AN ELECTRIC FIELD exists at any point in space where a test charge, if placed at that point, would experience an electrical force. The direction of the electric field at a point is the same as the direction of the force experienced by a positive test charge placed at the point.

Electric field lines can be used to sketch electric fields. The line through a point has the same direction at that point as the electric field. Where the field lines are closest together, the electric field force is largest. Field lines come out of positive charges (because a positive charge repels a positive test charge) and come into negative charges (because they attract the positive test charge).

THE ELECTRIC INTENSITY (E) (also called the *electric field strength*) at a point is equal to the force experienced by a unit positive test charge placed at that point. Because the electric intensity is a force (per unit charge), it is a vector. The units of **E** are N/C or (see Chapter 25) V/m.

If a charge q is placed at a point where the electric field due to other charges is **E**, the charge will experience a force **F** given by

$$\mathbf{F} = q\mathbf{E}$$

If q is negative, then **F** will be opposite in direction to **E**.

ELECTRIC INTENSITY DUE TO A POINT CHARGE: To find the electric intensity E (the signed magnitude of **E**) due to a point charge q, we make use of Coulomb's law. If a point charge q' is placed at a distance r from the charge q, it will experience a force

$$F = \frac{1}{4\pi\epsilon} \frac{qq'}{r^2} = q'\left(\frac{1}{4\pi\epsilon} \frac{q}{r^2} \right)$$

But if a point charge q' is placed at a position where the electric field is E, then the force on q' is

$$F = q'E$$

Comparing these two expressions for F, we see that

$$E = \frac{1}{4\pi\epsilon} \frac{q}{r^2}$$

This is the electric field intensity at a distance r from a point charge q. The same relation applies at points outside a finite spherical charge q. For q positive, E is positive and **E** is directed radially outwards from q; for q negative, E is negative and **E** is radially inwards.

SUPERPOSITION PRINCIPLE: The force experienced by a charge due to other charges is the vector sum of the Coulomb forces acting on it due to these other charges. Similarly, the electric intensity **E** at a point due to several charges is the vector sum of the intensities due to the individual charges.

Solved Problems

24.1. Two coins lie 1.5 m apart on a table. They carry identical charges. How large is the charge on each if a coin experiences a force of 2.0 N?

The diameter of a coin is small compared to the 1.5 m separation. We may therefore approximate the coins as point charges. Coulomb's law, $F = kq_1q_2/r^2$, gives

$$q_1q_2 = q^2 = \frac{Fr^2}{k} = \frac{(2\ \text{N})(1.5\ \text{m})^2}{9 \times 10^9\ \text{N} \cdot \text{m}^2/\text{C}^2} = 5 \times 10^{-10}\ \text{C}^2$$

from which $q = 2.2 \times 10^{-5}\ \text{C} = 22\ \mu\text{C}$.

24.2. Repeat Problem 24.1 if the coins are separated by a distance of 1.5 m in a large vat of water. The dielectric constant of water is about 80.

From Coulomb's law,

$$F = \frac{1}{K} \frac{1}{4\pi\epsilon_0} \frac{q^2}{r^2} = \frac{k}{K} \frac{q^2}{r^2}$$

where K, the dielectric constant, is 80 in this case. (In Problem 24.1, $K = 1.0006 \approx 1.00$ for air.) Then,

$$q = \sqrt{\frac{Fr^2K}{k}} = \sqrt{\frac{(2)(1.5)^2(80)}{9 \times 10^9}} = 2 \times 10^{-4} \ C = 200 \ \mu C$$

24.3. A helium nucleus has charge $+2e$ and a neon nucleus $+10e$, where e is the charge quantum, 1.60×10^{-19} C. Find the repulsion force exerted on one by the other when they are 3 nanometers (1 nm $= 10^{-9}$ m) apart. Assume them to be in vacuum.

Nuclei have radii of order 10^{-15} m. We can assume them to be point charges in this case. Then

$$F = k \frac{qq'}{r^2} = (9 \times 10^9 \ N \cdot m^2/C^2) \frac{(2)(10)(1.6 \times 10^{-19} \ C)^2}{(3 \times 10^{-9} \ m)^2} = 5.1 \times 10^{-10} \ N$$

24.4. In the Bohr model of the hydrogen atom, an electron ($q = -e$) circles a proton ($q' = e$) in an orbit of radius 5.3×10^{-11} m. The attraction of the proton for the electron furnishes the centripetal force needed to hold the electron in orbit. Find (a) the force of electrical attraction between the particles and (b) the electron's speed. The electron mass is 9.1×10^{-31} kg.

(a)
$$F = k \frac{qq'}{r^2} = (9 \times 10^9 \ N \cdot m^2/C^2) \frac{(1.6 \times 10^{-19} \ C)^2}{(5.3 \times 10^{-11} \ m)^2} = 8.2 \times 10^{-8} \ N$$

(b) The force found in (a) is the centripetal force. Therefore,

$$8.2 \times 10^{-8} \ N = \frac{mv^2}{r}$$

from which

$$v = \sqrt{\frac{(8.2 \times 10^{-8} \ N)(r)}{m}} = \sqrt{\frac{(8.2 \times 10^{-8} \ N)(5.3 \times 10^{-11} \ m)}{9.1 \times 10^{-31} \ kg}} = 2.2 \times 10^6 \ m/s$$

24.5. Find the ratio of the Coulomb electrical force, F_e, to the gravitational force, F_g, between two electrons.

From Coulomb's law and Newton's law of gravitation,

$$F_e = k \frac{q^2}{r^2} \qquad \text{and} \qquad F_g = G \frac{m^2}{r^2}$$

Therefore

$$\frac{F_e}{F_g} = \frac{kq^2/r^2}{Gm^2/r^2} = \frac{kq^2}{Gm^2}$$

$$= \frac{(9 \times 10^9 \ N \cdot m^2/C^2)(1.6 \times 10^{-19} \ C)^2}{(6.67 \times 10^{-11} \ N \cdot m^2/kg^2)(9.1 \times 10^{-31} \ kg)^2} = 4.2 \times 10^{42}$$

As we see, the electric force is much stronger than the gravitational force.

24.6. As shown in Fig. 24-1, two identical balls, each of mass 0.10 g, carry identical charges and are suspended by two threads of equal length. At equilibrium they position themselves as shown. Find the charge on either ball.

Consider the ball on the left. It is in equilibrium under three forces: (1) the tension T in the thread; (2) the force of gravity,

$$mg = (1 \times 10^{-4} \ kg)(9.8 \ m/s^2) = 9.8 \times 10^{-4} \ N$$

(3) the Coulomb repulsion F.

Writing $\sum F_x = 0$ and $\sum F_y = 0$ for the ball on the left, we obtain

$$T \cos 60° - F = 0 \qquad T \sin 60° - mg = 0$$

From the second equation,

$$T = \frac{mg}{\sin 60°} = \frac{9.8 \times 10^{-4} \text{ N}}{0.866} = 1.13 \times 10^{-3} \text{ N}$$

Substituting in the first equation,

$$F = T \cos 60° = (1.13 \times 10^{-3} \text{ N})(0.50) = 5.7 \times 10^{-4} \text{ N}$$

But this is the Coulomb force, kqq'/r^2. Therefore

$$qq' = q^2 = \frac{Fr^2}{k} = \frac{(5.7 \times 10^{-4} \text{ N})(0.4 \text{ m})^2}{9 \times 10^9 \text{ N} \cdot \text{m}^2/\text{C}^2}$$

from which $q = 0.1 \ \mu\text{C}$.

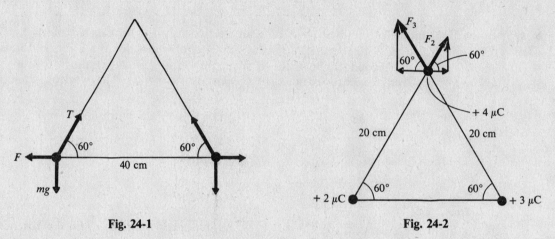

Fig. 24-1   Fig. 24-2

24.7. The charges shown in Fig. 24-2 are stationary. Find the force on the 4 μC charge due to the other two.

From Coulomb's law we have

$$F_2 = k \frac{qq'}{r^2} = (9 \times 10^9 \text{ N} \cdot \text{m}^2/\text{C}^2) \frac{(2 \times 10^{-6} \text{ C})(4 \times 10^{-6} \text{ C})}{(0.20 \text{ m})^2} = 1.8 \text{ N}$$

$$F_3 = k \frac{qq'}{r^2} = (9 \times 10^9 \text{ N} \cdot \text{m}^2/\text{C}^2) \frac{(3 \times 10^{-6} \text{ C})(4 \times 10^{-6} \text{ C})}{(0.20 \text{ m})^2} = 2.7 \text{ N}$$

The resultant force on the 4 μC charge has components

$$F_x = F_2 \cos 60° - F_3 \cos 60° = (1.8 - 2.7)(0.5) \text{ N} = -0.45 \text{ N}$$
$$F_y = F_2 \sin 60° + F_3 \sin 60° = (1.8 + 2.7)(0.866) \text{ N} = 3.9 \text{ N}$$

Then

$$F = \sqrt{F_x^2 + F_y^2} = \sqrt{(0.45)^2 + (3.9)^2} \text{ N} = 3.9 \text{ N}$$

The resultant makes an angle of $\arctan(0.45/3.9) = 7°$ with the positive y-axis, i.e. $\theta = 97°$.

24.8. Two charges are placed on the x-axis: $+3$ μC at $x = 0$ and -5 μC at $x = 40$ cm. Where must a third charge, q, be placed if the force it experiences is to be zero?

The situation is shown in Fig. 24-3. We know that q must be placed somewhere on the x-axis. (Why?) Suppose that q is positive. When it is placed in interval BC, the two forces on it are in the same direction and cannot cancel. When it is placed to the right of C, the attractive force from the -5 μC is always larger than the repulsion of the $+3$ μC. Therefore, the force on q cannot be zero in

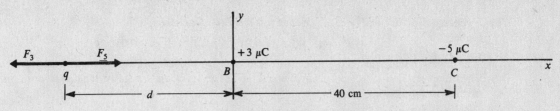

Fig. 24-3

this region.   Only in the region to the left of B can cancellation occur.   (Can you show that this is also true if q is negative?)

For q placed as shown, when the net force on it is zero, we have $F_3 = F_5$ and so, for distances in meters,

$$k \frac{q(3 \times 10^{-6}\,\text{C})}{d^2} = k \frac{q(5 \times 10^{-6}\,\text{C})}{(0.4 + d)^2}$$

After canceling q, k, and 10^{-6} C from each side, we cross multiply to obtain

$$5d^2 = 3(0.4 + d)^2 \qquad \text{or} \qquad d^2 - 1.2\,d - 0.24 = 0$$

Using the quadratic formula,

$$d = \frac{-b \pm \sqrt{b^2 - 4ac}}{2a} = \frac{1.2 \pm \sqrt{1.44 + 0.96}}{2} = 0.6 \pm 0.775 \text{ m}$$

Two answers, 1.375 m and -0.175 m, are therefore found for d.   The first is the correct one; the second gives the point in BC where the two forces have the same magnitude (but do not cancel).

24.9.   Compute (a) the electric field intensity E in air at a distance of 30 cm from a point charge $q_1 = 5 \times 10^{-9}$ C,   (b) the force on a charge $q_2 = 4 \times 10^{-10}$ C   placed 30 cm from q_1, and (c) the force on a charge $q_3 = -4 \times 10^{-10}$ C   placed 30 cm from q_1 (in the absence of q_2).

(a)
$$E = k \frac{q_1}{r^2} = (9 \times 10^9 \text{ N} \cdot \text{m}^2/\text{C}^2) \frac{5 \times 10^{-9}\,\text{C}}{(0.3 \text{ m})^2} = 500 \text{ N/C}$$

directed away from q_1.

(b)
$$F = Eq_2 = (500 \text{ N/C})(4 \times 10^{-10} \text{ C}) = 2 \times 10^{-7} \text{ N}$$

directed away from q_1.

(c)
$$F = Eq_3 = (500 \text{ N/C})(-4 \times 10^{-10} \text{ C}) = -2 \times 10^{-7} \text{ N}$$

This force is directed toward q_1.

24.10.   For the situation shown in Fig. 24-4, find (a) the electric field intensity E at point P, (b) the force on a -4×10^{-8} C charge placed at P, (c) where in the region the electric field would be zero.

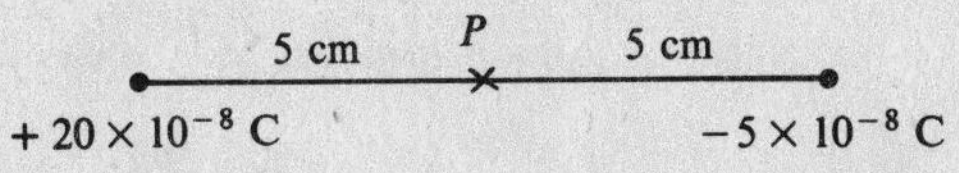

Fig. 24-4

(a)   A positive test charge placed at P will be repelled to the right by the positive charge and attracted to the right by the negative charge.   Because $\mathbf{E}_1$ and $\mathbf{E}_2$ have the same direction, we can add their magnitudes to obtain the magnitude of the resultant field, i.e.

$$E = E_1 + E_2 = k \frac{|q_1|}{r_1^2} + k \frac{|q_2|}{r_2^2} = \frac{k}{r_1^2} (|q_1| + |q_2|)$$

where $r_1 = r_2 = 0.05$ m,   and $|q_1|$ and $|q_2|$ are the absolute values of q_1 and q_2.   Hence,

$$E = \frac{9 \times 10^9}{(0.05)^2} (25 \times 10^{-8}) = 9 \times 10^5 \text{ N/C}$$

directed toward the right.

(b) A charge q placed at P will experience a force Eq. Therefore,

$$F = Eq = (9 \times 10^5 \text{ N/C})(-4 \times 10^{-8} \text{ C}) = -0.036 \text{ N}$$

The negative sign tells us the force is directed toward the left.   This is correct because the electric field represents the force on a positive charge.   The force on a negative charge is opposite in direction to the field.

(c) Reasoning as in Problem 24.8, we conclude that the field will be zero somewhere to the right of the -5×10^{-8} C charge. Represent the distance to that point from the -5×10^{-8} C charge by d. At that point,

$$E_1 - E_2 = 0$$

because the field due to the positive charge is to the right, while the field due to the negative charge is to the left.   Thus

$$k\left(\frac{|q_1|}{r_1^2} - \frac{|q_2|}{r_2^2} \right) = (9 \times 10^9)\left[\frac{20 \times 10^{-8}}{(d+0.1)^2} - \frac{5 \times 10^{-8}}{d^2} \right] = 0$$

Simplifying, we obtain

$$3d^2 - 0.2\,d - 0.01 = 0$$

which gives $d = 0.10$ m   and -0.03 m.   Only the plus sign has meaning here and therefore $d = 0.10$ m.   The point in question is 10 cm to the right of the negative charge.

24.11. Two charged metal plates in vacuum are 15 cm apart as shown in Fig. 24-5.   The electric field between the plates is uniform and has an intensity $E = 3000$ N/C.   An electron ($q = -e$, $m = 9.1 \times 10^{-31}$ kg) is released from rest at point P just outside the negative plate.   (a) How long will it take to reach the other plate?   (b) How fast will it be going just before it hits?

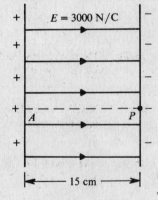

Fig. 24-5

The electric field lines show the force on a positive charge.   (A positive charge would be repelled to the right by the positive plate and attracted to the right by the negative plate.)   An electron, being negative, will experience a force in the opposite direction, toward the left, of magnitude

$$F = |q|E = (1.6 \times 10^{-19} \text{ C})(3000 \text{ N/C}) = 4.8 \times 10^{-16} \text{ N}$$

Because of this force, the electron experiences an acceleration toward the left given by

$$a = \frac{F}{m} = \frac{4.8 \times 10^{-16} \text{ N}}{9.1 \times 10^{-31} \text{ kg}} = 5.3 \times 10^{14} \text{ m/s}^2$$

In the motion problem for the electron released at the negative plate and traveling to the positive plate,

$$v_0 = 0 \qquad x = 0.15 \text{ m} \qquad a = 5.3 \times 10^{14} \text{ m/s}^2$$

(a) Then $x = v_0 t + \frac{1}{2} at^2$ gives

$$t = \sqrt{\frac{2x}{a}} = \sqrt{\frac{(2)(0.15 \text{ m})}{5.3 \times 10^{14} \text{ m/s}^2}} = 2.4 \times 10^{-8} \text{ s}$$

(b) $$v = v_0 + at = 0 + (5.3 \times 10^{14} \text{ m/s}^2)(2.4 \times 10^{-8} \text{ s}) = 1.30 \times 10^7 \text{ m/s}$$

As we shall see in Chapter 43, relativistic effects begin to become important at speeds above this. Therefore, this approach must be modified for very fast particles.

24.12. Suppose in Fig. 24-5 an electron is shot straight upward from point P with a speed of 5×10^6 m/s.   How far above A will it strike the positive plate?

This is a projectile problem. (Since the gravitational force is so small compared to the electrical force, we ignore gravity.) The only force acting on the electron after its release is the electric field, a horizontal force. We found in Problem 24.11(a) that under this force the electron has a time of flight of 2.4×10^{-8} s. The vertical displacement in this time is

$$(5 \times 10^6 \text{ m/s})(2.4 \times 10^{-8} \text{ s}) = 0.12 \text{ m}$$

The electron strikes the positive plate 12 cm above point A.

24.13. In Fig. 24-5 a proton ($q = +e$, $m = 1.67 \times 10^{-27}$ kg) is shot with speed 2×10^5 m/s toward P from A. What will be its speed just before hitting the plate at P?

$$a = \frac{F}{m} = \frac{qE}{m} = \frac{(1.6 \times 10^{-19} \text{ C})(3000 \text{ N/C})}{1.67 \times 10^{-27} \text{ kg}} = 2.9 \times 10^{11} \text{ m/s}^2$$

For the motion problem in the horizontal direction,

$$v_0 = 2 \times 10^5 \text{ m/s} \qquad x = 0.15 \text{ m} \qquad a = 2.9 \times 10^{11} \text{ m/s}^2$$

We use $v_f^2 = v_0^2 + 2ax$ to find

$$v_f = \sqrt{v_0^2 + 2ax} = \sqrt{(2 \times 10^5)^2 + (2)(2.9 \times 10^{11})(0.15)} = 3.55 \times 10^5 \text{ m/s}$$

Supplementary Problems

24.14. How many electrons are contained in 1 C of charge? What are the mass and weight of the electrons in 1 C of charge?   *Ans.* 6.2×10^{18} electrons, 5.7×10^{-12} kg, 5.6×10^{-11} N

24.15. If two equal charges, each of 1 C, were separated in air by a distance of 1 km, what would be the force between them?   *Ans.*   9000 N repulsion

24.16. Determine the force between two free electrons spaced 1 angstrom (10^{-10} m) apart.   *Ans.* 2.3×10^{-8} N repulsion

24.17. What is the force of repulsion between two argon nuclei when separated by 1 nm (10^{-9} m)? The charge on an argon nucleus is $+18e$.   *Ans.* 7.5×10^{-8} N

24.18. Two equally charged pith balls are 3 cm apart in air and repel each other with a force of 4×10^{-5} N. Compute the charge on each ball.   *Ans.* 2×10^{-9} C

24.19. Three point charges are placed at the following points on the x-axis: $+2$ μC at $x = 0$, -3 μC at $x = 40$ cm, -5 μC at $x = 120$ cm. Find the force (a) on the -3 μC charge, (b) on the -5 μC charge. *Ans.* (a) -0.55 N; (b) 0.15 N

24.20. Four equal point charges, $+3$ μC, are placed at the four corners of a square that is 40 cm on a side. Find the force on any one of the charges.   *Ans.*   0.97 N outward along the diagonal

24.21. Four equal-magnitude point charges (3 μC) are placed at the corners of a square that is 40 cm on a side. Two, diagonally opposite each other, are positive and the other two are negative. Find the force on either negative charge.   *Ans.*   0.46 N inward along the diagonal

24.22.   Charges of $+2$, $+3$, and -8 μC are placed at the vertices of an equilateral triangle of side 10 cm. Calculate the magnitude of the force acting on the -8 μC charge due to the other two charges.
Ans.   31.4 N

24.23.   One charge ($+5$ μC) is placed at $x = 0$, and a second charge ($+7$ μC) at $x = 100$ cm. Where can a third be placed and experience zero net force due to the other two?   *Ans.* $x = 45.8$ cm

24.24.   Two identical tiny metal balls carry charges of $+3$ nC and -12 nC. They are 3 cm apart. (*a*) Compute the force of attraction. (*b*) The balls are now touched together and then separated to 3 cm. Describe the forces on them now.   *Ans.*   (*a*) 3.6×10^{-4} N attraction;   (*b*) 2×10^{-4} N repulsion

24.25.   A charge, $+6$ μC, experiences a force of 2×10^{-3} N in the $+x$-direction at a certain point in space. (*a*) What was the electric field there before the charge was placed there?   (*b*) Describe the force a -2 μC charge would experience if it were used in place of the $+6$ μC.
Ans.   (*a*) 333 N/C in $+x$-direction;   (*b*) 6.7×10^{-4} N in $-x$-direction

24.26.   A point charge, -3×10^{-5} C, is placed at the origin of coordinates. Find the electric field at the point $x = 5$ m on the x-axis.   *Ans.*   10 800 N/C in $-x$-direction

24.27.   Four equal-magnitude (4 μC) charges are placed at the four corners of a square that is 20 cm on each side. Find the electric field intensity at the center of the square (*a*) if the charges are all positive, (*b*) if the charges alternate in sign as one goes around the perimeter of the square, (*c*) if the charges have the following sequence around the square: plus, plus, minus, minus.
Ans.   (*a*) zero;   (*b*) zero;   (*c*) 5.1×10^6 N/C toward the negative side

24.28.   Determine the acceleration of a proton ($q = +e$, $m = 1.67 \times 10^{-27}$ kg) in an electric field of intensity 500 N/C. How many times is this acceleration greater than that due to gravity?
Ans. 4.8×10^{10} m/s^2, 4.9×10^9

24.29.   A tiny, 0.60 g ball carries a charge of magnitude 8 μC. It is suspended by a thread in a downward electric field of intensity 300 N/C. What is the tension in the thread if the charge on the ball is (*a*) positive, (*b*) negative?   *Ans.*   (*a*) 8.3×10^{-3} N;   (*b*) 3.5×10^{-3} N

24.30.   The tiny ball at the end of the thread shown in Fig. 24-6 has a mass of 0.60 g and is in a horizontal electric field of intensity 700 N/C. It is in equilibrium in the position shown. What are the magnitude and sign of the charge on the ball?   *Ans.* -3.1×10^{-6} C

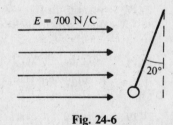

$E = 700$ N/C

20°

24.31.   An electron ($q = -e$, $m = 9.1 \times 10^{-31}$ kg) is projected out along the $+x$-axis with an initial speed of 3×10^6 m/s. It goes 45 cm and stops due to a uniform electric field in the region. Find the magnitude and direction of the field.   *Ans.*   57 N/C in $+x$-direction

Fig. 24-6

24.32.   A particle of mass m and charge $-e$ is projected with horizontal speed v into an electric field of intensity E directed downward. Find: (*a*) the horizontal and vertical components of its acceleration, a_x and a_y; (*b*) its horizontal and vertical displacements, x and y, after time t; (*c*) the equation of its trajectory.

$$Ans. \quad (a)\ a_x = 0, \quad a_y = \frac{Ee}{m}\ ; \quad (b)\ x = vt, \quad y = \frac{1}{2}\,a_y t^2 = \frac{1}{2}\left(\frac{Ee}{m}\right)t^2;$$

$$(c)\ y = \frac{1}{2}\left(\frac{Ee}{mv^2}\right)x^2 \quad \text{(a parabola)}$$

Chapter 25

Potential; Capacitance

THE POTENTIAL DIFFERENCE from one point, A, to another point, B, is the work done against electrical forces in carrying a unit positive test charge from A to B. We represent the potential difference from A to B by $V_B - V_A$ or by V. Its units are those of work per charge (joules/coulomb), called *volts*.

$$1 \text{ volt (V)} = 1 \text{ J/C}$$

Because work is a scalar quantity, so too is potential difference. Like work, potential difference may be positive or negative.

The work W done in transporting a charge q from one point, A, to a second point, B, is

$$W = q(V_B - V_A) = qV$$

ABSOLUTE POTENTIAL: The absolute potential at a point is the work done against electrical forces in carrying a unit positive test charge from infinity to that point. Hence the absolute potential at a point B is the difference in potential from $A = \infty$ to B.

Consider a point charge q in vacuum and a point P at a distance r from the point charge. The absolute potential at P due to the charge q is

$$V = k \frac{q}{r}$$

where $k = 9 \times 10^9 \text{ N} \cdot \text{m}^2/\text{C}^2$ is the Coulomb constant.

Because of the superposition principle and the scalar nature of potential difference, the absolute potential at a point due to a number of point charges is

$$V = k \sum \frac{q_i}{r_i}$$

where the r_i are the distances of the charges q_i from the point in question. Negative q's contribute negative terms to the potential, while positive q's contribute positive terms.

ELECTRICAL POTENTIAL ENERGY (EPE): In order to carry a charge q from infinity to a point where the absolute potential is V, work in the amount qV must be done on the charge. This work appears as electrical potential energy (EPE) stored in the charge.

When a charge q is carried through a *potential difference* V, work in the amount qV must be done on the charge. This work results in a change qV in the EPE of the charge. For a *potential rise*, V will be positive and the EPE will increase if q is positive. But for a *potential drop*, V will be negative and the EPE of the charge will decrease if q is positive.

V RELATED TO E: Because E is force per unit charge and V is work per unit charge, E and V are interrelated in the same way as force and work. Suppose points A and B have the small separation $\Delta x = x_B - x_A$ in a region where the x-component of the field intensity is E_x. Then

$$V_B - V_A = (-E_x)\Delta x$$

The minus sign arises because V is defined as the work done *against* the field E.

In the special case of a uniform field, the magnitude of the potential difference between two points whose separation in the direction of the field is d is $|V| = Ed$. The field between two large, parallel, oppositely charged plates is uniform. If the separation of the plates is d, then the potential difference between the plates is Ed.

ELECTRON VOLT (eV) ENERGY UNIT: The work done in carrying a charge $+e$ (coulombs) through a potential rise of 1 volt is defined to be 1 *electron volt* (eV).   Therefore,

$$1 \text{ eV} = (1.602 \times 10^{-19} \text{ C})(1 \text{ V}) = 1.602 \times 10^{-19} \text{ J}$$

Equivalently,

$$\text{work or energy (in eV)} = \frac{\text{work (in joules)}}{e}$$

A CAPACITOR or *condenser* consists of two conductors separated by an insulator or dielectric.   The *capacitance* of a capacitor is defined by

$$\text{capacitance } C = \frac{\text{magnitude of charge } q \text{ on either conductor}}{\text{magnitude of potential difference } V \text{ between conductors}}$$

For q in coulombs and V in volts, C will be in *farads* (F).   Convenient submultiples of the farad are:

$$1 \ \mu\text{F} = 1 \text{ microfarad} = 10^{-6} \text{ F}$$

$$1 \text{ pF} = 1 \text{ picofarad} = 10^{-12} \text{ F}$$

PARALLEL PLATE CAPACITOR: The capacitance of a parallel plate capacitor with two large plates, the area of one side of one plate which is opposed by the other plate being A square meters and the distance between them d meters, is

$$C = K\epsilon_0 \frac{A}{d}$$

where K (dimensionless) is the dielectric constant (see Chapter 24) of the nonconducting material (the dielectric) between the plates, and

$$\epsilon_0 = 8.85 \times 10^{-12} \text{ C}^2/\text{N} \cdot \text{m}^2 = 8.85 \times 10^{-12} \text{ F/m}$$

For vacuum, $K = 1$, so that a dielectric-filled, parallel plate capacitor has a capacitance K times larger than the same capacitor with vacuum between its plates.   This result holds for a capacitor of arbitrary shape.

CAPACITORS IN PARALLEL AND SERIES: As shown in Fig. 25-1, capacitances add for capacitors in parallel, whereas reciprocal capacitances add for capacitors in series.

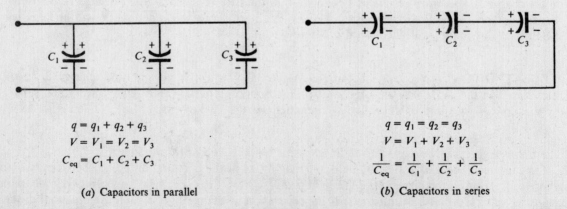

$$q = q_1 + q_2 + q_3$$
$$V = V_1 = V_2 = V_3$$
$$C_{eq} = C_1 + C_2 + C_3$$

(a) Capacitors in parallel

$$q = q_1 = q_2 = q_3$$
$$V = V_1 + V_2 + V_3$$
$$\frac{1}{C_{eq}} = \frac{1}{C_1} + \frac{1}{C_2} + \frac{1}{C_3}$$

(b) Capacitors in series

Fig. 25-1

ENERGY STORED IN A CAPACITOR: In a capacitor the potential difference is proportional to the charge ($V = q/C$). While a capacitor is being charged, the charge builds up from an initial value zero to a final value q. Hence the potential difference builds up from zero to a final value V, the average value during the process being $\frac{1}{2}V$. Now the work W required to transfer a total charge q through an average potential difference $\frac{1}{2}V$ is $W = q(\frac{1}{2}V)$. Thus the electrical energy W stored in a charged capacitor is

$$W = \tfrac{1}{2}qV = \tfrac{1}{2}CV^2 = \tfrac{1}{2}q^2/C$$

using $q = CV$.

Solved Problems

25.1. How much work is required to carry an electron from the positive terminal of a 12 V battery to the negative terminal?

Going from the positive to the negative terminal, one passes through a potential drop. In this case it is $V = -12$ V. Then

$$\text{work} = qV = (-1.6 \times 10^{-19}\text{ C})(-12\text{ V}) = 1.92 \times 10^{-18}\text{ J}$$

As a check, we notice that an electron, if left to itself, will move from negative to positive because it is a negative charge. Hence positive work must be done to carry it in the reverse direction as required here.

25.2. How much electrical potential energy does a proton lose as it falls through a potential drop of 5 kV?

The proton carries a positive charge. It will therefore move from regions of high potential to low potential if left free to do so. Its change in potential energy as it moves through a potential difference V is Vq. In our case, $V = -5$ kV. Therefore,

$$\text{change in EPE} = Vq = (-5000\text{ V})(1.6 \times 10^{-19}\text{ C}) = -8 \times 10^{-16}\text{ J}$$

25.3. An electron starts from rest and falls through a potential rise of 80 V. What is its final speed?

Positive charges tend to fall through potential drops; negative charges, such as electrons, tend to fall through potential rises.

$$\text{change in EPE} = Vq = (80\text{ V})(-1.6 \times 10^{-19}\text{ C}) = -1.28 \times 10^{-17}\text{ J}$$

This lost EPE appears as KE of the electron.

$$\text{EPE lost} = \text{KE gained}$$

$$1.28 \times 10^{-17}\text{ J} = \tfrac{1}{2}mv_f^2 - \tfrac{1}{2}mv_0^2 = \tfrac{1}{2}mv_f^2 - 0$$

$$v_f = \sqrt{\frac{(1.28 \times 10^{-17}\text{ J})(2)}{9.1 \times 10^{-31}\text{ kg}}} = 5.3 \times 10^6\text{ m/s}$$

25.4. (a) What is the absolute potential at each of the following radii from a charge of 2 μC: $r = 10$ cm and $r = 50$ cm? (b) How much work is required to carry a 0.05 μC charge from the point at $r = 50$ cm to $r = 10$ cm?

(a)
$$V_{10} = k\frac{q}{r} = (9 \times 10^9\text{ N} \cdot \text{m}^2/\text{C}^2)\frac{2 \times 10^{-6}\text{ C}}{0.10\text{ m}} = 1.8 \times 10^5\text{ V}$$

$$V_{50} = \frac{10}{50}V_{10} = 0.36 \times 10^5\text{ V}$$

(b)
$$\text{work} = q(V_{10} - V_{50}) = (5 \times 10^{-8}\text{ C})(1.44 \times 10^5\text{ V}) = 7.2 \times 10^{-3}\text{ J}$$

25.5. Suppose, in Problem 25.4(a), that a proton is released at $r = 10$ cm. How fast will it be moving as it passes a point at $r = 50$ cm?

Going from one point to the other, there is a potential drop of

$$\text{potential drop} = 1.80 \times 10^5 \text{ V} - 0.36 \times 10^5 \text{ V} = 1.44 \times 10^5 \text{ V}$$

The proton acquires KE as it falls through this potential drop.

$$\text{KE gained} = \text{EPE lost}$$

$$\tfrac{1}{2}mv_f^2 - \tfrac{1}{2}mv_0^2 = qV$$

$$\tfrac{1}{2}(1.67 \times 10^{-27} \text{ kg})v_f^2 - 0 = (1.6 \times 10^{-19} \text{ C})(1.44 \times 10^5 \text{ V})$$

from which $v_f = 5.3 \times 10^6$ m/s.

25.6. A tin nucleus has a charge $+50e$. (a) Find the absolute potential V at a radius 10^{-12} m from the nucleus. (b) If a proton is released from this point, how fast will it be moving when it is 1 m from the nucleus?

(a)
$$V = k\frac{q}{r} = (9 \times 10^9)\frac{(50)(1.6 \times 10^{-19})}{10^{-12}} = 72\,000 \text{ V} = 72 \text{ kV}$$

(b) The proton is repelled by the nucleus and flies out to infinity. The absolute potential at a point is the potential difference between the point in question and infinity. Hence there is a potential drop of 72 kV as the proton flies to infinity.

Usually we would simply assume that 1 m is far enough from the nucleus to consider it to be at infinity. But, as a check, let us compute V at $r = 1$ m.

$$V_{1 \text{ m}} = k\frac{q}{r} = (9 \times 10^9)\frac{(50)(1.6 \times 10^{-19})}{1} = 7.2 \times 10^{-8} \text{ V}$$

which is essentially zero in comparison with 72 kV.

As the proton falls through 72 kV,

$$\text{gain in KE} = \text{loss in EPE}$$

$$\tfrac{1}{2}mv_f^2 - \tfrac{1}{2}mv_0^2 = qV$$

$$\tfrac{1}{2}(1.67 \times 10^{-27} \text{ kg})v_f^2 - 0 = (1.6 \times 10^{-19} \text{ C})(72\,000 \text{ V})$$

from which $v_f = 3.7 \times 10^6$ m/s.

25.7. The following point charges are placed on the x-axis: $+2$ μC at $x = 20$ cm, -3 μC at $x = 30$ cm, -4 μC at $x = 40$ cm. Find the absolute potential on the axis at $x = 0$.

Potential is a scalar and so

$$V = k\sum \frac{q_i}{r_i}$$

$$= (9 \times 10^9)\left[\frac{2 \times 10^{-6}}{0.20} + \frac{-3 \times 10^{-6}}{0.30} + \frac{-4 \times 10^{-6}}{0.40}\right]$$

$$= (9 \times 10^9)[10 \times 10^{-6} - 10 \times 10^{-6} - 10 \times 10^{-6}] = -90\,000 \text{ V}$$

25.8. Two point charges, $+q$ and $-q$, are separated by a distance d. Where, besides at infinity, is the absolute potential zero?

At the point (or points) in question,

$$0 = k\frac{q}{r_1} + k\frac{-q}{r_2} \qquad \text{or} \qquad r_1 = r_2$$

This condition holds everywhere on a plane which is the perpendicular bisector of the line joining the two charges. Therefore the absolute potential is zero everywhere on that plane.

25.9.   Four point charges are placed at the four corners of a square that is 30 cm on each side. Find the potential at the center of the square if (a) the four charges are each $+2$ μC and (b) two of the four charges are $+2$ μC and two are -2 μC.

(a) $$V = k \sum \frac{q_i}{r_i} = k \frac{\sum q_i}{r} = (9 \times 10^9 \text{ N} \cdot \text{m}^2/\text{C}^2) \frac{(4)(2 \times 10^{-6} \text{ C})}{(0.3 \text{ m}) \cos 45°} = 3.4 \times 10^5 \text{ V}$$

(b) $$V = (9 \times 10^9 \text{ N} \cdot \text{m}^2/\text{C}^2) \frac{(2 + 2 - 2 - 2) \times 10^{-6} \text{ C}}{(0.3 \text{ m}) \cos 45°} = 0$$

25.10.   In Fig. 25-2, the charge at A is $+200$ pC, while the charge at B is -100 pC. (a) Find the absolute potentials at points C and D. (b) How much work must be done to transfer a charge of $+500$ μC from point C to point D?

(a) $$V_C = k \sum \frac{q_i}{r_i} = (9 \times 10^9 \text{ N} \cdot \text{m}^2/\text{C}^2) \left[\frac{2 \times 10^{-10} \text{ C}}{0.8 \text{ m}} - \frac{1 \times 10^{-10} \text{ C}}{0.2 \text{ m}} \right] = -2.25 \text{ V}$$

$$V_D = (9 \times 10^9) \left[\frac{2 \times 10^{-10}}{0.2} - \frac{1 \times 10^{-10}}{0.8} \right] = +7.88 \text{ V}$$

(b)   There is a potential rise from C to D of $V = V_D - V_C = 7.88 \text{ V} - (-2.25 \text{ V}) = 10.13 \text{ V}$.

work $= Vq = (10.13 \text{ V})(5 \times 10^{-4} \text{ C}) = 5.06 \times 10^{-3}$ J

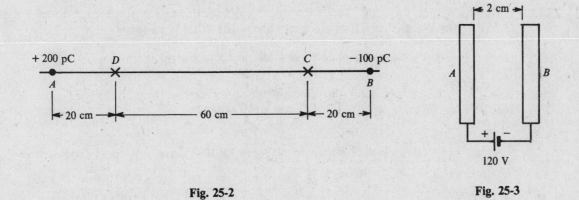

Fig. 25-2   Fig. 25-3

25.11.   In Fig. 25-3 we show two large metal plates connected to a 120 V battery.   Assume the plates to be in vacuum and to be much larger than shown.   Find: (a) E between the plates, (b) force experienced by an electron between the plates, (c) EPE lost by an electron as it moves from plate B to plate A, (d) speed of the electron released from plate B just before striking plate A.

(a) E will be directed from the positive plate A to the negative plate B. It is uniform between large parallel plates and is given by

$$E = \frac{V}{d} = \frac{120 \text{ V}}{0.02 \text{ m}} = 6000 \text{ V/m}$$

directed from left to right.   The units V/m are the same as N/C; very often electric field strengths are specified in V/m.

(b) $$F = qE = (-1.6 \times 10^{-19} \text{ C})(6000 \text{ V/m}) = -9.6 \times 10^{-16} \text{ N}$$

The negative sign tells us that $\mathbf{F}$ is directed oppositely to $\mathbf{E}$.   Since plate A is positive, the electron is attracted by it.   The force on the electron is towards the left.

(c)   change in EPE = Vq = (120 V)(-1.6×10^{-19} C) = -1.92×10^{-17} J

Notice that V is a potential rise when going from B to A.

(d)   loss in EPE = gain in KE

$$1.92 \times 10^{-17} \text{ J} = \tfrac{1}{2}mv_f^2 - \tfrac{1}{2}mv_0^2$$

$$1.92 \times 10^{-17} \text{ J} = \tfrac{1}{2}(9.1 \times 10^{-31} \text{ kg})v_f^2 - 0$$

from which $v_f = 6.5 \times 10^6$ m/s.

25.12. As shown in Fig. 25-4, a charged particle remains stationary between the two charged plates. The plate separation is 2 cm and $m = 4 \times 10^{-13}$ kg and $q = 2.4 \times 10^{-18}$ C for the particle.   Find the potential difference between the plates.

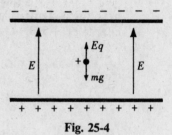

Fig. 25-4

Since the particle is in equilibrium,

weight of particle = upward electrical force

$mg = qE$

$$E = \frac{mg}{q} = \frac{(4 \times 10^{-13} \text{ kg})(9.8 \text{ m/s}^2)}{2.4 \times 10^{-18} \text{ C}} = 1.63 \times 10^6 \text{ V/m}$$

But for a parallel plate system,

$$V = Ed = (1.63 \times 10^6 \text{ V/m})(0.02 \text{ m}) = 32.7 \text{ kV}$$

25.13. An alpha particle ($q = 2e$, $m = 6.7 \times 10^{-27}$ kg) falls through a potential drop of 3×10^6 V (3 MV).   (a) What is its KE in electron volts?   (b) What is its speed?

(a)   energy in eV = $\dfrac{qV}{e} = \dfrac{(2e)(3 \times 10^6)}{e} = 6 \times 10^6$ eV

or 6 MeV.

(b)   EPE lost = KE gained

$$qV = \tfrac{1}{2}mv_f^2 - \tfrac{1}{2}mv_0^2$$

$$(2)(1.6 \times 10^{-19} \text{ C})(3 \times 10^6 \text{ V}) = \tfrac{1}{2}(6.7 \times 10^{-27} \text{ kg})v_f^2 - 0$$

from which $v_f = 1.7 \times 10^7$ m/s.

25.14. A capacitor with air between its plates has a capacitance of 8 μF.   Determine its capacitance when a dielectric with dielectric constant 6.0 is between its plates.

C with dielectric = K (C with air) = (6.0)(8 μF) = 48 μF

25.15. What is the charge on a 300 pF capacitor when it is charged to a voltage of 1 kV?

$q = CV = (300 \times 10^{-12}$ F$)(1000$ V$) = 3 \times 10^{-7}$ C = 0.3 μC

25.16. A metal sphere mounted on an insulating rod carries a charge of 6 nC when its potential is 200 V higher than its surroundings.   What is the capacitance of the capacitor formed by the sphere and its surroundings?

$$C = \frac{q}{V} = \frac{6 \times 10^{-9} \text{ C}}{200 \text{ V}} = 30 \text{ pF}$$

25.17. A 1.2 μF capacitor is charged to 3 kV.   Compute the energy stored in the capacitor.

energy = $\tfrac{1}{2}qV = \tfrac{1}{2}CV^2 = \tfrac{1}{2}(1.2 \times 10^{-6}$ F$)(3000$ V$)^2 = 5.4$ J

25.18. The series combination of two capacitors shown in Fig. 25-5 is connected across 1000 V. Compute (a) the equivalent capacitance C_{eq} of the combination, (b) the magnitudes of the charges on the capacitors, (c) the potential differences across the capacitors, (d) the energy stored in the capacitors.

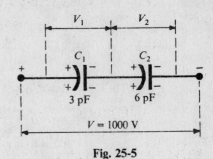

(a) $$\frac{1}{C_{eq}} = \frac{1}{C_1} + \frac{1}{C_2} = \frac{1}{3 \text{ pF}} + \frac{1}{6 \text{ pF}} = \frac{1}{2 \text{ pF}}$$

from which $C = 2$ pF.

Fig. 25-5

(b) In a series combination, each capacitor carries the same charge, which is the charge on the combination. Thus, using the result of (a),

$$q_1 = q_2 = q = C_{eq}V = (2 \times 10^{-12} \text{ F})(1000 \text{ V}) = 2 \text{ nC}$$

(c) $$V_1 = \frac{q_1}{C_1} = \frac{2 \times 10^{-9} \text{ C}}{3 \times 10^{-12} \text{ F}} = 667 \text{ V}$$

$$V_2 = \frac{q_2}{C_2} = \frac{2 \times 10^{-9} \text{ C}}{6 \times 10^{-12} \text{ F}} = 333 \text{ V}$$

(d)   energy in $C_1 = \frac{1}{2}q_1V_1 = \frac{1}{2}(2 \times 10^{-9} \text{ C})(667 \text{ V}) = 6.7 \times 10^{-7} \text{ J}$

energy in $C_2 = \frac{1}{2}q_2V_2 = \frac{1}{2}(2 \times 10^{-9} \text{ C})(333 \text{ V}) = 3.3 \times 10^{-7} \text{ J}$

energy in combination $= (6.7 + 3.3) \times 10^{-7} \text{ J} = 10 \times 10^{-7} \text{ J}$

The last result is also directly given by $\frac{1}{2}qV$ or $\frac{1}{2}C_{eq}V^2$.

25.19. The parallel capacitor combination shown in Fig. 25-6 is connected across a 120 V source. Determine the equivalent capacitance C_{eq} and the charge on each capacitor.

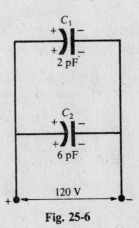

For a parallel combination,

$$C_{eq} = C_1 + C_2 = 6 \text{ pF} + 2 \text{ pF} = 8 \text{ pF}$$

Each capacitor has a 120 V potential difference impressed on it. Therefore,

$$q_1 = C_1V_1 = (2 \times 10^{-12} \text{ F})(120 \text{ V}) = 2.4 \times 10^{-10} \text{ C}$$
$$q_2 = C_2V_2 = (6 \times 10^{-12} \text{ F})(120 \text{ V}) = 7.2 \times 10^{-10} \text{ C}$$

The charge on the combination is $q_1 + q_2 = 9.6 \times 10^{-10}$ C. Or, we could write

$$q = C_{eq}V = (8 \times 10^{-12} \text{ F})(120 \text{ V}) = 9.6 \times 10^{-10} \text{ C}$$

Fig. 25-6

25.20. A certain parallel plate capacitor consists of two plates, each with area 200 cm², separated by a 0.4 cm air gap. (a) Compute its capacitance. (b) If the capacitor is connected across a 500 V source, find the charge on it, the energy stored in it, and the value of E between the plates. (c) If a liquid with $K = 2.60$ is poured between the plates so as to fill the air gap, how much additional charge will flow onto the capacitor from the 500 V source?

(a) For a parallel plate capacitor,

$$C = K\epsilon_0 \frac{A}{d} = (1)(8.85 \times 10^{-12} \text{ F/m}) \frac{200 \times 10^{-4} \text{ m}^2}{4 \times 10^{-3} \text{ m}} = 4.4 \times 10^{-11} \text{ F}$$

or **44 pF**.

(b)
$$q = CV = (4.4 \times 10^{-11} \text{ F})(500 \text{ V}) = 2.2 \times 10^{-8} \text{ C}$$
$$\text{energy} = \tfrac{1}{2}qV = \tfrac{1}{2}(2.2 \times 10^{-8} \text{ C})(500 \text{ V}) = 5.5 \times 10^{-6} \text{ J}$$
$$E = \frac{V}{d} = \frac{500 \text{ V}}{4 \times 10^{-3} \text{ m}} = 1.25 \times 10^5 \text{ V/m}$$

(c) The capacitor will now have a capacitance K times larger than before.   Therefore,

$$q = CV = (2.60 \times 4.4 \times 10^{-11} \text{ F})(500 \text{ V}) = 5.7 \times 10^{-8} \text{ C}$$

The capacitor already had a charge of 2.2×10^{-8} C and so $(5.7 - 2.2) \times 10^{-8}$ or 3.5×10^{-8} C must have been added to it.

25.21. Two capacitors, 3 μF and 4 μF, are individually charged across a 6 V battery.   After being disconnected from the battery, they are connected together with the negative plate of one attached to the positive plate of the other.   What is the final charge on each capacitor?

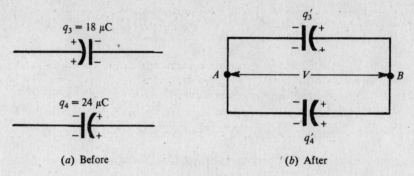

(a) Before   (b) After

Fig. 25-7

The situation is shown in Fig. 25-7.   Before being connected, their charges are

$$q_3 = CV = (3 \times 10^{-6} \text{ F})(6 \text{ V}) = 18 \ \mu\text{C}$$
$$q_4 = CV = (4 \times 10^{-6} \text{ F})(6 \text{ V}) = 24 \ \mu\text{C}$$

As seen in the figure, the charges will partly cancel when the capacitors are connected together.   Their final charges are given by

$$q_3' + q_4' = q_4 - q_3 = 6 \ \mu\text{C}$$

Also, the potential across each is now the same, so that $V = q/C$ gives

$$\frac{q_3'}{3 \times 10^{-6} \text{ F}} = \frac{q_4'}{4 \times 10^{-6} \text{ F}} \qquad \text{or} \qquad q_3' = 0.75 \ q_4'$$

Substitution of this in the previous equation gives

$$0.75 \ q_4' + q_4' = 6 \ \mu\text{C} \qquad \text{or} \qquad q_4' = 3.43 \ \mu\text{C}$$

Then $q_3' = 0.75 \ q_4' = 2.57 \ \mu$C.

Supplementary Problems

25.22. Two metal plates are attached to the two terminals of a 1.50 V battery.   How much work is required to carry a $+5 \ \mu$C charge (a) from the negative to the positive plate, (b) from the positive to the negative plate?   *Ans.*   (a) 7.5×10^{-6} J;   (b) -7.5×10^{-6} J

25.23. The plates described in Problem 25.22 are in vacuum. An electron ($q = -e$, $m = 9.1 \times 10^{-31}$ kg) is released at the negative plate and falls freely to the positive plate. How fast is it going just before it strikes the plate?   *Ans.* 7.3×10^5 m/s

25.24. A proton ($q = e$, $m = 1.67 \times 10^{-27}$ kg) is accelerated from rest through a potential difference of 1 MV. What is its final speed?   *Ans.* 1.4×10^7 m/s

25.25. An electron gun shoots electrons ($q = -e$, $m = 9.1 \times 10^{-31}$ kg) at a metal plate that is 4 mm away in vacuum. The plate is 5.0 V lower in potential than the gun. How fast must the electrons be moving as they leave the gun if they are to reach the plate?   *Ans.* 1.33×10^6 m/s

25.26. The potential difference between two large parallel metal plates is 120 V. The plate separation is 3.0 mm. Find the electric field intensity between the plates.   *Ans.*   40 kV/m toward negative plate

25.27. An electron ($q = -e$, $m = 9.1 \times 10^{-31}$ kg) is shot with speed 5×10^6 m/s parallel to a uniform electric field of intensity 3000 V/m. How far will the electron go before it stops?   *Ans.*   2.4 cm

25.28. A potential difference of 24 kV maintains a downward-directed electric field between two horizontal parallel plates separated by 1.8 cm. Find the charge on an oil droplet of mass 2.2×10^{-13} kg that remains stationary in the field between the plates.   *Ans.* 1.6×10^{-18} C $= 10e$

25.29. Determine the absolute potential in air at a distance of 3 cm from a point charge of 5×10^{-8} C.
Ans.   15 000 V

25.30. Compute the electric intensity and absolute potential at a distance of 1 nm from a helium nucleus of charge $+2e$. What is the potential energy (relative to infinity) of a proton at this position?
Ans. 2.9×10^9 N/C, 2.9 V, 4.6×10^{-19} J

25.31. What work is required to bring a charge of 0.2 μC from a point in vacuum 30 cm from a point charge of 3 μC to a point 12 cm from it?   *Ans.*   0.027 J

25.32. A point charge, $+2$ μC, is placed at the origin of coordinates. A second, -3 μC, is placed on the x-axis at $x = 100$ cm. At what point (or points) on the x-axis will the absolute potential be zero?
Ans. $x = 40$ cm   and $x = -200$ cm

25.33. In Problem 25.32, what is the difference in potential between the following two points on the x-axis: point A at $x = 10$ cm   and point B at $x = 90$ cm? Which point is at the higher potential?
Ans. 4×10^5 V, point A

25.34. A capacitor with air between its plates has capacitance 3 μF. What is the capacitance when wax of dielectric constant 2.8 is between the plates?   *Ans.*   8.4 μF

25.35. Determine the charge on each plate of a 0.05 μF capacitor when the potential difference between the plates is 200 V.   *Ans.*   10 μC

25.36. A capacitor is charged with 9.6 nC and has a 120 V potential difference between its terminals. Compute its capacitance and the energy stored in it.   *Ans.*   80 pF, 5.76×10^{-7} J

25.37. Compute the energy stored in a 60 pF capacitor (*a*) when charged to a potential difference of 2 kV, (*b*) when the charge on each plate is 30 nC.   *Ans.*   (*a*) 1.2×10^{-4} J;   (*b*) 7.5×10^{-6} J

25.38.   Three capacitors, each of capacitance 120 pF, are charged each to 500 V and then connected in series. Determine (a) the potential difference between the end plates, (b) the charge on each capacitor, and (c) the energy stored in the system.   Ans.   (a) 1500 V;   (b) 60 nC;   (c) 4.5×10^{-5} J

25.39.   Three capacitors (2, 5, and 7 μF) are connected in series.   What is their equivalent capacitance?
Ans.   1.19 μF

25.40.   Three capacitors (2, 5, and 7 μF) are connected in parallel.   What is their equivalent capacitance?
Ans.   14 μF

25.41.   The capacitor combination in Problem 25.39 is connected in series with the combination in Problem 25.40.   What is the capacitance of this new combination?   Ans.   1.09 μF

25.42.   Two capacitors (0.30 and 0.50 μF) are connected in parallel.   (a) What is their equivalent capacitance? A charge of 200 μC is now placed on the combination capacitor.   (b) What is the potential difference across it?   (c) What are the charges on the capacitors?
Ans.   (a) 0.8 μF;   (b) 250 V;   (c) 75 μC, 125 μC

25.43.   A 2 μF capacitor is charged to 50 V and then connected in parallel (positive plate to positive plate) with a 4 μF capacitor charged to 100 V.   (a) What are the final charges on the capacitors?   (b) What is the potential difference across each?   Ans.   (a) 167 μC, 333 μC;   (b) 83 V

25.44.   Repeat Problem 25.43 if the capacitors are connected positive plate of one to negative plate of the other.
Ans.   (a) 100 μC, 200 μC;   (b) 50 V

25.45.   (a) Calculate the capacitance of a capacitor consisting of two parallel plates separated by a layer of paraffin wax 0.5 cm thick, the area of each plate being 80 cm^2.   The dielectric constant for the wax is 2. (b) If the capacitor is connected to a 100 V source, calculate the charge on the capacitor and the energy stored in the capacitor.   Ans.   (a) 28 pF;   (b) 2.8 nC, 1.4×10^{-7} J

Chapter 26

Current, Resistance, and Ohm's Law

A CURRENT (I) of electricity exists in a region whenever electric charge is being transported from one point to another in that region. Suppose that charge is moving through a wire. If a charge q is transported through a given cross section of the wire in time t, then the current through the wire is

$$I \text{ (current)} = \frac{q \text{ (charge transported)}}{t \text{ (time taken to transport this charge)}}$$

Here, q is in coulombs, t is in seconds, and I is in *amperes* (1 A = 1 C/s).

A BATTERY is a source of electrical energy. If the battery has no internal energy losses occurring in it, then the potential difference (see Chapter 25) between its terminals is called the *electromotive force* (emf) of the battery. Unless otherwise stated, it will be assumed that the terminal potential difference (t.p.d.) of a battery is equal to its emf. The unit for emf is the same as the unit for potential difference, the volt.

THE RESISTANCE of a wire or other object is a measure of the potential difference that must be impressed across the object to cause unit current to flow through it.

$$R \text{ (resistance)} = \frac{V \text{ (potential difference)}}{I \text{ (current)}}$$

The unit of resistance is the *ohm*, for which the symbol Ω (Greek omega) is used. $1\,\Omega = 1\,\text{V/A}$.

OHM'S LAW originally contained two parts. Its first part was simply the defining equation for resistance, $V = IR$. We often refer to this equation as being Ohm's law. However, Ohm also stated that R was a constant independent of V and I. This latter part of the law is only approximately correct.

The relation $V = IR$ can be applied to any resistor, where V is the potential difference (p.d.) between the two ends of the resistor, I is the current through the resistor, and R is the resistance of the resistor under those conditions.

MEASUREMENT OF RESISTANCE BY AMMETER AND VOLTMETER: The current is measured by inserting in series a (low-resistance) ammeter into the circuit. The potential difference is measured by connecting the terminals of a (high-resistance) voltmeter across the resistance being measured, i.e. in parallel. The resistance is computed by dividing the voltmeter reading by the ammeter reading, according to Ohm's law, $R = V/I$. (If an exact value of the resistance is required, the resistances of the voltmeter and ammeter must be considered parts of the circuit.)

THE TERMINAL VOLTAGE of a battery or generator when it delivers a current I is equal to the total electromotive force (emf or $\mathcal{E}$) minus the potential drop (or voltage drop) in its *internal resistance*, r.

(1) When delivering current (*on discharge*):

terminal voltage = emf − (voltage drop in internal resistance) = $\mathcal{E} - Ir$

(2) When receiving current (*on charge*):

$$\text{terminal voltage} = \text{emf} + (\text{voltage drop in internal resistance}) = \mathcal{E} + Ir$$

(3) When no current exists:

$$\text{terminal voltage} = \text{emf of battery or generator}$$

RESISTIVITY: The resistance R of a wire of length L and cross-sectional area A is

$$R = \rho \frac{L}{A}$$

where ρ is a constant called the *resistivity* and is a characteristic of the material from which the wire is made.   For L in m, A in m^2, and R in Ω, the units of ρ are $\Omega \cdot m$.

RESISTANCE VARIES WITH TEMPERATURE: If a wire has a resistance R_0 at a temperature T_0, then its resistance R at a temperature T is

$$R = R_0 + \alpha R_0 (T - T_0)$$

where α is the *temperature coefficient of resistance* of the material of the wire.   Usually α varies with temperature and so this relation is applicable only over a small temperature range.   The units of α are K^{-1} or $°C^{-1}$.

A similar relation applies to the variation of resistivity with temperature.   If ρ_0 and ρ are the resistivities at T_0 and T respectively, then

$$\rho = \rho_0 + \alpha \rho_0 (T - T_0)$$

POTENTIAL CHANGES: The potential difference across a resistor R through which a current I flows is, by Ohm's law, IR.   The end of the resistor at which the current enters is the high-potential end of the resistor.   Current always flows "downhill," from high to low potential, through a resistor.

The positive terminal of a battery is always the high-potential terminal if internal resistance of the battery is negligible or small.   This is true irrespective of the direction of the current through the battery.

Solved Problems

26.1.   A steady current of 0.5 A flows through a wire.   How much charge passes through the wire in one minute?

Because $I = q/t$,   we have $q = It = (0.5 \text{ A})(60 \text{ s}) = 30 \text{ C}$.   (Recall that $1 \text{ A} = 1 \text{ C/s}$.)

26.2.   How many electrons flow through a light bulb each second if the current through the light bulb is 0.75 A?

From $I = q/t$,

$$q = \text{charge through bulb in 1 s} = (0.75 \text{ A})(1 \text{ s}) = 0.75 \text{ C}$$

But the magnitude of the charge on each electron is $e = 1.6 \times 10^{-19} \text{ C}$.   Therefore

$$\text{number} = \frac{\text{charge}}{\text{charge/electron}} = \frac{0.75 \text{ C}}{1.6 \times 10^{-19} \text{ C}} = 4.7 \times 10^{18}$$

26.3.   A certain light bulb has a resistance of 240 Ω when lighted.   How much current will flow through it when connected across 120 V, its normal operating voltage?

$$I = \frac{V}{R} = \frac{120 \text{ V}}{240 \text{ }\Omega} = 0.50 \text{ A}$$

26.4. An electric heater uses 5 A when connected across 110 V. Determine its resistance.

$$R = \frac{V}{I} = \frac{110 \text{ V}}{5 \text{ A}} = 22 \ \Omega$$

26.5. What is the potential drop across an electric hot plate which draws 5 A when its hot resistance is 24 Ω?

$$V = IR = (5 \text{ A})(24 \ \Omega) = 120 \text{ V}$$

26.6. In the Bohr model, the electron of a hydrogen atom moves in a circular orbit of radius 5.3×10^{-11} m with a speed of 2.2×10^6 m/s. Determine its frequency f and the current I in the orbit.

$$f = \frac{v}{2\pi r} = \frac{2.2 \times 10^6 \text{ m/s}}{2\pi(5.3 \times 10^{-11} \text{ m})} = 6.6 \times 10^{15} \text{ rev/s}$$

Each time the electron goes around the orbit, it carries a charge e around the loop. The charge passing a point on the loop each second is

$$\text{current} = I = ef = (1.6 \times 10^{-19} \text{ C})(6.6 \times 10^{15} \text{ s}^{-1}) = 1.06 \times 10^{-3} \text{ A}$$

26.7. A current of 3 A flows through the wire shown in Fig. 26-1. What would a voltmeter read when connected from (a) A to B, (b) A to C, (c) A to D?

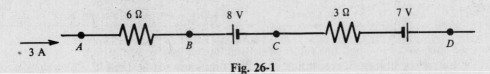

Fig. 26-1

(a) Point A is at the higher potential because current always flows "downhill" through a resistor. There is a potential drop of $IR = (3 \text{ A})(6 \ \Omega) = 18 \text{ V}$ as one goes from A to B. The voltmeter will read -18 V.

(b) In going from B to C, one goes from the positive to the negative side of the battery; hence there is a potential drop of 8 V in going from B to C. This drop adds to the drop of 18 V from A to B, found in (a), to give a 26 V drop from A to C. The voltmeter will read -26 V from A to C.

(c) In going from C to D, there is first a drop of $IR = (3 \text{ A})(3 \ \Omega) = 9 \text{ V}$ in going through the resistor. Then, because one goes from the negative to the positive terminal of the 7 V battery, there is a 7 V rise in going through the battery. The voltmeter connected from A to D will read

$$-18 \text{ V} - 8 \text{ V} - 9 \text{ V} + 7 \text{ V} = -28 \text{ V}$$

26.8. What is the potential difference between points A and D in Fig. 26-2? Which point is at the higher potential?

Notice first that the current is the same in each part of this series circuit. Recall also that the current goes through a resistor from high to low potential.

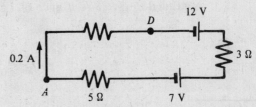

Start at A and proceed counterclockwise. The potential changes are as follows from A to D:

Fig. 26-2

$$+ (5 \ \Omega)(0.2 \text{ A}) + 7 \text{ V} + (3 \ \Omega)(0.2 \text{ A}) - 12 \text{ V} = -3.4 \text{ V}$$

Therefore, point D is 3.4 V lower in potential than point A.

Check: Proceeding clockwise from D gives

$$+12 - (3)(0.2) - 7 - (5)(0.2) = 3.4 \text{ V}$$

and so A is 3.4 V higher than D.

26.9. A dry cell has an emf of 1.52 V. Its terminal potential drops to zero when a current of 25 A passes through it. What is its internal resistance?

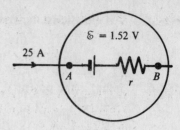

As is shown in Fig. 26-3, the battery acts like a pure emf, $\mathcal{E}$, in series with a resistor, r. We are told that, under the conditions shown, the potential difference from A to B is zero. Therefore

$$0 = +\mathcal{E} - Ir \qquad \text{or} \qquad 0 = 1.52\ \text{V} - (25\ \text{A})r$$

from which the internal resistance is $r = 0.061\ \Omega$.

Fig. 26-3

26.10. A direct-current generator has an emf of 120 V, i.e. its terminal voltage is 120 V when no current is flowing from it. At an output of 20 A the terminal potential is 115 V. (a) What is the internal resistance r of the generator? (b) What will be the terminal voltage at an output of 40 A?

The situation is much like that shown in Fig. 26-3. Now, however, $\mathcal{E} = 120\ \text{V}$ and I is no longer 25 A.

(a) In this case, $I = 20\ \text{A}$ and the p.d. from A to B is 115 V. Therefore

$$115\ \text{V} = +120\ \text{V} - (20\ \text{A})r$$

from which $r = 0.25\ \Omega$.

(b) Now $I = 40\ \text{A}$.

$$\text{terminal p.d.} = \mathcal{E} - Ir = 120\ \text{V} - (40\ \text{A})(0.25\ \Omega) = 110\ \text{V}$$

26.11. As shown in Fig. 26-4, the ammeter-voltmeter method is used to measure an unknown resistance R. The ammeter reads 0.3 A and the voltmeter reads 1.50 V. Compute the value of R if the ammeter and voltmeter are ideal.

$$R = \frac{V}{I} = \frac{1.50\ \text{V}}{0.3\ \text{A}} = 5.0\ \Omega$$

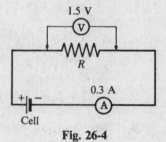

26.12. A metal rod is 2 m long and 8 mm in diameter. Compute its resistance if the resistivity of the metal is $1.76 \times 10^{-8}\ \Omega \cdot \text{m}$.

$$R = \rho \frac{L}{A} = (1.76 \times 10^{-8}\ \Omega \cdot \text{m}) \frac{2\ \text{m}}{\pi (4 \times 10^{-3}\ \text{m})^2} = 7 \times 10^{-4}\ \Omega$$

Fig. 26-4

26.13. Number 10 wire has a diameter of 2.59 mm. How many meters of number 10 aluminum wire is needed to give a resistance of 1 Ω? ρ for aluminum is $2.8 \times 10^{-8}\ \Omega \cdot \text{m}$.

From $R = \rho L / A$, we have

$$L = \frac{RA}{\rho} = \frac{(1\ \Omega)(\pi)(2.59 \times 10^{-3}\ \text{m})^2 / 4}{2.8 \times 10^{-8}\ \Omega \cdot \text{m}} = 188\ \text{m}$$

26.14. Number 24 copper wire has diameter 0.0201 in. Compute (a) the cross-sectional area of the wire in circular mils and (b) the resistance of 100 ft of the wire. The resistivity of copper is $10.4\ \Omega \cdot \text{circular mils/ft}$.

The area of a circle in circular mils is defined as the square of the diameter of the circle expressed in mils, where 1 mil = 0.001 in.

(a)   area in circular mils $= (20.1\ \text{mil})^2 = 404$ circular mils

(b) $R = \rho \dfrac{L}{A} = (10.4\ \Omega \cdot \text{circular mil/ft}) \dfrac{100\ \text{ft}}{404\ \text{circular mils}} = 2.57\ \Omega$

26.15.   The resistance of a coil of copper wire is 3.35 Ω at 0 °C. What is its resistance at 50 °C? For copper, $\alpha = 4.3 \times 10^{-3}$ °C^{-1}.

$$R = R_0 + \alpha R_0(T - T_0) = 3.35 \ \Omega + (4.3 \times 10^{-3} \ °C^{-1})(3.35 \ \Omega)(50 \ °C) = 4.07 \ \Omega$$

26.16.   A resistor is to have a constant resistance of 30 Ω, independent of temperature. For this, an aluminum resistor with resistance R_{01} at 0 °C is used in series with a carbon resistor with resistance R_{02} at 0 °C. Evaluate R_{01} and R_{02}, given that $\alpha_1 = 3.9 \times 10^{-3}$ °C^{-1} for aluminum and $\alpha_2 = -0.5 \times 10^{-3}$ °C^{-1} for carbon.

The combined resistance at temperature T will be

$$R_1 + R_2 = [R_{01} + \alpha_1 R_{01}(T - T_0)] + [R_{02} + \alpha_2 R_{02}(T - T_0)]$$
$$= (R_{01} + R_{02}) + (\alpha_1 R_{01} + \alpha_2 R_{02})(T - T_0)$$

We thus have the two conditions

$$R_{01} + R_{02} = 30 \ \Omega \qquad \text{and} \qquad \alpha_1 R_{01} + \alpha_2 R_{02} = 0$$

Substituting the given values of α_1 and α_2, and then solving for R_{01} and R_{02}, we find

$$R_{01} = 3.41 \ \Omega \qquad R_{02} = 26.6 \ \Omega$$

Supplementary Problems

26.17.   How many electrons per second pass through a section of wire carrying a current of 0.7 A?
Ans. 4.4×10^{18} electrons/s

26.18.   An electron gun in a TV set shoots out a beam of electrons. The beam current is 1.0×10^{-5} A. How many electrons strike the TV screen each second? How much charge strikes the screen in a minute?
Ans. 6.3×10^{13} electrons/s, -6×10^{-4} C/min

26.19.   What is the current through an 8 Ω toaster when it is operating on 120 V?   *Ans.*   15 A

26.20.   What potential difference is required to pass 3 A through 28 Ω?   *Ans.*   84 V

26.21.   Determine the potential difference between the ends of a wire of resistance 5 Ω if 720 C passes through it per minute.   *Ans.*   60 V

26.22.   A copper bus bar carrying 1200 A has a potential drop of 1.2 mV along 24 cm of its length. What is the resistance per m of the bar?   *Ans.*   4.2 $\mu\Omega$/m

26.23.   An ammeter is connected in series with an unknown resistance, and a voltmeter is connected across the terminals of the resistance. If the ammeter reads 1.2 A and the voltmeter reads 18 V, compute the value of the resistance. Assume ideal meters.   *Ans.*   15 Ω

26.24.   An electric utility company runs two 100 ft copper wires from the street mains up to a customer's premises. If the wire resistance is 0.100 ohm per 1000 ft, calculate the line voltage drop for an estimated load current of 120 A.   *Ans.*   2.4 V

26.25.   When testing the insulation resistance between a motor winding and the motor frame, the value obtained is one megohm ($10^6 \ \Omega$). How much current passes through the insulation of the motor if the test voltage is 1000 V?   *Ans.*   1 mA

26.26.   Compute the internal resistance of an electric generator which has an emf of 120 V and a terminal voltage of 110 V when supplying 20 A.   *Ans.*   0.5 Ω

26.27.   A dry cell delivering 2 A has terminal voltage 1.41 V. What is the internal resistance of the cell if its open-circuit voltage is 1.59 V?   *Ans.*   0.09 Ω

26.28.   A cell has emf 1.54 V. When it is in series with a 1 Ω resistance, the reading of a voltmeter connected across the cell terminals is 1.40 V. Determine the cell's internal resistance.   *Ans.*   0.10 Ω

26.29.   The internal resistance of a 6.4 V storage battery is 4.8 mΩ. What is the theoretical maximum current on short circuit? (In practice the leads and connections have some resistance and this value would not be attained.)   *Ans.*   1300 A

26.30.   A battery has emf 13.2 V and internal resistance 24 mΩ. If the load current is 20.0 A, find the terminal voltage.   *Ans.*   12.7 V

26.31.   What is the maximum allowable discharge current of a 55-cell storage battery having emf 121 V and total internal resistance 0.100 Ω if the terminal voltage must not fall below 110 V?   *Ans.*   110 A

26.32.   A storage battery has emf 25 V and internal resistance 0.20 Ω. Compute its terminal voltage (*a*) when it is delivering 8 A, (*b*) when it is being charged with 8 A.   *Ans.*   (*a*) 23.4 V;   (*b*) 26.6 V

26.33.   A battery charger supplies a current of 10 A to charge a storage battery which has an open-circuit voltage of 5.6 V. If the voltmeter connected across the charger reads 6.8 V, what is the internal resistance of the battery at this time?   *Ans.*   0.12 Ω

26.34.   Find the potential difference between points *A* and *B* in Fig. 26-5 if *R* is 0.7 Ω. Which point is at the higher potential?   *Ans.*   −5.1 V, point *A*

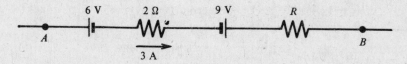

Fig. 26-5

26.35.   Repeat Problem 26.34 if the current flows in the opposite direction and *R* = 0.7 Ω.
          Ans.   11.1 V, point *B*

26.36.   In Fig. 26-5, how large must *R* be if the potential drop from *A* to *B* is 12 V?   *Ans.*   3 Ω

26.37.   Compute the resistance of 180 m of silver wire having a cross section of 0.3 mm². The resistivity of silver is 1.6×10^{-8} Ω·m.   *Ans.*   9.6 Ω

26.38.   The resistivity of aluminum is 2.8×10^{-8} Ω·m. How long a piece of aluminum wire 1 mm in diameter is needed to give a resistance of 4 Ω?   *Ans.*   112 m

26.39.   Number 6 copper wire has a diameter of 0.162 in. (*a*) Calculate its area in circular mils. (*b*) If ρ = 10.4 Ω·circular mils/ft, find the resistance of 1000 ft of the wire.
          Ans.   (*a*) 26 000 circular mils;   (*b*) 0.40 Ω

26.40.   A coil of wire has a resistance of 25.00 Ω at 20 °C and a resistance of 25.17 Ω at 35 °C. What is its temperature coefficient of resistance?   *Ans.* 4.5×10^{-4} °C^{-1}

Chapter 27

Electrical Power

THE ELECTRICAL WORK (in joules) required to transfer a charge q (in coulombs) through a potential difference V (in volts) is given by

$$\text{work} = qV$$

As discussed in Chapter 25, when q and V are given their proper signs, the work will have its proper sign. Thus, to carry a positive charge through a potential rise, a positive amount of work must be done on the charge.

THE ELECTRICAL POWER (in watts) delivered by an energy source as it carries a charge q (in coulombs) through a potential rise V (in volts) in a time t (in seconds) is

$$\text{power furnished} = \frac{\text{work}}{\text{time}} = \frac{Vq}{t}$$

Because $q/t = I$, this can be rewritten as

$$\text{power furnished} = VI$$

where I is in amperes.

THE POWER LOSS IN A RESISTOR is found by replacing V in VI by IR, or by replacing I in VI by V/R, to give

$$\text{power loss in resistor} = VI = I^2R = \frac{V^2}{R}$$

THE HEAT GENERATED per second in a resistor is equal to the power loss in the resistor.

$$\text{heat (in joules) generated per second} = VI = I^2R \quad = watts$$

Because $1\ \text{J} = 0.239\ \text{cal}$,

$$\text{heat (in calories) generated per second} = 0.239\ VI = 0.239\ I^2R$$

CONVENIENT CONVERSIONS

$1\ \text{W} = 1\ \text{J/s} = 0.239\ \text{cal/s} = 0.738\ \text{ft} \cdot \text{lb/s}$
$1\ \text{kW} = 1.341\ \text{hp} = 56.9\ \text{Btu/min}$
$1\ \text{hp} = 746\ \text{W} = 33\,000\ \text{ft} \cdot \text{lb/min} = 42.4\ \text{Btu/min}$
$1\ \text{kW} \cdot \text{h} = 3.6 \times 10^6\ \text{J} = 3.6\ \text{MJ}$

Solved Problems

27.1. Compute the work and the average power required to transfer 96 000 C of charge in one hour through a potential rise of 50 V.

$$\text{work} = qV = (96\,000\ \text{C})(50\ \text{V}) = 4.8 \times 10^6\ \text{J}$$

$$\text{power} = \frac{\text{work}}{\text{time}} = \frac{4.8 \times 10^6\ \text{J}}{3600\ \text{s}} = 1.33\ \text{kW}$$

27.2. How much current does a 60 W light bulb draw when connected to its proper voltage, 120 V?

From $P = VI$,

$$I = \frac{P}{V} = \frac{60\ \text{W}}{120\ \text{V}} = 0.50\ \text{A}$$

27.3. An electric motor takes 5 A from a 110 V line. Determine the power input and the energy, in J and kW·h, supplied to the motor in 2 h.

$$\text{power} = P = VI = (110\ \text{V})(5\ \text{A}) = 550\ \text{W} = 0.55\ \text{kW}$$

$$\text{energy} \quad = Pt = (550\ \text{W})(7200\ \text{s}) = 3.96 \times 10^6\ \text{J}$$

$$= (0.55\ \text{kW})(2\ \text{hr}) = 1.10\ \text{kW·h}$$

27.4. An electric iron of resistance 20 Ω takes a current of 5 A. Calculate the heat, in joules and in calories, developed in 30 seconds.

$$\text{heat in joules} = I^2Rt = (5\ \text{A})^2(20\ \Omega)(30\ \text{s}) = 15 \times 10^3\ \text{J}$$

$$\text{heat in calories} = (0.239)(I^2Rt) = (0.239\ \text{cal/J})(15 \times 10^3\ \text{J}) = 3.6 \times 10^3\ \text{cal}$$

27.5. An electric heater of resistance 8 Ω draws 15 A from the service mains. At what rate is heat developed, in W and in cal/s? What is the cost of operating the heater for a period of 4 h at 10¢/kW·h?

$$\text{heat rate in W} = I^2R = (15\ \text{A})^2(8\ \Omega) = 1800\ \text{W} = 1.8\ \text{kW}$$

$$\text{heat rate in cal/s} = (0.239)I^2R = (0.239\ \text{cal/J})(1800\ \text{J/s}) = 430\ \text{cal/s}$$

$$\text{cost} = (1.8\ \text{kW})(4\ \text{h})(10¢/\text{kW·h}) = 72¢$$

27.6. A coil develops 800 cal/s when 20 V is supplied across its ends. Compute its resistance.

$$800\ \text{cal/s} = (0.239\ \text{cal/J})\frac{V^2}{R} = (0.239\ \text{cal/J})\frac{(20\ \text{V})^2}{R}$$

Solving, $R = 0.12\ \Omega$.

27.7. A line having a total resistance of 0.2 Ω delivers 10 kW at 250 V to a small factory. What is the efficiency of the transmission?

$$\text{power lost in line} = I^2R = \left(\frac{IV}{V}\right)^2 R = \left(\frac{10\,000\ \text{W}}{250\ \text{V}}\right)^2(0.2\ \Omega) = 320\ \text{W}$$

$$\text{efficiency} = \frac{\text{power delivered by line}}{\text{power supplied to line}} = \frac{10\ \text{kW}}{(10 + 0.32)\ \text{kW}} = 0.97 = 97\%$$

27.8. A hoist motor supplied by 240 V requires 12 A to lift a ton weight at the rate of 25 ft/min. Determine the power input to the motor and the power output, both in horsepower, and the overall efficiency of the system.

$$\text{power input} = IV = (12 \text{ A})(240 \text{ V}) = 2880 \text{ W} = (2.880 \text{ kW})(1.34 \text{ hp/kW}) = 3.86 \text{ hp}$$

$$\text{power output} = Fv = (2000 \text{ lb})(25 \text{ ft/min})\left(\frac{1 \text{ hp}}{33\,000 \text{ ft} \cdot \text{lb/min}}\right) = 1.52 \text{ hp}$$

$$\text{overall efficiency} = \frac{1.52 \text{ hp output}}{3.86 \text{ hp input}} = 0.394 = 39.4\%$$

27.9. What is the cost of electrically heating 50 liters of water from 40 °C to 100 °C at 8¢/kW · h?

$$\text{heat gained by water} = (\text{mass}) \times (\text{specific heat}) \times (\text{temperature rise})$$

$$= (50 \times 10^3 \text{ g}) \times (1 \text{ cal/g} \cdot °C) \times (60 °C)$$

$$= 3 \times 10^6 \text{ cal}$$

$$\text{cost} = (3 \times 10^6 \text{ cal})\left(\frac{4.184 \text{ J}}{1 \text{ cal}}\right)\left(\frac{1 \text{ kW} \cdot \text{h}}{3.6 \times 10^6 \text{ J}}\right)\left(\frac{8¢}{1 \text{ kW} \cdot \text{h}}\right) = 28¢$$

Supplementary Problems

27.10. A heater is labeled 1600 W/120 V. How much current does the heater draw from a 120 V source?
Ans. 13.3 A

27.11. A bulb is stamped 40 W/120 V. What is its resistance when lighted by a 120 V source?
Ans. 360 Ω

27.12. A spark of man-made 10 MV lightning had an energy output of 0.125 MW · s. How many coulombs of charge flowed? *Ans.* 0.0125 C

27.13. A current of 1.5 A exists in a conductor whose terminals are connected across a potential difference of 100 V. Compute the total charge transferred in one minute, the work done in transferring this charge, and the power expended in heating the conductor if all the electrical energy is converted into heat. *Ans.* 90 C, 9000 J, 150 W

27.14. An electric motor takes 15 A at 110 V. Determine (*a*) the power input, (*b*) the cost of operating the motor for 8 h at 10¢/kW · h. *Ans.* (*a*) 1.65 kW; (*b*) $1.32

27.15. A current of 10 A exists in a line of 0.15 Ω resistance. Compute the rate of production of heat in watts. *Ans.* 15 W

27.16. An electric broiler develops 400 cal/s when the current through it is 8 A. Determine the resistance of the broiler. *Ans.* 26 Ω

27.17. A 25 W, 120 V bulb has a cold resistance of 45 Ω and a hot resistance of 575 Ω. When the voltage is switched on, what is the instantaneous current? What is the current under normal operation?
Ans. 2.67 A, 0.209 A

27.18. At a rated current of 400 A, a defective switch becomes overheated due to faulty surface contact. A millivoltmeter connected across the switch shows a 100 mV drop. What is the power loss due to the contact resistance? *Ans.* 40 W

27.19. A 10 Ω electric heater operates on a 110 V line. Compute the rate at which heat is developed, in W and in cal/s. *Ans.* 1210 W = 290 cal/s

27.20. An electric motor, which has 95% efficiency, uses 20 A at 110 V. What is the horsepower output of the motor? How many watts are lost in heat? How many calories of heat are developed per second? If the motor operates for 3 h, what energy, in MJ and in kW · h, is consumed?
Ans. 2.8 hp, 110 W, 26.4 cal/s, 23.8 MJ = 6.6 kW · h

27.21. An electric crane uses 8 A at 150 V to raise 1000 lb at the rate of 22 ft/min. Determine the efficiency of the system. *Ans.* 41.4%

27.22. What should be the resistance of a heating coil which will be used to raise the temperature of 500 g of water from 28 °C to the boiling point in 2 minutes, assuming that 25% of the heat is lost? The heater operates on a 110 V line. *Ans.* 7.23 Ω

27.23. Compute the cost per hour at 8¢/kW · h of electrically heating a room, if it requires 1 kg/h of anthracite coal having a heat of combustion of 8000 kcal/kg. *Ans.* 74¢/h

27.24. Power is transmitted at 80 kV between two stations. If the voltage can be increased to 160 kV without change in cable size, how much additional power can be transmitted for the same current? What effect does the power increase have on the line heating loss?
Ans. additional power = original power, no effect

27.25. A storage battery, of emf 6.4 V and internal resistance 0.08 Ω, is being charged by a current of 15 A. Calculate (*a*) the power loss in internal heating of the battery, (*b*) the rate at which energy is stored in the battery, (*c*) its terminal voltage. *Ans.* (*a*) 18 W; (*b*) 96 W; (*c*) 7.6 V

27.26. A tank containing 200 liters of water was used as a constant-temperature bath. How long would it take to heat the bath from 20 °C to 25 °C with a 250 W immersion heater? Neglect the heat capacity of the tank frame and any heat losses to the air. *Ans.* 4.6 h

Chapter 28

Equivalent Resistance; Simple Circuits

RESISTORS IN SERIES: When the current can follow only one path as it flows through two or more resistors connected in line, the resistors are *in series*. A typical case is seen in Fig. 28-1(*a*). For several resistors in series, their equivalent resistance, R_{eq}, is given by

$$R_{eq} = R_1 + R_2 + R_3 + \cdots \qquad \text{(series)}$$

where $R_1, R_2, R_3, \ldots$, are the resistances of the several resistors. Observe that resistances in series combine like capacitances in parallel (see Chapter 25).

In a series combination, the current through each resistance is the same. The potential drop (p.d.) across the combination is equal to the sum of the individual potential drops.

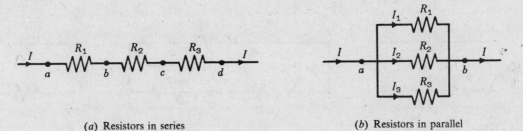

(*a*) Resistors in series   (*b*) Resistors in parallel

Fig. 28-1

RESISTORS IN PARALLEL: Several resistors are connected *in parallel* between A and B if one end of each is connected by low-resistance wires to point A and the other end of each is connected to point B. A typical case is shown in Fig. 28-1(*b*). Their equivalent resistance, R_{eq}, is given by

$$\frac{1}{R_{eq}} = \frac{1}{R_1} + \frac{1}{R_2} + \frac{1}{R_3} + \cdots \qquad \text{(parallel)}$$

The equivalent resistance is always less than the smallest of the individual resistances. Connecting additional resistances in parallel decreases R_{eq} for the combination. Observe that resistances in parallel combine like capacitances in series (see Chapter 25).

The potential drop V across one resistor in a parallel combination is the same as the potential drop across each of the others. The current through the nth resistor is $I_n = V/R_n$ and the total current entering the combination is equal to the sum of these individual branch currents.

Solved Problems

28.1. Resistors R_1, R_2, and R_3 are (*a*) in series and (*b*) in parallel, as shown in Fig. 28-1(*a*) and (*b*). Derive the formula for the equivalent resistance R_{eq} of each network.

(*a*) For the series network,

$$V_{ad} = V_{ab} + V_{bc} + V_{cd} = IR_1 + IR_2 + IR_3$$

since the current I is the same in each resistor. Dividing by I,

$$\frac{V_{ad}}{I} = R_1 + R_2 + R_3 \qquad \text{or} \qquad R_{eq} = R_1 + R_2 + R_3$$

since V_{ad}/I is by definition the equivalent resistance R_{eq} of the network.

(b) The p.d. across each resistor is the same, whence

$$I_1 = \frac{V_{ab}}{R_1} \qquad I_2 = \frac{V_{ab}}{R_2} \qquad I_3 = \frac{V_{ab}}{R_3}$$

Since the line current I is the sum of the branch currents,

$$I = I_1 + I_2 + I_3 = \frac{V_{ab}}{R_1} + \frac{V_{ab}}{R_2} + \frac{V_{ab}}{R_3}$$

Dividing by V_{ab},

$$\frac{I}{V_{ab}} = \frac{1}{R_1} + \frac{1}{R_2} + \frac{1}{R_3} \qquad \text{or} \qquad \frac{1}{R_{eq}} = \frac{1}{R_1} + \frac{1}{R_2} + \frac{1}{R_3}$$

since V_{ab}/I is by definition the equivalent resistance R_{eq} of the network.

28.2. As shown in Fig. 28-2(a), a battery (internal resistance 1 Ω) is connected in series with two resistors. Compute (a) the current in the circuit, (b) the p.d. across each resistor, (c) the terminal p.d. of the battery.

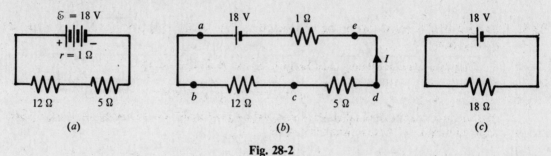

Fig. 28-2

The circuit is redrawn in Fig. 28-2(b) so as to show the battery resistance. We have
$$R_{eq} = 5 \ \Omega + 12 \ \Omega + 1 \ \Omega = 18 \ \Omega$$

Hence the circuit is equivalent to the one shown in Fig. 28-2(c). Applying $V = IR$ to it, we have:

(a)
$$I = \frac{V}{R} = \frac{18 \ \text{V}}{18 \ \Omega} = 1.00 \ \text{A}$$

(b) Since $I = 1$ A, we can find the p.d. from b to c by
$$V_{bc} = IR_{bc} = (1 \ \text{A})(12 \ \Omega) = 12 \ \text{V}$$

and from c to d by

$$V_{cd} = IR_{cd} = (1 \ \text{A})(5 \ \Omega) = 5 \ \text{V}$$

Notice that I is the same at all points in a series circuit.

(c) The terminal p.d. of the battery is the p.d. from a to e. Therefore
$$\text{terminal p.d.} = V_{bc} + V_{cd} = 12 + 5 = 17 \ \text{V}$$

Or, we could start at e and keep track of the voltage changes as we go through the battery from e to a. Taking voltage drops as negative, we have

$$\text{terminal p.d.} = -Ir + \mathcal{E} = -(1 \ \text{A})(1 \ \Omega) + 18 \ \text{V} = 17 \ \text{V}$$

28.3. A 120 V house circuit has the following light bulbs turned on: 40 W, 60 W, and 75 W. Find the equivalent resistance of these lights.

House circuits are so constructed that each device is connected in parallel with the others. From $P = VI = V^2/R$, we have for the first bulb

$$R_1 = \frac{V^2}{P_1} = \frac{(120)^2}{40} = 360 \ \Omega$$

Similarly, $R_2 = 240\ \Omega$ and $R_3 = 192\ \Omega$.   Because they are in parallel,

$$\frac{1}{R_{eq}} = \frac{1}{360\ \Omega} + \frac{1}{240\ \Omega} + \frac{1}{192\ \Omega} \qquad \text{or} \qquad R_{eq} = 82\ \Omega$$

As a check, we note that the total power drawn from the line is $40 + 60 + 75 = 175$ W.   Then, using $P = V^2/R$,

$$R_{eq} = \frac{V^2}{(\text{total power})} = \frac{(120)^2}{175} = 82\ \Omega$$

28.4. What resistance must be placed in parallel with $12\ \Omega$ to obtain a combined resistance of $4\ \Omega$?

$$\frac{1}{R_{eq}} = \frac{1}{R_1} + \frac{1}{R_2}$$

$$\frac{1}{4\ \Omega} = \frac{1}{12\ \Omega} + \frac{1}{R_2}$$

$$R_2 = 6\ \Omega$$

28.5. Several $40\ \Omega$ resistors are to be connected so that 15 A flows from a 120 V source.   How can this be done?

The equivalent resistance must be such that 15 A flows from 120 V.

$$R_{eq} = \frac{V}{I} = \frac{120\ \text{V}}{15\ \text{A}} = 8\ \Omega$$

The resistors must be in parallel, since the combined resistance is to be smaller than any of them.   If the required number of $40\ \Omega$ resistors is n, then

$$\frac{1}{8\ \Omega} = n\left(\frac{1}{40\ \Omega}\right) \qquad \text{or} \qquad n = 5$$

28.6. For each circuit shown in Fig. 28-3, determine the current I through the battery.

(a) The 3 and $7\ \Omega$ are in parallel; their joint resistance, R_1, is found from

$$\frac{1}{R_1} = \frac{1}{3} + \frac{1}{7} = \frac{10}{21} \qquad \text{or} \qquad R_1 = 2.1\ \Omega$$

Then the equivalent resistance of the entire circuit is

$$R_{eq} = 2.1 + 5 + 0.4 = 7.5\ \Omega$$

and the battery current is

$$I = \frac{\mathcal{E}}{R_{eq}} = \frac{30}{7.5} = 4\ \text{A}$$

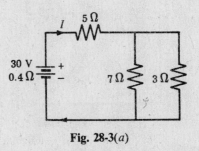

Fig. 28-3(a)

(b) The 7, 1 and $10\ \Omega$ are in series; their joint resistance is $18\ \Omega$.   Then $18\ \Omega$ is in parallel with $6\ \Omega$; their combined resistance, R_1, is given by

$$\frac{1}{R_1} = \frac{1}{18} + \frac{1}{6} \qquad \text{or} \qquad R_1 = 4.5\ \Omega$$

Hence, the equivalent resistance of the entire circuit is

$$R_{eq} = 4.5 + 2 + 8 + 0.3 = 14.8\ \Omega$$

and the battery current is

$$I = \frac{\mathcal{E}}{R_{eq}} = \frac{20}{14.8} = 1.35\ \text{A}$$

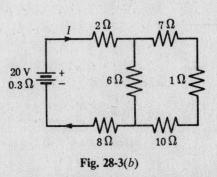

Fig. 28-3(b)

(c)   The 5 and 19 Ω are in series; their joint resistance is 24 Ω.   Then 24 Ω is in parallel with 8 Ω; their joint resistance, R_1, is given by

$$\frac{1}{R_1} = \frac{1}{24} + \frac{1}{8} \qquad \text{or} \qquad R_1 = 6 \ \Omega$$

Now $R_1 = 6 \ \Omega$ is in series with 15 Ω; their joint resistance is $6 + 15 = 21 \ \Omega$.   Thus 21 Ω is in parallel with 9 Ω; their combined resistance is found from

$$\frac{1}{R_2} = \frac{1}{21} + \frac{1}{9} \qquad \text{or} \qquad R_2 = 6.3 \ \Omega$$

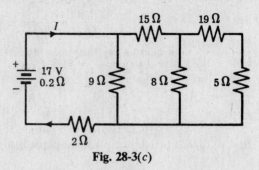

Fig. 28-3(c)

Hence the equivalent resistance of the entire circuit is

$$R_{eq} = 6.3 + 2 + 0.2 = 8.5 \ \Omega$$

and the battery current is

$$I = \frac{\mathcal{E}}{R_{eq}} = \frac{17}{8.5} = 2 \ \text{A}$$

28.7.   For the situation shown in Fig. 28-4, find the current in each resistor and the current drawn from the battery.

Notice that the p.d. from a to b is 40 V.   Therefore, the p.d. across each resistor is 40 V.   Then,

$$I_2 = \frac{40 \ \text{V}}{2 \ \Omega} = 20 \ \text{A} \qquad I_5 = \frac{40 \ \text{V}}{5 \ \Omega} = 8 \ \text{A} \qquad I_8 = \frac{40 \ \text{V}}{8 \ \Omega} = 5 \ \text{A}$$

Because I splits into the three currents,

$$I = I_2 + I_5 + I_8 = 20 + 8 + 5 = 33 \ \text{A}$$

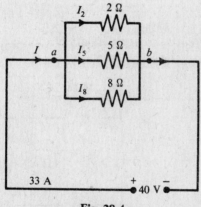

Fig. 28-4

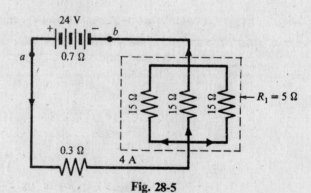

Fig. 28-5

28.8.   In Fig. 28-5, the battery has an internal resistance of 0.7 Ω.   Find (a) the current drawn from the battery, (b) the current in each 15 Ω resistor, (c) the terminal voltage of the battery.

(a)   Parallel group resistance R_1:

$$\frac{1}{R_1} = \frac{1}{15} + \frac{1}{15} + \frac{1}{15} = \frac{3}{15} \qquad \text{or} \qquad R_1 = 5 \ \Omega$$

$$R_{eq} = 5 + 0.3 + 0.7 = 6 \ \Omega$$

$$I = \frac{\mathcal{E}}{R_{eq}} = \frac{24 \ \text{V}}{6 \ \Omega} = 4 \ \text{A}$$

(b)   **Method 1**

The three-resistor combination is equivalent to $R_1 = 5 \ \Omega$.   A current of 4 A flows through it.   Hence, the p.d. across the combination is

$$IR_1 = (4 \ \text{A})(5 \ \Omega) = 20 \ \text{V}$$

This is also the p.d. across each 15 Ω resistor.   Therefore, the current through each 15 Ω resistor is

$$I_{15} = \frac{V}{R} = \frac{20\,\text{V}}{15\,\Omega} = 1.33\,\text{A}$$

Method 2

In this special case, we know that one-third of the current will go through each 15 Ω resistor. Hence

$$I_{15} = \frac{4\,\text{A}}{3} = 1.33\,\text{A}$$

(c)   Start at a and go to b outside the battery:

$$V \text{ from } a \text{ to } b = -(4\,\text{A})(0.3\,\Omega) - (4\,\text{A})(R_1)$$

$$= -(4\,\text{A})(0.3\,\Omega) - (4\,\text{A})(5\,\Omega) = -21.2\,\text{V}$$

The terminal p.d. of the battery is 21.2 V.   Or, we could write for this case of a discharging battery,

$$\text{terminal p.d.} = \mathcal{E} - Ir = 24\,\text{V} - (4\,\text{A})(0.7\,\Omega) = 21.2\,\text{V}$$

28.9.   Find the equivalent resistance between points a and b for the combination shown in Fig. 28-6(a).

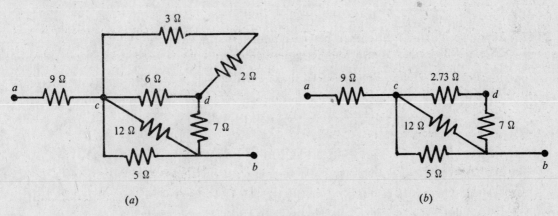

(a)   (b)

Fig. 28-6

The 3 and 2 Ω resistors are in series and are equivalent to a 5 Ω resistor.   The equivalent 5 Ω is in parallel with the 6 Ω, and their equivalent, R_1, is

$$\frac{1}{R_1} = \frac{1}{5} + \frac{1}{6} = 0.20 + 0.167 = 0.367 \qquad \text{or} \qquad R_1 = 2.73\,\Omega$$

The circuit thus far reduced is shown in Fig. 28-6(b).

The 7 and 2.73 Ω are equivalent to 9.73 Ω.   Now the 5, 12, and 9.73 Ω are in parallel and their equivalent, R_2, is

$$\frac{1}{R_2} = \frac{1}{5} + \frac{1}{12} + \frac{1}{9.73} = 0.386 \qquad \text{or} \qquad R_2 = 2.6\,\Omega$$

This 2.6 Ω is in series with the 9 Ω resistor.   Therefore, the equivalent resistance of the combination is $9 + 2.6 = 11.6\,\Omega$.

28.10.   A current of 5 A flows into the circuit of Fig. 28-6 at point a and out at point b.   (a) What is the potential difference from a to b?   (b) How much current flows through the 12 Ω resistor?

In Problem 28.9, we found that the equivalent resistance for this combination is 11.6 Ω, and the current through it is 5 A.

(a)   voltage drop from a to $b = IR_{\text{eq}} = (5\,\text{A})(11.6\,\Omega) = 58\,\text{V}$

(b) The voltage drop from a to c is   (5 A)(9 Ω) = 45 V.   Hence, from part (a), the voltage drop from c to b is

$$58 \text{ V} - 45 \text{ V} = 13 \text{ V}$$

and the current in the 12 Ω resistor is

$$I_{12} = \frac{V}{R} = \frac{13 \text{ V}}{12 \text{ Ω}} = 1.08 \text{ A}$$

28.11. As shown in Fig. 28-7, the current I divides into I_1 and I_2. Find I_1 and I_2 in terms of I, R_1, and R_2.

The potential drops across R_1 and R_2 are the same, so

$$I_1 R_1 = I_2 R_2$$

But, $I = I_1 + I_2$ and so $I_2 = I - I_1$. Substituting in the first equation gives

$$I_1 R_1 = (I - I_1)R_2 = IR_2 - I_1 R_2 \qquad \text{or} \qquad I_1 = \frac{R_2}{R_1 + R_2} I$$

Using this result together with the first equation gives

$$I_2 = \frac{R_1}{R_2} I_1 = \frac{R_1}{R_1 + R_2} I$$

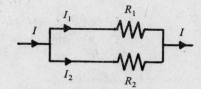

Fig. 28-7

28.12. Find the potential difference between points P and Q in Fig. 28-8.   Which point is at the higher potential?

From the result of Problem 28.11, the currents through P and Q are

$$I_P = \frac{2 \text{ Ω} + 18 \text{ Ω}}{10 \text{ Ω} + 5 \text{ Ω} + 2 \text{ Ω} + 18 \text{ Ω}} (7 \text{ A}) = 4 \text{ A}$$

$$I_Q = \frac{10 \text{ Ω} + 5 \text{ Ω}}{10 \text{ Ω} + 5 \text{ Ω} + 2 \text{ Ω} + 18 \text{ Ω}} (7 \text{ A}) = 3 \text{ A}$$

Start at point P and go through point a to point Q:

voltage change from P to Q = +(4 A)(10 Ω) − (3 A)(2 Ω) = +34 V

(Notice that we go through a potential rise from P to a because we are going against the current.   From a to Q there is a drop.)   Therefore, the voltage difference between P and Q is 34 V, with Q being at the higher potential.

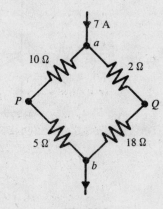

Fig. 28-8

28.13. For the circuit of Fig. 28-9(a), find (a) I_1, I_2, and I_3; (b) the current in the 12 Ω resistor.

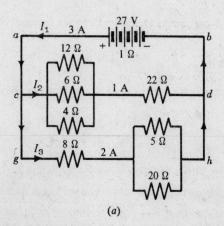

(a)

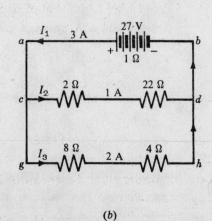

(b)

Fig. 28-9

(a) The circuit reduces at once to that shown in Fig. 28-9(b). There we have 24 Ω in parallel with 12 Ω, so the equivalent resistance below points a and b is

$$\frac{1}{R_{ab}} = \frac{1}{24} + \frac{1}{12} = \frac{3}{24} \qquad \text{or} \qquad R_{ab} = 8\ \Omega$$

Adding to this the 1 Ω internal resistance of the battery gives a total equivalent resistance of 9 Ω.
To find the current from the battery,

$$I_1 = \frac{\mathcal{E}}{R_{eq}} = \frac{27\ \text{V}}{9\ \Omega} = 3\ \text{A}$$

This same current flows through the equivalent resistance below a and b, and so

$$\text{p.d. from } a \text{ to } b = \text{p.d. from } c \text{ to } d = I_1 R_{ab} = (3\ \text{A})(8\ \Omega) = 24\ \text{V}$$

Applying $V = IR$ to branch cd gives

$$I_2 = \frac{V_{cd}}{R_{cd}} = \frac{24\ \text{V}}{24\ \Omega} = 1.0\ \text{A}$$

Similarly,

$$I_3 = \frac{V_{gh}}{R_{gh}} = \frac{24\ \text{V}}{12\ \Omega} = 2.0\ \text{A}$$

As a check, we note that $I_2 + I_3 = 3\ \text{A} = I_1$,   as should be.

(b) Because $I_2 = 1$ A,   the p.d. across the 2 Ω resistor in Fig. 28-9(b) is   (1 A)(2 Ω) = 2 V. But this is also the p.d. across the 12 Ω resistor in Fig. 28-9(a). Applying $V = IR$ to the 12 Ω gives

$$I_{12} = \frac{V_{12}}{R} = \frac{2\ \text{V}}{12\ \Omega} = 0.167\ \text{A}$$

28.14. A certain galvanometer has a resistance of 400 Ω and deflects full scale for a current of 0.2 mA through it. How large a shunt resistance is required to change it to a 3 A ammeter?

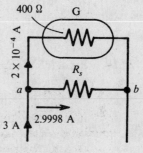

Fig. 28-10

In Fig. 28-10 we label the galvanometer G and the shunt resistance R_s. At full scale deflection, the currents are as shown.
The voltage drop from a to b across G is the same as across R_s. Therefore

$$(2.9998\ \text{A})R_s = (2 \times 10^{-4}\ \text{A})(400\ \Omega)$$

from which $R_s = 0.027\ \Omega$.

28.15. A voltmeter which deflects full scale for a potential difference of 5.0 V across it is to be made by connecting a resistance R_x in series with a galvanometer. The 80 Ω galvanometer deflects full scale for a potential of 20 mV across it. Find R_x.

When deflecting full scale, the current through the galvanometer is

$$I = \frac{V}{R} = \frac{20 \times 10^{-3}\ \text{V}}{80\ \Omega} = 2.5 \times 10^{-4}\ \text{A}$$

When R_x is connected in series with the galvanometer, we wish I to be 2.5×10^{-4} A for a potential difference of 5 V across the combination. Hence $V = IR$ becomes

$$5\ \text{V} = (2.5 \times 10^{-4}\ \text{A})(80\ \Omega + R_x)$$

from which $R_x = 19\,920\ \Omega$.

28.16. The currents in the circuit of Fig. 28-11 are steady. Find I_1, I_2, I_3, and the charge on the capacitor.

When a capacitor has a constant charge, as it does here, the current flowing to it is zero. Therefore, $I_2 = 0$ and the circuit behaves just as though the center wire were missing.

With the center wire missing, the remaining circuit is simply 12 Ω connected across a 15 V battery. Therefore,

$$I_1 = \frac{\mathcal{E}}{R} = \frac{15\ V}{12\ \Omega} = 1.25\ A$$

In addition, because $I_2 = 0$, we have $I_3 = I_1 = 1.25\ A$.

To find the charge on the capacitor, first find the voltage difference between points a and b. Start at a and go around the upper path.

$$\text{voltage change from } a \text{ to } b = -(5\ \Omega)I_3 + 6\ V + (3\ \Omega)I_2$$
$$= -(5\ \Omega)(1.25\ A) + 6\ V + (3\ \Omega)(0) = -0.25\ V$$

Therefore b is at the lower potential and the capacitor plate at b is negative. To find the charge on the capacitor,

$$Q = CV_{ab} = (2 \times 10^{-6}\ F)(0.25\ V) = 0.5\ \mu C$$

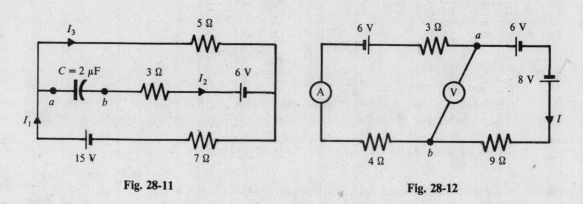

Fig. 28-11   Fig. 28-12

28.17. Find the ammeter reading and the voltmeter reading in the circuit of Fig. 28-12. Assume both meters to be ideal.

The ideal voltmeter has infinite resistance and so its wire can be removed without altering the circuit. The ideal ammeter has zero resistance. It can be shown (see Chapter 29) that batteries in series simply add or subtract. The two 6 V batteries cancel each other because they tend to push current in opposite directions. As a result, the circuit behaves as though it had a single 8 V battery that causes a clockwise current.

The equivalent resistance is $3 + 4 + 9 = 16\ \Omega$ and the equivalent battery is 8 V. Therefore,

$$I = \frac{\mathcal{E}}{R} = \frac{8\ V}{16\ \Omega} = 0.50\ A$$

and this is what the ammeter will read.

Adding up the voltage changes from a to b around the right-hand side of the circuit gives

$$\text{voltage change from } a \text{ to } b = -6\ V + 8\ V - (0.5\ A)(9\ \Omega) = -2.5\ V$$

Therefore, a voltmeter connected from a to b will read 2.5 V, with b being at the lower potential.

Supplementary Problems

28.18. Compute the equivalent resistance of 4 and 8 Ω (a) in series, (b) in parallel.
Ans. (a) 12 Ω; (b) 2.67 Ω

28.19. Compute the equivalent resistance of (*a*) 3, 6, and 9 Ω in parallel; (*b*) 3, 4, 7, 10, and 12 Ω in parallel; (*c*) three 33 Ω heating elements in parallel; (*d*) twenty 100 Ω lamps in parallel.
Ans.   (*a*) 1.64 Ω;   (*b*) 1.10 Ω;   (*c*) 11 Ω;   (*d*) 5 Ω

28.20. Show that if two resistors, R_1 and R_2, are connected in parallel, with current I_1 in R_1 and I_2 in R_2, then $I_1/I_2 = R_2/R_1$.

28.21. What resistance must be placed in parallel with 20 Ω to make the combined resistance 15 Ω?
Ans.   60 Ω

28.22. How many 160 Ω resistors (in parallel) are required to carry 5 A on a 100 V line?   *Ans.*   8

28.23. Three resistors, of 8, 12 and 24 Ω, are in parallel, and a current of 20 A is drawn by the combination. Determine (*a*) the potential difference across the combination and (*b*) the current through each resistance.   *Ans.*   (*a*) 80 V;   (*b*) 10, 6.7, 3.3 A

28.24. Two resistors, of 4 and 12 Ω, are connected in parallel across a 22 V battery having internal resistance 1 Ω.   Compute (*a*) the battery current, (*b*) the current in the 4 Ω resistor, (*c*) the terminal voltage of the battery, (*d*) the current in the 12 Ω resistor.
Ans.   (*a*) 5.5 A;   (*b*) 4.12 A;   (*c*) 16.5 V;   (*d*) 1.38 A

28.25. Three resistors, of 40, 60 and 120 Ω, are connected in parallel, and this parallel group is connected in series with 15 Ω in series with 25 Ω.   The whole system is then connected to a 120 V source.   Determine (*a*) the current in the 25 Ω, (*b*) the potential drop across the parallel group, (*c*) the potential drop across the 25 Ω, (*d*) the current in the 60 Ω, (*e*) the current in the 40 Ω.
Ans.   (*a*) 2 A;   (*b*) 40 V;   (*c*) 50 V;   (*d*) 0.67 A;   (*e*) 1 A

28.26. What shunt resistance should be connected in parallel with an ammeter having a resistance of 0.04 Ω so that 25% of the total current would pass through the ammeter?   *Ans.*   0.0133 Ω

28.27. A 36 Ω galvanometer is shunted by a resistance of 4 Ω.   What part of the total current will pass through the instrument?   *Ans.*   1/10

28.28. A relay of resistance 6 Ω operates with a minimum current of 0.03 A.   It is required that the relay operate when the current in the line attains 0.24 A.   What resistance should be used to shunt the relay?
Ans.   0.857 Ω

28.29. Show that if two resistors are connected in parallel, the rates of heat production in each vary inversely as their resistances.

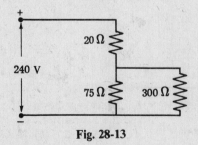

28.30. For the circuit shown in Fig. 28-13, find the current through each resistance and the potential drop across each resistance.
Ans.   for 20 Ω, 3 A and 60 V; for 75 Ω, 2.4 A and 180 V; for 300 Ω, 0.6 A and 180 V

Fig. 28-13

28.31. For the circuit shown in Fig. 28-14, find (*a*) its equivalent resistance; (*b*) the current drawn from the power source; (*c*) the potential difference across *ab*, *cd*, and *de*; (*d*) the current in each resistance.
Ans.   (*a*) 15 Ω;   (*b*) 20 A;   (*c*) V_{ab} = 80 V, V_{cd} = 120 V, V_{de} = 100 V;   (*d*) I_4 = 20 A, I_{10} = 12 A, I_{15} = 8 A, I_9 = 11.1 A, I_{18} = 5.56 A, I_{30} = 3.3 A

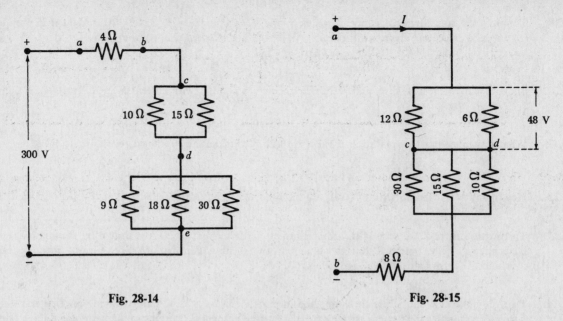

Fig. 28-14   Fig. 28-15

28.32. It is known that the potential difference across the 6 Ω resistance in Fig. 28-15 is 48 V. Determine (*a*) the entering current I, (*b*) the potential difference across the 8 Ω resistance, (*c*) the potential difference across the 10 Ω resistance, (*d*) the potential difference from *a* to *b*. (*Hint:* The wire connecting *c* and *d* can be shrunk to zero length without altering the currents or potentials.)
Ans. (*a*) 12 A;   (*b*) 96 V;   (*c*) 60 V;   (*d*) 204 V

28.33. In the circuit shown in Fig. 28-16, 23.9 cal of heat is produced each second in the 4 Ω resistance. Assuming the ammeter A and the voltmeters V_1 and V_2 to be ideal, what will be their readings?
Ans.   5.8 A, 8 V, 58 V

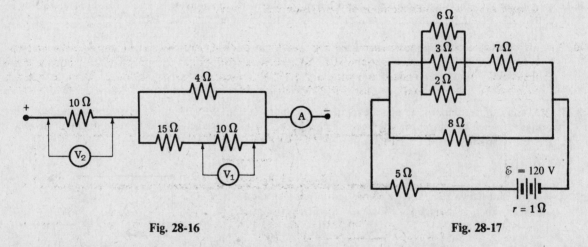

Fig. 28-16   Fig. 28-17

28.34. For the circuit shown in Fig. 28-17, find (*a*) equivalent resistance; (*b*) currents through the 5 Ω, the 7 Ω and the 3 Ω resistors; (*c*) total power output of battery.
Ans. (*a*) 10 Ω;   (*b*) 12 A, 6 A, 2 A;   (*c*) 1296 W

28.35. In the circuit shown in Fig. 28-18, the ideal ammeter A registers 2 A. (*a*) Assuming XY to be a resistance, find its value. (*b*) Assuming XY to be a battery (with 2 Ω internal resistance) that is being charged, find its emf. (*c*) Under the conditions of part (*b*), what is the potential change from point Y to point X?   *Ans.* (*a*) 5 Ω;   (*b*) 6 V;   (*c*) −10 V

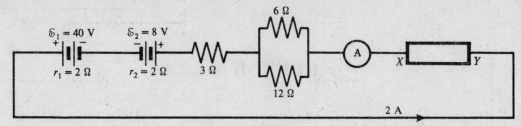

Fig. 28-18

28.36. The *Wheatstone bridge* shown in Fig. 28-19 is being used to measure resistance X. At balance, the current through the galvanometer G is zero and resistances L, M, and N are 3, 2, and 10 Ω respectively. Find the value of X.   *Ans.* 15 Ω

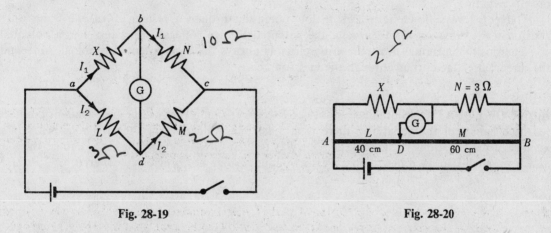

Fig. 28-19   **Fig. 28-20**

28.37. The slidewire Wheatstone bridge shown in Fig. 28-20 is balanced when the uniform slide wire AB is divided as shown. Find the value of the resistance X.   *Ans.* 2 Ω

28.38. The slidewire potentiometer shown in Fig. 28-21 has a steady current flowing through the uniform-resistance wire. A balance is found at a slider position of 45 cm, as shown, when an unknown emf is measured. When a standard cell with emf 1.018 V is used, the slider position is 30 cm at balance. What is the unknown emf?   *Ans.* 1.527 V

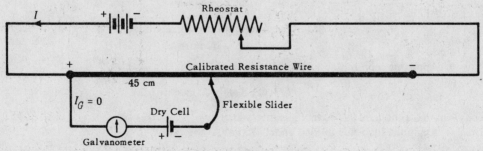

Fig. 28-21

Chapter 29

Kirchhoff's Laws

KIRCHHOFF'S POINT (OR JUNCTION) RULE: The sum of all the currents coming into a point must equal the sum of all the currents leaving the point.

KIRCHHOFF'S LOOP (OR CIRCUIT) RULE: As one traces out a closed circuit, the algebraic sum of the potential changes encountered is zero. In this sum, a potential rise is positive and a potential drop is negative.

Current always flows from high to low potential through a resistor. As one traces through a resistor in the direction of the current, the potential change is negative because it is a potential drop.

The positive terminal of a pure emf source is always the high-potential terminal, independent of the direction of the current through the emf source.

A SET OF EQUATIONS obtained by use of Kirchhoff's loop rule will be independent provided that, for each subset of the equations, there is at least one voltage-change term that appears in an *odd number* of the equations in the subset. One way to guarantee independence is to choose the loops one at a time, such that each new loop covers a voltage change not previously covered.

Solved Problems

29.1. Find the currents in the circuit shown in Fig. 29-1.

This circuit cannot be reduced further because it contains no resistors in simple series or parallel. We therefore revert to Kirchhoff's rules. If the currents had not been labeled and shown by arrows, we would do that first. No special care need be taken in assigning the current directions, since those chosen incorrectly will simply give negative numerical values.

Apply the point rule to point b in Fig. 29-1:

$$\text{current into } b = \text{current out of } b$$
$$I_1 + I_2 + I_3 = 0 \qquad\qquad (1)$$

Apply the loop rule to loop *adba*. In volts,

$$-7I_1 + 6 + 4 = 0 \qquad \text{or} \qquad I_1 = \frac{10}{7} \text{ A}$$

(Why must the term $7I_1$ have a negative sign?)

Apply the loop rule to loop *abca*. In volts,

$$-4 - 8 + 5I_2 = 0 \qquad \text{or} \qquad I_2 = \frac{12}{5} \text{ A}$$

(Why must the signs be as written?)

Now return to (1) to find

$$I_3 = -I_1 - I_2 = -\frac{10}{7} - \frac{12}{5} = \frac{-50 - 84}{35} = -3.83 \text{ A}$$

The negative sign tells us that I_3 is opposite in direction to that shown in the figure.

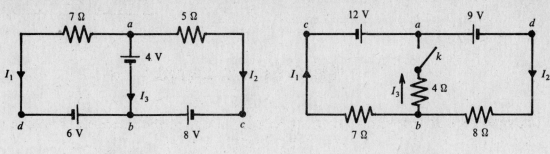

Fig. 29-1 ·   Fig. 29-2

29.2. In Fig. 29-2, find I_1, I_2, and I_3 if switch k is (a) open, (b) closed.

(a) When k is open, $I_3 = 0$, because no current can flow through the open switch. Apply the point rule to point a.

$$I_1 + I_3 = I_2 \qquad \text{or} \qquad I_2 = I_1 + 0 = I_1$$

Apply the loop rule to loop $acbda$. In volts,

$$-12 + 7I_1 + 8I_2 + 9 = 0 \tag{1}$$

To understand the signs used, remember that current always flows from high to low potential through a resistor.

Because $I_2 = I_1$, (1) becomes

$$15I_1 = 3 \qquad \text{or} \qquad I_1 = 0.20 \text{ A}$$

Also, $I_2 = I_1 = 0.20$ A. Notice that this is the same result that one would obtain by replacing the two batteries by a single 3 V battery.

(b) With k closed, I_3 is no longer known to be zero. Applying the point rule to point a gives

$$I_1 + I_3 = I_2 \tag{2}$$

Applying the loop rule to loop $acba$ gives, in volts,

$$-12 + 7I_1 - 4I_3 = 0 \tag{3}$$

and to loop $adba$ gives

$$-9 - 8I_2 - 4I_3 = 0 \tag{4}$$

Applying the loop rule to the remaining loop, $acbda$, would yield a redundant equation, because, among the three loop equations, each voltage change would appear twice.

We must now solve (2), (3), and (4) for I_1, I_2, and I_3. From (4),

$$I_3 = -2I_2 - 2.25$$

Substituting this in (3) gives

$$-12 + 7I_1 + 9 + 8I_2 = 0 \qquad \text{or} \qquad 7I_1 + 8I_2 = 3$$

Also substituting for I_3 in (2) gives

$$I_1 - 2I_2 - 2.25 = I_2 \qquad \text{or} \qquad I_1 = 3I_2 + 2.25$$

Substituting this value in the previous equation,

$$21I_2 + 15.75 + 8I_2 = 3 \qquad \text{or} \qquad I_2 = -0.44 \text{ A}$$

Using this in the equation for I_1 gives

$$I_1 = 3(-0.44) + 2.25 = -1.32 + 2.25 = 0.93 \text{ A}$$

Notice that the minus sign is a part of the value we have found for I_2. It must be carried along with its numerical value. Now we can use (2) to find

$$I_3 = I_2 - I_1 = (-0.44) - 0.93 = -1.37 \text{ A}$$

29.3. Each of the cells shown in Fig. 29-3 has an emf of 1.50 V and a 0.075 Ω internal resistance. Find I_1, I_2, and I_3.

Applying the point rule to point a,

$$I_1 = I_2 + I_3 \qquad\qquad (1)$$

Applying the loop rule to loop $abcea$ gives, in volts,

$$-(0.075)I_2 + 1.5 - (0.075)I_2 + 1.5 - 3I_1 = 0$$

or

$$3I_1 + 0.15\,I_2 = 3 \qquad\qquad (2)$$

Also, for loop $adcea$,

$$-(0.075)I_3 + 1.5 - (0.075)I_3 + 1.5 - 3I_1 = 0$$

or

$$3I_1 + 0.15\,I_3 = 3 \qquad\qquad (3)$$

Solve (2) for $3I_1$ and substitute in (3), yielding

$$3 - 0.15\,I_3 + 0.15\,I_2 = 3 \qquad \text{or} \qquad I_2 = I_3$$

as we might have guessed from the symmetry of the problem.   Then (1) yields

$$I_1 = 2I_2$$

Substituting this in (2) gives

$$6I_2 + 0.15\,I_2 = 3 \qquad \text{or} \qquad I_2 = 0.488 \text{ A}$$

Then, $I_3 = I_2 = 0.488$ A   and $I_1 = 2I_2 = 0.976$ A.

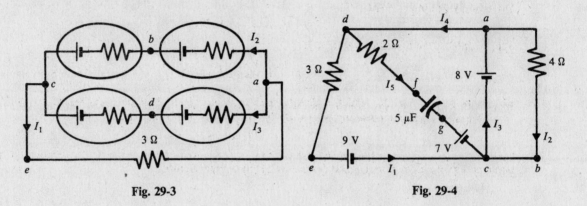

Fig. 29-3   Fig. 29-4

29.4.   The currents are steady in the circuit of Fig. 29-4.   Find I_1, I_2, I_3, I_4, I_5, and the charge on the capacitor.

The capacitor passes no current when charged and so $I_5 = 0$.   Consider loop $acba$.   The loop rule gives

$$-8 + 4I_2 = 0 \qquad \text{or} \qquad I_2 = 2 \text{ A}$$

Using loop $adeca$ gives

$$-3I_1 - 9 + 8 = 0 \qquad \text{or} \qquad I_1 = -0.33 \text{ A}$$

Apply the point rule at point c:

$$I_1 + I_5 + I_2 = I_3 \qquad \text{or} \qquad I_3 = 1.67 \text{ A}$$

and at point a:

$$I_3 = I_4 + I_2 \qquad \text{or} \qquad I_4 = -0.33 \text{ A}$$

(We should have realized this at once, because $I_5 = 0$ and so $I_4 = I_1$.)

To find the charge on the capacitor, we need the voltage across it, V_{fg}.   Applying the loop rule to loop $dfgced$ gives

$$-2I_5 + V_{fg} - 7 + 9 + 3I_1 = 0 \qquad \text{or} \qquad 0 + V_{fg} - 7 + 9 - 1.0 = 0$$

from which $V_{fg} = -1$ V.   The negative sign tells us that plate g is negative.   The capacitor's charge is

$$Q = CV = (5 \ \mu\text{F})(1 \text{ V}) = 5 \ \mu\text{C}$$

29.5. For the circuit shown in Fig. 29-5, the resistance R is $5\ \Omega$ and $\mathcal{E} = 20$ V. Find the readings of the ammeter and the voltmeter. Assume the meters to be ideal.

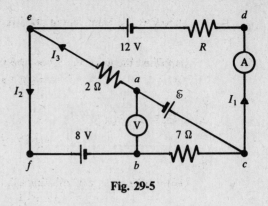

Fig. 29-5

The ideal voltmeter has infinite resistance and so it can be removed from the circuit with no effect. Let us write the loop equation for loop $cdefc$:

$$-RI_1 + 12 - 8 - 7I_2 = 0$$

which becomes

$$5I_1 + 7I_2 = 4 \qquad\qquad (1)$$

Next write the loop equation for loop $cdeac$. It is

$$-5I_1 + 12 + 2I_3 + 20 = 0$$

or

$$5I_1 - 2I_3 = 32 \qquad\qquad (2)$$

But the point rule applied at e gives

$$I_1 + I_3 = I_2 \qquad\qquad (3)$$

Substituting (3) in (1) gives

$$5I_1 + 7I_1 + 7I_3 = 4$$

Solve this for I_3 and substitute in (2):

$$5I_1 - 2\left(\frac{4 - 12I_1}{7}\right) = 32$$

which yields $I_1 = 3.93$ A, which is the ammeter reading. Then (1) gives $I_2 = -2.24$ A.

To find the voltmeter reading, V_{ab}, write the loop equation for loop $abca$:

$$V_{ab} - 7I_2 - \mathcal{E} = 0$$

Substituting the known values of I_2 and $\mathcal{E}$, then solving, we obtain $V_{ab} = 4.3$ V. Since this is the potential change from a to b, point b must be at the higher potential.

29.6. In the circuit of Fig. 29-5, $I_1 = 0.20$ A and $R = 5\ \Omega$. Find $\mathcal{E}$.

Write the loop equation for loop $cdefc$.

$$-RI_1 + 12 - 8 - 7I_2 = 0 \qquad \text{or} \qquad -(5)(0.2) + 12 - 8 - 7I_2 = 0$$

from which $I_2 = 0.43$ A. We can now find I_3 by applying the point rule at e.

$$I_1 + I_3 = I_2 \qquad \text{or} \qquad I_3 = I_2 - I_1 = 0.23 \text{ A}$$

Now apply the loop rule to loop $cdeac$.

$$-(5)(0.2) + 12 + (2)(0.23) + \mathcal{E} = 0$$

from which $\mathcal{E} = -11.5$ V. The negative sign tells us that the polarity of the battery is actually the reverse of that shown.

Supplementary Problems

29.7. For the circuit shown in Fig. 29-6, find the current in the 0.96 Ω resistor and the terminal voltages of the batteries.   *Ans.*   5 A, 4.8 V, 4.8 V

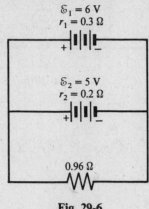

Fig. 29-6

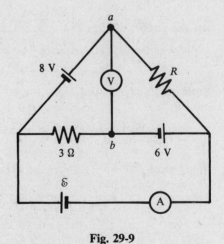

Fig. 29-7

29.8. For the network shown in Fig. 29-7, determine (*a*) the three currents I_1, I_2, and I_3, and (*b*) the terminal voltages of the three batteries.
Ans.   (*a*) $I_1 = 2$ A, $I_2 = 1$ A, $I_3 = -3$ A;   (*b*) $V_{16} = 14$ V, $V_4 = 3.8$ V, $V_{10} = 8.5$ V

29.9. Refer back to Fig. 29-5. If the voltmeter reads 16 V (with point *b* at the higher potential) and $I_2 = 0.2$ A,   find $\mathcal{E}$, *R*, and the ammeter reading.   *Ans.*   14.6 V, 0.21 Ω, 12.2 A

29.10. Find I_1, I_2, I_3, and the potential difference from point *b* to point *e* in Fig. 29-8.
Ans.   2 A, −8 A, 6 A, −13 V

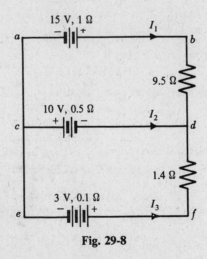

Fig. 29-8

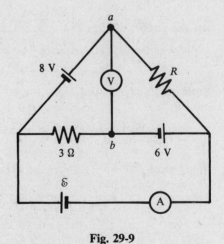

Fig. 29-9

29.11. In Fig. 29-9, $R = 10$ Ω   and $\mathcal{E} = 13$ V.   Find the readings of the ideal ammeter and voltmeter.
Ans.   8.4 A, 27 V with point *a* positive

29.12. In Fig. 29-9, the voltmeter reads 14 V (with point *a* at the higher potential) and the ammeter reads 4.5 A.
Find $\mathcal{E}$ and *R*.   *Ans.* $\mathcal{E} = 0$, $R = 3.2$ Ω

Chapter 30

Electrolysis

FARADAY'S LAWS OF ELECTROLYSIS:

(1) The mass of a substance liberated or deposited at an electrode is proportional to the quantity of electricity (i.e. to the number of coulombs) that has passed through the electrolyte.

(2) The masses of different substances liberated or deposited by the same quantity of electricity are proportional to their equivalent weights.

The *equivalent weight* of an element is its atomic weight divided by its valence. Thus the equivalent weight of copper (atomic weight 63.54) is 63.54/2 for the electrolysis of solutions containing Cu^{++}, because the reaction at the cathode (the negative electrode) is

$$Cu^{++} + 2e^- \rightarrow Cu$$

But if a solution of Cu^+ were electrolyzed, the equivalent weight of copper would be 63.54/1, because only 1 electron would be involved in the electrode reaction

$$Cu^+ + e^- \rightarrow Cu$$

When the equivalent weight of a substance is expressed in grams, it is called the *gram-equivalent weight*.

One *faraday* ($\mathscr{F}$), which is 96 500 C, is the quantity of electricity that will liberate or deposit 1 gram-equivalent weight of any substance. Thus the mass m in grams of any substance liberated in electrolysis is

$$m = \text{gram-equivalent weight} \times \text{number of faradays transferred}$$

ELECTROCHEMICAL EQUIVALENT of a substance is the mass of that substance liberated by 1 C of electric charge. Thus the mass m of a substance liberated in electrolysis is

$$m = \text{electrochemical equivalent} \times \text{number of coulombs transferred}$$

Solved Problems

30.1. The charge on the electron is 1.602×10^{-19} C. How many electronic charges are contained in 1000 $\mathscr{F}$?

$$\text{number} = \frac{(1000)(96\,500 \text{ C})}{1.602 \times 10^{-19} \text{ C/electron}} = 6.02 \times 10^{26} \text{ electrons}$$

We recognize this number as Avogadro's number, N_A. Therefore, Faraday's laws may be summarized as follows: Exactly 1000 $\mathscr{F}$ of electricity is required to provide 1 kmol of electrons and thereby liberate or deposit $(1/v)$ kmol of a substance whose valence is v.

30.2. A charge of 0.2 $\mathscr{F}$ is passed through an electrolytic cell containing ferric iron (Fe^{+++}). Assuming that the only cathode reaction is

$$Fe^{+++} + 3e^- \rightarrow Fe$$

what mass of iron will be deposited?

$$\text{equivalent weight of iron} = \frac{\text{atomic weight}}{\text{valence}} = \frac{55.85}{3} = 18.62$$

Since 1 $\mathscr{F}$ deposits 1 gram-equivalent weight, or 18.62 g of iron, 0.2 $\mathscr{F}$ will deposit

$$(0.2)(18.62 \text{ g}) = 3.72 \text{ g}$$

of iron.

30.3. How many grams of zinc would be deposited by the same number of coulombs that deposits 2 grams of silver?

The equivalent weights of silver (Ag^+) and zinc (Zn^{++}) are

$$\frac{107.88}{1} = 107.88 \qquad \frac{65.37}{2} = 32.69$$

Then, by Faraday's second law,

$$\text{mass of zinc} = \frac{32.69}{107.88} (2 \text{ g}) = 0.61 \text{ g}$$

30.4. A steady current of 5 A maintained for 30 min deposits 3.048 g of zinc at the cathode. Determine the equivalent weight of zinc.

One faraday, or 96 500 C, deposits 1 gram-equivalent weight of a substance.   Given that

$$(5 \text{ A})(1800 \text{ s}) = 9000 \text{ C}$$

deposits 3.048 g of zinc, 96 500 C would deposit

$$\frac{96\,500}{9000} (3.048 \text{ g}) = 32.7 \text{ g Zn}$$

The equivalent weight of Zn is 32.7.

30.5. A current of 15 A is employed to plate nickel in a nickel sulfate bath.   Both Ni and H_2 are formed at the cathode.   The current efficiency with respect to the formation of Ni is 60%. (a) How many grams of nickel are plated on the cathode per hour?   (b) What is the thickness of the plating if the cathode consists of a sheet of metal 4 cm square which is coated on both faces?   Nickel has specific gravity 8.9, atomic weight 58.71, and valence 2.

(a) $$\text{equivalent weight of Ni} = \frac{\text{atomic weight}}{\text{valence}} = \frac{58.71}{2} = 29.4$$

The amount of charge used to form Ni is $(0.60)(15 \text{ A})(3600 \text{ s}) = 3.24 \times 10^4$ C.   Since 9.65×10^4 C deposits 1 gram-equivalent weight, or 29.4 g, of Ni, 3.24×10^4 C deposits

$$\frac{3.24 \times 10^4 \text{ C}}{9.65 \times 10^4 \text{ C}} (29.4 \text{ g Ni}) = 9.9 \text{ g Ni}$$

(b) $$\text{volume of 9.9 g Ni} = \frac{\text{mass}}{\text{density}} = \frac{9.9 \text{ g}}{8.9 \text{ g/cm}^3} = 1.11 \text{ cm}^3$$

$$\text{thickness of plating} = \frac{\text{volume of plating}}{\text{area of plating}} = \frac{1.11 \text{ cm}^3}{2(4 \text{ cm})^2} = 0.035 \text{ cm}$$

30.6. (a) Compute the electrochemical equivalent of gold.   Gold has atomic weight 197.0 and valence 3.   (b) What constant current is required to deposit on the cathode 5 g of gold per hour?

(a) $$\text{electrochemical equiv} = \frac{\text{gram-equivalent wt}}{96\,500 \text{ C}} = \frac{(197.0/3) \text{ g}}{96\,500 \text{ C}} = 0.681 \text{ mg/C}$$

(b) $$\text{mass deposited} = (\text{electrochemical equivalent})(\text{number of coulombs transferred})$$

$$5 \text{ g} = (6.81 \times 10^{-4} \text{ g/C})(I \times 3600 \text{ s})$$
$$I = 2.04 \text{ A}$$

Supplementary Problems

30.7. What current is required to pass 1 $\mathscr{F}$/h through an electroplating bath? How many grams of aluminum and of cadmium will be liberated by 1 $\mathscr{F}$? Equivalent weight of aluminum is 8.99; of cadmium, 56.2. *Ans.* 26.8 A, 8.99 g Al, 56.2 g Cd

30.8. How many grams of aluminum are deposited electrolytically in 30 min by a current of 40 A? Equivalent weight of Al is 8.99. *Ans.* 6.71 g

30.9. A certain current liberates 0.504 g of hydrogen in 2 h. How many grams of oxygen and of copper (from Cu^{++} solution) can be liberated by the same current maintained for the same time? Equivalent weights are: 1.008 for hydrogen, 8.00 for oxygen, 31.8 for copper. *Ans.* 4.00 g oxygen, 15.9 g copper

30.10. A given current releases 2 g of oxygen in 12 min. How long will it take the same current to deposit 18 g of copper from a copper sulfate solution? Equivalent weight of oxygen is 8.00; of copper, 31.8. *Ans.* 27.2 min

30.11. An electrolytic cell contains a solution of $CuSO_4$ and an anode (positive electrode) of impure copper. How many pounds of copper will be refined (deposited on the cathode) by 150 A maintained for 12 h? Equivalent weight of copper is 31.8. Recall that 1 kg weighs 2.21 lb. *Ans.* 2.14 kg = 4.7 lb

30.12. How many hours will it take to produce 100 lb of electrolytic chlorine from NaCl in a cell carrying 1000 A? The anode efficiency for the chlorine reaction is 85%. Equivalent weight of chlorine is 35.46. *Ans.* 40.3 h

30.13. Compute the time required for a current of 3 A to decompose electrolytically 18 g of water. Equivalent weight of hydrogen is 1.0; of oxygen, 8.0. *Ans.* 18 h

30.14. Compute the electrochemical equivalent of magnesium. Magnesium has atomic weight 24.32 and valence 2. *Ans.* 0.126 mg/C

Chapter 31

Magnetic Fields

A MAGNETIC FIELD exists in a region of space if a moving charge there experiences a force (other than friction) due to its motion. More frequently, a magnetic field is detected by its effect on a compass needle. The compass needle lines up in the direction of the magnetic field.

Figure 31-1 shows a uniform magnetic field directed toward the right. As shown, a compass needle orients along the field lines. The force **F** on a charge moving with velocity **v** through the field is also shown.

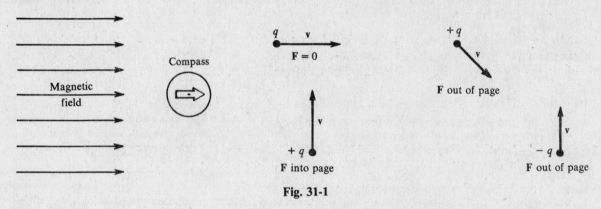

Fig. 31-1

THE DIRECTION OF THE FORCE on a charge $+q$ moving in a magnetic field can be found from a *right-hand rule* (Fig. 31-2).

Hold the right hand flat. Point its fingers in the direction of the field. Orient the thumb along the direction of the velocity of the positive charge. Then the palm of the hand pushes in the direction of the force on the charge. The force direction on a negative charge is opposite to that on a positive charge.

It is often helpful to note that the field line through the particle and the velocity vector of the particle determine a plane (the plane of the page in Fig. 31-2). The force vector is always perpendicular to this plane.

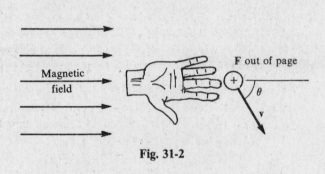

Fig. 31-2

THE MAGNITUDE OF THE FORCE (F) on a charge moving in a magnetic field depends upon the product of four factors:
   (1) q, the charge magnitude (in C)
   (2) v, the magnitude of the velocity of the charge (in m/s)
   (3) B, the strength of the magnetic field
   (4) $\sin \theta$, where θ is the angle between the field lines and the velocity **v**
Therefore, $F \propto qvB \sin \theta$.

THE MAGNETIC FIELD AT A POINT is represented by a vector, **B**. Its direction is that of the magnetic field. The names given to **B** are the *magnetic induction*, the *magnetic flux density*, and, colloquially, the *magnetic field strength*.

We define the magnitude of **B** and its units by making the proportionality $F \propto qvB \sin\theta$ into an equality. Thus,

$$F = qvB \sin\theta$$

where F is in newtons, q is in coulombs, v is in m/s, and B is in a unit called the *tesla* (T). This unit is also called the *weber per square meter*: $1\,\text{T} = 1\,\text{Wb/m}^2$ (see Chapter 32). Still encountered is the cgs unit of B, the *gauss* (G), where

$$1\,\text{G} = 10^{-4}\,\text{T}$$

The earth's magnetic field has B somewhat less than 1 G. Also note that

$$1\,\text{T} = 1\,\text{Wb/m}^2 = 1\,\frac{\text{N}}{\text{C} \cdot (\text{m/s})} = 1\,\frac{\text{N}}{\text{A} \cdot \text{m}}$$

FORCE ON A CURRENT IN A MAGNETIC FIELD: Since a current is simply a stream of positive charges, the current experiences a force due to a magnetic field. The direction of the force is found by the right-hand rule shown in Fig. 31-2, with the direction of the current used in place of the velocity vector.

The magnitude of the force, ΔF, on a small length of wire, ΔL, is given by

$$\Delta F = I(\Delta L)B \sin\theta$$

where θ is the angle between the direction of the current I and the direction of the field. For a straight wire of length L in a uniform magnetic field, this becomes

$$F = ILB \sin\theta$$

Notice that the force is zero if the wire is in line with the field lines. The force is maximum if the field lines are perpendicular to the wire. In analogy to the case of a moving charge, the force is perpendicular to the plane defined by the wire and the field lines.

TORQUE ON A FLAT COIL in a uniform magnetic field: The torque τ on a coil of N loops, each carrying a current I, in an external magnetic field B is

$$\tau = NIAB \sin\theta$$

where A is the area of the coil and θ is the angle between the field lines and a perpendicular to the plane of the coil. For the direction of rotation of the coil, we have the following right-hand rule:

Orient the right thumb perpendicular to the plane of the coil, such that the fingers run in the direction of the current flow. Then the torque acts to rotate the thumb into alignment with the external field (in which orientation the torque will be zero).

SOURCES OF MAGNETIC FIELDS: To the best of our present knowledge, all magnetic fields are caused by the motion of charges. Even the fields produced by permanent magnets are the result of charge motion within the atoms composing the magnet.

The magnetic fields produced by several simple current configurations are shown in Fig. 31-3. Below each is given the value of B at the indicated point P in that configuration. The constant $\mu_0 = 4\pi \times 10^{-7}\,\text{T} \cdot \text{m/A}$ is called the *permeability of free space*. It is assumed that the coil is in vacuum or air.

THE DIRECTION OF THE MAGNETIC FIELD of a current-carrying wire can be found using a right-hand rule. As shown in Fig. 31-3(*a*),

Grasp the wire in the right hand, with the thumb pointing in the direction of the current. The fingers then circle the wire in the same direction as the magnetic field does.

This same rule can be used to find the general direction of the field for a current loop such as that shown in Fig. 31-3(*b*). Then, however, only the field due to the grasped portion of the loop is found.

current I

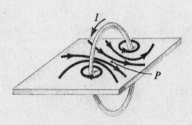

(a)   Long straight wire:

$$B = \frac{\mu_o I}{2\pi r}$$

where r is the distance to
P from the axis of the wire

(b)   Center of a circular coil
with radius a and N loops:

$$B = \frac{\mu_o NI}{2a}$$

(c)   Interior point of
long solenoid with
n loops per meter:

$$B = \mu_0 nI$$

It is constant in
the interior

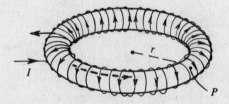

(d)   Interior point
of toroid having
N loops:

$$B = \frac{\mu_0 NI}{2\pi r}$$

where r is the
radius of the circle
on which P lies

Fig. 31-3

MAGNETIC FIELD OF A CURRENT ELEMENT: The
current element of length ΔL shown in Fig. 31-4 contributes
$\Delta \mathbf{B}$ to the field at P.   The magnitude of $\Delta \mathbf{B}$ is

$$\Delta B = \frac{\mu_0 I \, \Delta L}{4\pi r^2} \sin\theta$$

where r and θ are defined in the figure.   The direction of $\Delta \mathbf{B}$
is perpendicular to the plane determined by ΔL and r (the
plane of the page).   In the case shown, the right-hand rule
tells us that $\Delta \mathbf{B}$ is out of the page.

When r is directed along ΔL, then $\theta = 0$ and thus
$\Delta B = 0$.   This means that the field due to a straight wire at a
point on the line of the wire is zero.

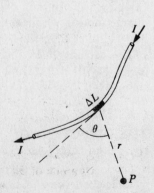

Fig. 31-4

Solved Problems

31.1. A uniform magnetic field, $B = 3$ G, exists in the $+x$-direction. A proton $(q = +e)$ shoots through the field in the $+y$-direction with a speed of 5×10^6 m/s. (*a*) Find the magnitude and direction of the force on the proton. (*b*) Repeat if the proton is replaced by an electron.

(*a*) The situation is shown in Fig. 31-5. We have

$$F = qvB \sin \theta = (1.6 \times 10^{-19} \text{ C})(5 \times 10^6 \text{ m/s})(3 \times 10^{-4} \text{ T}) \sin 90° = 2.4 \times 10^{-16} \text{ N}$$

The force is perpendicular to the xy-plane, the plane defined by the field lines and **v**. The right-hand rule tells us that the force is directed into the page, in the $-z$-direction.

(*b*) The magnitude of the force is the same as in (*a*), 2.4×10^{-16} N. But, because the electron is negative, the force direction is reversed. The force is in the $+z$-direction.

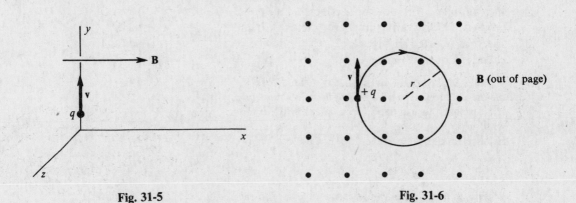

Fig. 31-5   Fig. 31-6

31.2. The charge shown in Fig. 31-6 is a proton $(q = +e,\ m = 1.67 \times 10^{-27}$ kg) with speed 5×10^6 m/s. It is passing through a uniform magnetic field directed out of the page; B is 30 G. Describe the path followed by the proton.

Because the proton's velocity is perpendicular to **B**, the force on the proton is

$$qvB \sin 90° = qvB$$

This force is perpendicular to **v** and so it does no work on the proton. It simply deflects the proton and causes it to follow the circular path shown, as you can verify using the right-hand rule. The force qvB is radially inward and supplies the centripetal force for the circular motion:

$$qvB = \frac{mv^2}{r} \qquad \text{or} \qquad r = \frac{mv}{qB} \qquad\qquad (1)$$

For the data,

$$r = \frac{(1.67 \times 10^{-27} \text{ kg})(5 \times 10^6 \text{ m/s})}{(1.6 \times 10^{-19} \text{ C})(30 \times 10^{-4} \text{ T})} = 17.4 \text{ m}$$

Observe from (*1*) that the momentum of the charged particle is directly proportional to the radius of its circular orbit.

31.3. A proton enters a magnetic field of flux density 1.5 Wb/m² with a velocity of 2×10^7 m/s at an angle of 30° with the field. Compute the force on the proton.

$$F = qvB \sin \theta = (1.6 \times 10^{-19} \text{ C})(2 \times 10^7 \text{ m/s})(1.5 \text{ Wb/m}^2) \sin 30° = 2.4 \times 10^{-12} \text{ N}$$

31.4. A cathode ray beam (an electron beam; $m = 9.1 \times 10^{-31}$ kg, $q = -e$) is bent in a circle of radius 2 cm by a uniform field with $B = 4.5 \times 10^{-3}$ T. What is the speed of the electrons?

To describe a circle like this, the particle must be moving perpendicular to **B**. From (1) of Problem 31.2,

$$v = \frac{rqB}{m} = \frac{(0.02 \text{ m})(1.6 \times 10^{-19} \text{ C})(4.5 \times 10^{-3} \text{ T})}{9.1 \times 10^{-31} \text{ kg}} = 1.58 \times 10^7 \text{ m/s}$$

31.5. As shown in Fig. 31-7, a beam of particles of charge q enters a region where an electric field is uniform and directed downward. Its value is 80 kV/m. Perpendicular to **E** and directed into the page is a magnetic field $B = 0.4$ T. If the speed of the particles is properly chosen, the particles will not be deflected by these crossed electric and magnetic fields. What speed is selected in this case? (This device is called a *velocity selector*.)

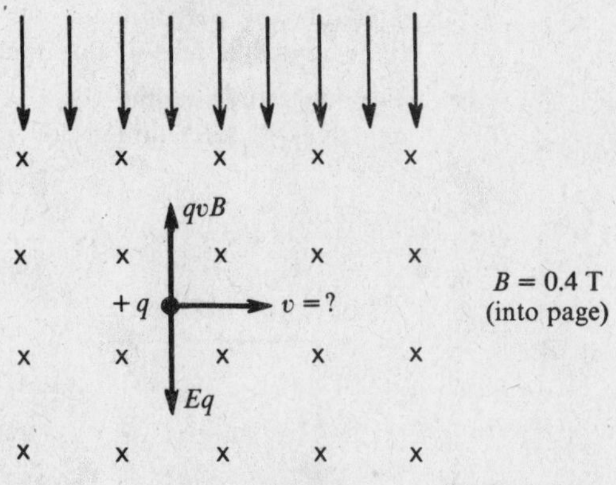

The electric field causes a downward force Eq on the charge if it is positive. The right-hand rule tells us that the magnetic force, $qvB \sin 90°$, is upward if q is positive. If these two forces are to balance so that the particle does not deflect, then

Fig. 31-7

$$Eq = qvB \sin 90° \qquad \text{or} \qquad v = \frac{E}{B} = \frac{80 \times 10^3 \text{ V/m}}{0.4 \text{ T}} = 2 \times 10^5 \text{ m/s}$$

When q is negative, both forces are reversed, so the result $v = E/B$ still holds.

31.6. In Fig. 31-8(a), a proton ($q = +e$, $m = 1.67 \times 10^{-27}$ kg) is shot with speed 8×10^6 m/s at an angle of 30° to an x-directed field $B = 0.15$ T. Describe the path followed by the proton.

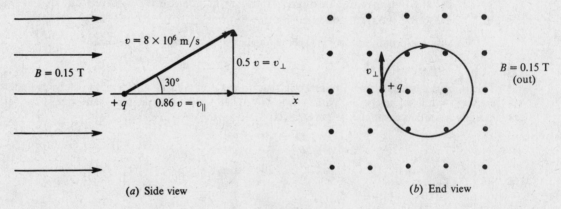

(a) Side view   (b) End view

Fig. 31-8

We resolve the particle velocity into components parallel to and perpendicular to the magnetic field. The magnetic force due to $v_\parallel$ is zero ($\sin \theta = 0$); the magnetic force due to $v_\perp$ has no x-component. Therefore, the x-motion is uniform, at speed

$$v_\parallel = (0.86)(8 \times 10^6 \text{ m/s}) = 6.88 \times 10^6 \text{ m/s}$$

while the transverse motion is circular (see Problem 31.2), with radius

$$r = \frac{mv_\perp}{qB} = \frac{(1.67 \times 10^{-27} \text{ kg})(0.5 \times 8 \times 10^6 \text{ m/s})}{(1.6 \times 10^{-19} \text{ C})(0.15 \text{ T})} = 0.28 \text{ m}$$

The proton will spiral along the *x*-axis; the radius of the spiral (or helix) will be 28 cm.

To find the *pitch* of the helix (the *x*-distance traveled during one revolution), we note that the time taken to complete one circle is

$$\text{period} = \frac{2\pi r}{v_\perp} = \frac{2\pi(0.28 \text{ m})}{(0.5)(8 \times 10^6 \text{ m/s})} = 4.4 \times 10^{-7} \text{ s}$$

During that time, the proton will travel an *x*-distance of

$$\text{pitch} = (v_\parallel)(\text{period}) = (6.88 \times 10^6 \text{ m/s})(4.4 \times 10^{-7} \text{ s}) = 3.0 \text{ m}$$

31.7. Alpha particles ($m = 6.68 \times 10^{-27}$ kg, $q = +2e$) are accelerated from rest through a p.d. of 1 kV. They then enter a magnetic field $B = 0.2$ T perpendicular to their direction of motion. Calculate the radius of their path.

Their final KE is equal to the electric potential energy they lose during acceleration, Vq.

$$\frac{1}{2}mv^2 = Vq \qquad \text{or} \qquad v = \sqrt{\frac{2Vq}{m}}$$

They follow a circular path in which

$$r = \frac{mv}{qB} = \frac{m}{qB}\sqrt{\frac{2Vq}{m}} = \frac{1}{B}\sqrt{\frac{2Vm}{q}}$$

$$= \frac{1}{0.2 \text{ T}}\sqrt{\frac{2(1000 \text{ V})(6.68 \times 10^{-27} \text{ kg})}{3.2 \times 10^{-19} \text{ C}}} = 0.032 \text{ m}$$

31.8. In Fig. 31-9, the magnetic field is out of the page and $B = 0.8$ T. The wire shown carries a current of 30 A. Find the magnitude and direction of the force on a 5 cm length of the wire.

We know that

$$\Delta F = I(\Delta L)B \sin\theta = (30 \text{ A})(0.05 \text{ m})(0.8 \text{ T})(1) = 1.2 \text{ N}$$

Using the right-hand rule, the force direction is perpendicular to both the wire and the field and is toward the bottom of the page.

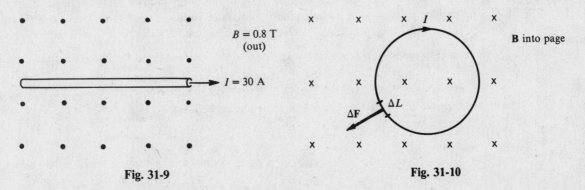

Fig. 31-9   Fig. 31-10

31.9. As shown in Fig. 31-10, a loop of wire carries a current *I* and its plane is perpendicular to a uniform magnetic field **B**. What are the resultant force and torque on the loop?

Consider the length ΔL shown. The force $\Delta \mathbf{F}$ on it has the direction indicated. A point directly opposite this on the loop has an equal, but opposite, force acting on it. Hence the forces on the loop cancel and the. resultant force on it is zero.

We see from the figure that the $\Delta \mathbf{F}$'s acting on the loop are trying to expand it, not rotate it. Therefore the torque on the loop is zero. Or, we can use the torque equation,

$$\text{torque} = NIAB \sin\theta$$

where θ is the angle between the field lines and the perpendicular to the plane of the loop. We see that $\theta = 0$. Therefore $\sin\theta = 0$ and the torque is zero.

31.10. The 40-loop coil shown in Fig. 31-11 carries a current of 2 A in a magnetic field $B = 0.25$ T. Find the torque on it. How will it rotate?

Method 1

$$\tau = NIAB \sin \theta = (40)(2 \text{ A})(0.10 \text{ m} \times 0.12 \text{ m})(0.25 \text{ T})(\sin 90°) = 0.24 \text{ N} \cdot \text{m}$$

(Remember that θ is the angle between the field lines and the perpendicular to the loop.) By the right-hand rule, the coil will turn about a vertical axis in such a way that side ad moves out of the page.

Method 2

Because sides dc and ab are in line with the field, the force on each of them is zero, while the force on each vertical wire is

$$f = ILB = (2 \text{ A})(0.12 \text{ m})(0.25 \text{ T}) = 0.060 \text{ N}$$

out of the page on side ab and into the page on side bc. Thus we have two forces, of magnitude $40f$ and separation 10 cm, comprising a couple (Problem 3.8). The torque is

$$\tau = (\text{force})(\text{separation}) = (40 \times 0.060 \text{ N})(0.10 \text{ m}) = 0.24 \text{ N} \cdot \text{m}$$

and tends to rotate side ad out of the page.

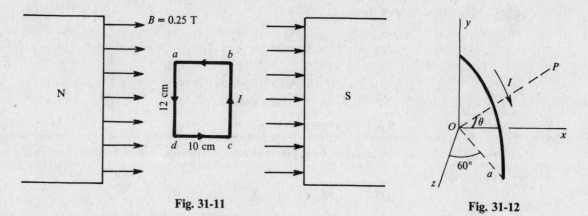

Fig. 31-11   Fig. 31-12

31.11. In Fig. 31-12 is shown one quarter of a single circular loop of wire that carries a current of 14 A. Its radius is $a = 5$ cm. A uniform magnetic field, $B = 300$ G, is directed in the $+x$-direction. Find the torque on the loop and the direction in which it will rotate.

The normal to the loop, OP, makes an angle $\theta = 60°$ with the $+x$-direction, the field direction. Hence,

$$\tau = NIAB \sin \theta = (1)(14 \text{ A})(\pi \times 25 \times 10^{-4} \text{ m}^2)(0.03 \text{ T}) \sin 60° = 2.9 \times 10^{-3} \text{ N} \cdot \text{m}$$

The right-hand rule shows that the loop will rotate about the y-axis so as to decrease the angle labeled 60°.

31.12. Compute the magnetic induction in air at a point 5 cm from a long straight wire carrying a current of 15 A.

$$B = \frac{\mu_0 I}{2\pi r} = \frac{(4\pi \times 10^{-7} \text{ T} \cdot \text{m/A})(15 \text{ A})}{2\pi (0.05 \text{ m})} = 6 \times 10^{-5} \text{ T}$$

31.13. A flat circular coil having 40 loops of wire on it has a diameter of 32 cm. What current must flow in its wires to produce a flux density of 3×10^{-4} Wb/m² at its center?

$$B = \frac{\mu_0 NI}{2r} \qquad \text{or} \qquad 3 \times 10^{-4} \text{ T} = \frac{(4\pi \times 10^{-7} \text{ T} \cdot \text{m/A})(40)I}{2(0.16 \text{ m})}$$

which gives $I = 1.9$ A.

31.14. An air core solenoid with 2000 loops on it is 60 cm long and has a diameter of 2 cm. If a current of 5 A is sent through it, what will be the flux density within it?

$$B = \mu_0 n I = (4\pi \times 10^{-7} \text{ T} \cdot \text{m/A}) \left(\frac{2000}{0.6 \text{ m}} \right) (5 \text{ A}) = 0.021 \text{ T}$$

31.15. In Bohr's model of the hydrogen atom, the electron travels with speed 2.2×10^6 m/s in a circle ($r = 5.3 \times 10^{-11}$ m) about the nucleus. Find the value of B at the nucleus because of the electron's motion.

In Problem 26.6 we found that the orbiting electron corresponds to a current loop with $I = 1.06$ mA. The field at the center of the current loop is

$$B = \frac{\mu_0 I}{2r} = \frac{(4\pi \times 10^{-7} \text{ T} \cdot \text{m/A})(1.06 \times 10^{-3} \text{ A})}{2(5.3 \times 10^{-11} \text{ m})} = 12.5 \text{ T}$$

31.16. A long wire carries a current of 20 A along the axis of a long solenoid. The field due to the solenoid is 4 mT. Find the resultant field at a point 3 mm from the solenoid axis.

The situation is shown in Fig. 31-13. The field of the solenoid, $\mathbf{B}_s$, is directed parallel to the wire. The field of the long straight wire, $\mathbf{B}_w$, circles the wire and is perpendicular to $\mathbf{B}_s$. We have $B_s = 4$ mT and

$$B_w = \frac{\mu_0 I}{2\pi r} = \frac{(4\pi \times 10^{-7} \text{ T} \cdot \text{m/A})(20 \text{ A})}{2\pi (3 \times 10^{-3} \text{ m})} = 1.33 \text{ mT}$$

Since $\mathbf{B}_s$ and $\mathbf{B}_w$ are perpendicular, their resultant, $\mathbf{B}$, has magnitude

$$B = \sqrt{(4)^2 + (1.33)^2} = 4.2 \text{ mT}$$

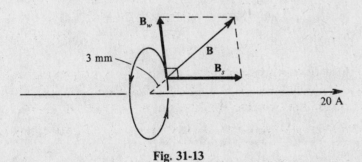

Fig. 31-13   Fig. 31-14

31.17. As shown in Fig. 31-14, two long parallel wires are 10 cm apart and carry currents of 6 A and 4 A. Find the force on a 1 m length of wire D if the currents are (a) parallel and (b) anti-parallel.

(a) This is the situation shown in Fig. 31-14. The field at wire D due to wire C is into the page and has the value

$$B = \frac{\mu_0 I}{2\pi r} = \frac{(4\pi \times 10^{-7} \text{ T} \cdot \text{m/A})(6 \text{ A})}{2\pi (0.10 \text{ m})} = 1.2 \times 10^{-5} \text{ T}$$

The force on 1 m of wire D due to this field is

$$F = ILB \sin \theta = (4 \text{ A})(1 \text{ m})(1.2 \times 10^{-5} \text{ T})(\sin 90°) = 4.8 \times 10^{-5} \text{ N}$$

The right-hand rule applied to wire D tells us the force on D is toward the left. The wires attract each other.

(b) If the current in D flows in the reverse direction, the force direction will be reversed. The wires will repel each other. The force per meter length is still 4.8×10^{-5} N.

31.18. Consider the three long, straight, parallel wires shown in Fig. 31-15. Find the force experienced by a 25 cm length of wire C.

The fields due to wires D and G at wire C are

$$B_D = \frac{\mu_0 I}{2\pi r} = \frac{(4\pi \times 10^{-7} \text{ T} \cdot \text{m/A})(30 \text{ A})}{2\pi (0.03 \text{ m})} = 2 \times 10^{-4} \text{ T}$$

into the page, and

$$B_G = \frac{(4\pi \times 10^{-7} \text{ T} \cdot \text{m/A})(20 \text{ A})}{2\pi (0.05 \text{ m})} = 0.8 \times 10^{-4} \text{ T}$$

out of the page. Therefore the field at the position of wire C is

$$B = 2 \times 10^{-4} - 0.8 \times 10^{-4} = 1.2 \times 10^{-4} \text{ T}$$

into the page. The force on a 25 cm length of C is

$$F = ILB \sin\theta = (10 \text{ A})(0.25 \text{ m})(1.2 \times 10^{-4} \text{ T})(\sin 90°) = 3 \times 10^{-4} \text{ N}$$

Using the right-hand rule at wire C tells us that the force on wire C is toward the right.

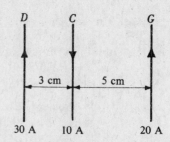

Fig. 31-15

31.19. A flat circular coil with 10 loops of wire on it has a diameter of 2 cm and carries a current of 0.5 A. It is mounted inside a long solenoid that has 200 loops on its 25 cm length. The current in the solenoid is 2.4 A. Compute the torque required to hold the coil with its axis perpendicular to that of the solenoid.

Let the subscripts s and c refer to the solenoid and coil respectively. Then

$$\tau = N_c I_c A_c B_s \sin 90°$$

But $B_s = \mu_0 n I_s = \mu_0 (N_s/L_s) I_s$,   which gives

$$\tau = \mu_0 N_c N_s I_c I_s (\pi r_c^2)/L_s$$

$$= (4\pi \times 10^{-7} \text{ T} \cdot \text{m/A})(10)(200)(0.5 \text{ A})(2.4 \text{ A})\pi(0.01 \text{ m})^2/(0.25 \text{ m})$$

$$= 3.8 \times 10^{-6} \text{ N} \cdot \text{m}$$

31.20. The wire shown in Fig. 31-16 carries a current of 40 A. Find the field at point P.

Since P lies on the lines of the straight wires, they contribute no field at P. A circular loop of radius r gives a field of $B = \mu_0 I/2r$ at its center point. Here we have only 3/4 of a loop and so

$$B \text{ at point } P = \frac{3}{4}\left(\frac{\mu_0 I}{2r}\right) = \frac{3 \times (4\pi \times 10^{-7} \text{ T} \cdot \text{m/A})(40 \text{ A})}{4 \times (2)(0.02 \text{ m})}$$

$$= 9.4 \times 10^{-4} \text{ T}$$

The field is out of the page.

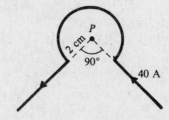

Fig. 31-16

Supplementary Problems

31.21. An ion ($q = +2e$) enters a magnetic field with flux density 1.2 Wb/m² at velocity of 2.5×10^5 m/s perpendicular to the field. Determine the force on the ion.   *Ans.* 9.6×10^{-14} N

31.22. Calculate the velocity of certain ions that pass undeflected through crossed E and B fields for which $E = 7.7$ kV/m   and $B = 0.14$ T.   *Ans.* 5.5×10^4 m/s

31.23. What might be the mass of a positive ion that is moving at 10^7 m/s and is bent into a circular path of radius 1.55 m by a magnetic field of 0.134 Wb/m²? There are several possible answers.
Ans. $n(3.3 \times 10^{-27}$ kg), where *ne* is the ion's charge.

31.24. An electron is accelerated from rest through a potential difference of 3750 V. It enters a region where $B = 4 \times 10^{-3}$ T perpendicular to its velocity. Calculate the radius of the path it will follow.
Ans. 5.2 cm

31.25. An electron is shot with speed 5×10^6 m/s out from the origin of coordinates. Its initial velocity makes an angle of 20° to the +x-axis. Describe its motion if a magnetic field $B = 0.0020$ T exists in the +x-direction. *Ans.* helix, $r = 0.49$ cm, pitch = 8.5 cm

31.26. The particle shown in Fig. 31-17 is positively charged. What is the direction of the force on it due to the magnetic field? Give its magnitude in terms of *B*, *q*, and *v*.
Ans. (a) into page, qvB; (b) out of page, $qvB \sin \theta$; (c) in plane of page at an angle $\theta + 90°$, qvB

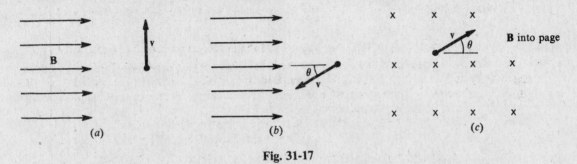

Fig. 31-17

31.27. A straight wire 15 cm long, carrying a current of 6 A, is in a uniform field of magnetic induction 0.4 T. What is the force on the wire when it is (a) at right angles to the field, (b) at 30° to the field?
Ans. (a) 0.36 N; (b) 0.18 N

31.28. What is the direction of the force, due to the earth's magnetic field, on a wire carrying current vertically downward? (*Hint:* The field is as if due to rotating negative charges.)
Ans. horizontally toward east

31.29. Find the force on each segment of the wire shown in Fig. 31-18 if $B = 0.15$ T. Assume the current in the wire to be 5 A.
Ans. In sections *AB* and *DE*, the force is zero; in section *BC*, 0.12 N into page; in section *CD*, 0.12 N out of page.

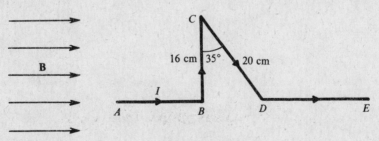

Fig. 31-18

31.30. A rectangular coil of 25 loops is suspended in a field of magnetic induction 0.2 Wb/m². The plane of the coil is parallel to the direction of the field. The dimensions of the coil are 15 cm perpendicular to the field lines and 12 cm parallel to them. What is the current in the coil if there is a torque of 5.4 N·m acting on it? *Ans.* 60 A

31.31. Compute the flux density in air at a point 6 cm from a long straight wire carrying a current of 9 A. *Ans.* 3×10^{-5} T

31.32. A closely wound, flat circular coil of 25 turns of wire has a diameter of 10 cm and carries a current of 4 A. Determine the magnetic induction at its center. *Ans.* 1.26×10^{-3} Wb/m^2

31.33. An air core solenoid 50 cm long has 4000 loops wound on it. Compute B in its interior when a current of 0.25 A exists in the winding. *Ans.* 2.5×10^{-3} T

31.34. A uniformly wound, air core toroid has 750 loops on it. The radius of the circle through the center of its windings is 5 cm. What current in the winding will produce a flux density of 1.8×10^{-3} T on this central circle? *Ans.* 0.6 A

31.35. Two long parallel wires are 4 cm apart and carry currents of 2 A and 6 A in the same direction. Compute the force between the wires per meter of wire length. *Ans.* 6×10^{-5} N/m attraction

31.36. Two long and fixed parallel wires, A and B, are 10 cm apart in air and carry currents of 40 A and 20 A, respectively, in opposite directions. Determine the resultant flux density (*a*) on a line midway between the wires and parallel to them, (*b*) on a line 8 cm from wire A and 18 cm from wire B. (*c*) What is the force per meter on a third long wire, midway between A and B and in their plane, when it carries a current of 5 A in the same direction as the current in A?
Ans. (*a*) 2.4×10^{-4} T; (*b*) 7.8×10^{-5} T; (*c*) 1.2×10^{-3} N/m toward A

31.37. A beam of electrons passes undeflected through two mutually perpendicular electric and magnetic fields. If the electric field is cut off and the same magnetic field maintained, the electrons move in the magnetic field in a circular path of radius 1.14 cm. Determine the ratio of the electronic charge to mass if $E = 8$ kV/m and the magnetic field has flux density 2×10^{-3} T. *Ans.* $e/m = 1.75 \times 10^{11}$ C/kg

Chapter 32

Induced EMF; Magnetic Flux

MAGNETIC EFFECTS OF MATTER: Most materials have only a very slight effect on a steady magnetic field. The effect of a material is best described in terms of an experiment.

Suppose that a very long solenoid or a toroid is used in vacuum. With a fixed current in the coil, the magnetic induction at a certain point in the solenoid or toroid is B_v, where the subscript v stands for vacuum. If now the solenoid or toroid core is filled with a material, the field at that point will change to a new value B. We define:

$$\text{relative permeability of the material} = k_m = B/B_v$$
$$\text{permeability of the material} = \mu = k_m \mu_0$$

Recall that μ_0 is the permeability of free space, $4\pi \times 10^{-7} \, \text{T} \cdot \text{m/A}$.

Diamagnetic materials have values for k_m slightly less than unity (0.999 984 for solid lead, for example). They slightly decrease the value of B in the solenoid or toroid.

Paramagnetic materials have values for k_m slightly larger than unity (1.000 021 for solid aluminum, for example). They slightly increase the value of B in the solenoid or toroid.

Ferromagnetic materials, such as iron and its alloys, have k_m-values of about 50 or larger. They greatly increase the value of B in the toroid or solenoid.

MAGNETIC FLUX LINES: A magnetic field may be represented by field lines, to which **B** is everywhere tangent. If, furthermore, these lines are constructed in such a way that the number of lines piercing a unit area perpendicular to them is equal to the local value of B, then the lines are called *lines of magnetic flux*.

THE MAGNETIC FLUX (Φ) through an area A is defined to be the number of flux lines that pass through the area. If $B_\perp$ is the component of **B** perpendicular to the surface of area A, then

$$\text{flux through area} = \Phi = B_\perp A$$

The flux is expressed in *webers* (Wb).

AN INDUCED EMF exists in a loop whenever there is a change in the flux through the area of the loop. The induced emf exists only during the time that the flux through the area is changing.

FARADAY'S LAW FOR INDUCED EMF: Suppose that a coil with N loops is subject to a changing magnetic flux through the coil. If a change in flux through the coil, $\Delta\Phi$, occurs in a time Δt, then the average emf induced between the two terminals of the coil is given by

$$\mathcal{E} = -N\frac{\Delta\Phi}{\Delta t}$$

The emf $\mathcal{E}$ is in V if $\Delta\Phi/\Delta t$ is in Wb/s. The minus sign indicates that the induced emf opposes the change which produces it, as stated generally in

LENZ'S LAW: An induced emf is always in such a direction as to oppose the change in magnetic flux that produced it. For example, if the flux is increasing through a coil, the current produced by the induced emf will generate a flux that tends to cancel the increasing flux. Or, if the flux is decreasing through the coil, the induced emf's current will produce a flux that tends to restore the decreasing flux.

MOTIONAL EMF: When a conductor moves through a magnetic field so as to cut lines of flux, an induced emf will exist in it, in accordance with Faraday's law.   In this case, $\Delta\Phi$, the change in flux, is the number of flux lines cut by the moving conductor, and so

$$|\mathcal{E}| = \frac{\Delta\Phi}{\Delta t} = \text{number of lines cut per second}$$

The symbolism $|\mathcal{E}|$ means that we are concerned here only with the magnitude of the average induced emf; its direction will be considered below.

   The induced emf in a straight conductor of length L moving with velocity **v** perpendicular to a field **B** is given by

$$|\mathcal{E}| = BLv$$

where **B**, **v**, and the wire are mutually perpendicular.

   In this case Lenz's law still tells us that the induced emf opposes the process.   But now the opposition is produced by way of the force exerted by the magnetic field on the induced current in the conductor.   The current direction must be such that the force opposes the motion of the conductor. Knowing the current direction, we also know the direction of $\mathcal{E}$.

Solved Problems

32.1.   A solenoid is 40 cm long, has cross-sectional area 8 cm^2, and is wound with 300 turns of wire that carry a current of 1.2 A.   The relative permeability of its iron core is 600.   Compute (*a*) B for an interior point and (*b*) the flux through the solenoid.

   (*a*)   From Chapter 31, for a long solenoid,

$$B_v = \frac{\mu_0 NI}{L} = \frac{(4\pi \times 10^{-7}\ \text{T}\cdot\text{m/A})(300)(1.2\ \text{A})}{0.40\ \text{m}} = 1.13 \times 10^{-3}\ \text{T}$$

and so

$$B = k_m B_v = (600)(1.13 \times 10^{-3}\ \text{T}) = 0.68\ \text{T}$$

   (*b*)   Because the field lines are perpendicular to the cross-sectional area inside the solenoid,

$$\Phi = B_\perp A = BA = (0.68\ \text{T})(8 \times 10^{-4}\ \text{m}^2) = 5.4 \times 10^{-4}\ \text{Wb}$$

32.2.   The flux through a certain toroid changes from 0.65 mWb to 0.91 mWb when the air core is replaced by another material.   What are the relative permeability and the permeability of the material?

   The air core is essentially the same as a vacuum core.   Since $k_m = B/B_v$ and $\Phi = B_\perp A$,

$$k_m = \frac{\Phi}{\Phi_v} = \frac{0.91\ \text{mWb}}{0.65\ \text{mWb}} = 1.40$$

This is the relative permeability.   The magnetic permeability is

$$\mu = k_m \mu_0 = (1.40)(4\pi \times 10^{-7}\ \text{T}\cdot\text{m/A}) = 5.6\,\pi \times 10^{-7}\ \text{T}\cdot\text{m/A}$$

32.3.   The quarter-circle loop shown in Fig. 32-1 has an area of 15 cm^2.   A magnetic field, with $B = 0.16$ T,   exists in the $+x$-direction.   Find the flux through the loop in each orientation shown.

   We know that $\Phi = B_\perp A$.

   (*a*) $\Phi = B_\perp A = BA = (0.16\ \text{T})(15 \times 10^{-4}\ \text{m}^2) = 2.4 \times 10^{-4}\ \text{Wb}$

   (*b*) $\Phi = (B \cos 20°)A = (2.4 \times 10^{-4}\ \text{Wb})(\cos 20°) = 2.26 \times 10^{-4}\ \text{Wb}$

   (*c*) $\Phi = (B \sin 20°)A = (2.4 \times 10^{-4}\ \text{Wb})(\sin 20°) = 8.2 \times 10^{-5}\ \text{Wb}$

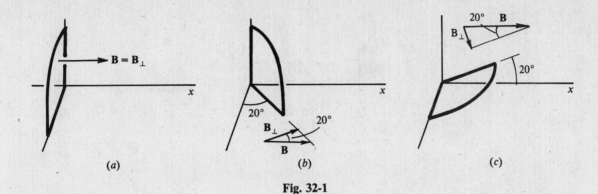

Fig. 32-1

32.4. A 50-loop circular coil has a radius of 3 cm. It is oriented so that the field lines of a magnetic field are parallel to a normal to the area of the coil. Suppose that the magnetic field is varied so that B increases from 0.10 T to 0.35 T in a time of 2 milliseconds. Find the average induced emf in the coil.

$$\Delta\Phi = B_{final}A - B_{initial}A = (0.25\ T)(\pi r^2) = (0.25\ T)\pi(0.03\ m)^2 = 7.1 \times 10^{-4}\ Wb$$

$$|\mathcal{E}| = N\left|\frac{\Delta\Phi}{\Delta t}\right| = (50)\frac{7.1 \times 10^{-4}\ Wb}{2 \times 10^{-3}\ s} = 17.7\ V$$

32.5. The magnet in Fig. 32-2 induces an emf in the coils as the magnet moves toward the right or the left. Find the directions of the induced currents through the resistors when the magnet is moving (a) toward the right and (b) toward the left.

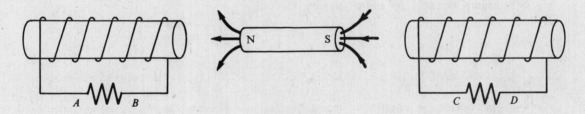

Fig. 32-2

(a) Consider first the coil on the left. As the magnet moves to the right, the flux through the coil, which is directed generally to the left, decreases. To compensate for this, the induced current in the coil will flow so as to produce a flux toward the left through itself. Apply the right-hand rule to the loop on the left end. For it to produce flux inside the coil directed toward the left, the current must flow through the resistor from B to A.

    Now consider the coil on the right. As the magnet moves toward the right, the flux inside the coil, also generally to the left, increases. The induced current in the coil will produce a flux toward the right to cancel this increased flux. Applying the right-hand rule to the loop on the right end, we find that the loop generates flux to the right inside itself if the current flows from C to D through the resistor.

(b) In this case the flux change caused by the magnet's motion is opposite to what it was in (a). Using the same type of reasoning, the induced currents flow through the resistors from A to B and from D to C.

32.6. In Fig. 32-3(a) there is a magnetic field in the $+x$-direction with $B = 0.20$ T. The loop has an area of 5 cm^2 and rotates about line CD as axis. Point A rotates toward positive x-values from the position shown. If line AE rotates through 50° from its indicated position in a time of 0.2 s, (a) what is the change in flux through the coil? (b) what is the average induced emf in it? (c) does the induced current flow from A to C or C to A in the upper part of the coil?

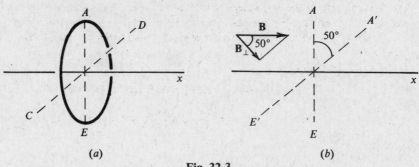

Fig. 32-3

(a)   initial flux $= B_{\perp}A = BA = (0.20 \text{ T})(5 \times 10^{-4} \text{ m}^2) = 1 \times 10^{-4} \text{ Wb}$

final flux $= (B \cos 50°)A = (1 \times 10^{-4} \text{ Wb})(\cos 50°) = 0.64 \times 10^{-4} \text{ Wb}$

$\Delta\Phi = 0.64 \times 10^{-4} \text{ Wb} - 1 \times 10^{-4} \text{ Wb} = -0.36 \times 10^{-4} \text{ Wb}$

(b) $|\mathscr{E}| = N\left|\dfrac{\Delta\Phi}{\Delta t}\right| = (1)\dfrac{0.36 \times 10^{-4} \text{ Wb}}{0.2 \text{ s}} = 1.8 \times 10^{-4} \text{ V}$

(c)   The flux through the loop from left to right decreased.   The induced current will tend to set up flux from left to right through the loop.   By the right-hand rule, the current flows from A to C. Alternatively, a torque must be set up that tends to rotate the loop back into its original position. The appropriate right-hand rule from Chapter 31 again gives a current flow from A to C.

32.7.   A coil of 50 loops is pulled in 0.02 s from between the poles of a magnet, where its area intercepts a flux of 3.1×10^{-4} Wb, to a place where the intercepted flux is 0.1×10^{-4} Wb. Determine the average emf induced in the coil.

$$|\mathscr{E}| = N\left|\frac{\Delta\Phi}{\Delta t}\right| = 50\,\frac{(3.1 - 0.1) \times 10^{-4} \text{ Wb}}{0.02 \text{ s}} = 0.75 \text{ V}$$

32.8.   A 5 Ω coil, of 100 turns and diameter 6 cm, is placed between the poles of a magnet so that the flux is maximum through its area.   When the coil is suddenly removed from the field of the magnet, a charge of 10^{-4} C flows through a 595 Ω galvanometer connected to the coil. Compute B between the poles of the magnet.

As the coil is removed, the flux changes from BA, where A is the coil area, to zero.   Therefore,

$$|\mathscr{E}| = N\left|\frac{\Delta\Phi}{\Delta t}\right| = N\,\frac{BA}{\Delta t}$$

We are told that $\Delta q = 10^{-4}$ C.   But, by Ohm's law,

$$|\mathscr{E}| = IR = \frac{\Delta q}{\Delta t}\,R$$

where $R = 600$ Ω, the total resistance.   If we now equate these two expressions for $|\mathscr{E}|$ and solve for B, we find

$$B = \frac{R\,\Delta q}{NA} = \frac{(600 \text{ Ω})(10^{-4} \text{ C})}{(100)(\pi \times 9 \times 10^{-4} \text{ m}^2)} = 0.212 \text{ T}$$

32.9.   A copper bar 30 cm long is perpendicular to a field of flux density 0.8 Wb/m^2 and moves at right angles to the field with a speed of 0.5 m/s.   Determine the emf induced in the bar.

$$|\mathscr{E}| = BLv = (0.8 \text{ Wb/m}^2)(0.3 \text{ m})(0.5 \text{ m/s}) = 0.12 \text{ V}$$

32.10.   As shown in Fig. 32-4, a metal rod makes contact with a partial circuit and completes the circuit.   The circuit area is perpendicular to a magnetic field with $B = 0.15$ T.   If the resistance of the total circuit is 3 Ω, how large a force is needed to move the rod as indicated with a constant speed of 2 m/s?

The induced emf in the rod causes a current to flow counterclockwise in the circuit. Because of this current in the rod, it experiences a force to the left due to the magnetic field. In order to pull the rod to the right with constant speed, this force must be balanced by the puller.

Method 1

The induced emf in the rod is

$$|\mathcal{E}| = BLv = (0.15\ \text{T})(0.50\ \text{m})(2\ \text{m/s}) = 0.15\ \text{V}$$

Then

$$I = \frac{|\mathcal{E}|}{R} = \frac{0.15\ \text{V}}{3\ \Omega} = 0.050\ \text{A}$$

from which

$$F = ILB \sin 90° = (0.050\ \text{A})(0.50\ \text{m})(0.15\ \text{T})(1) = 3.75 \times 10^{-3}\ \text{N}$$

Method 2

The emf induced in the loop is

$$|\mathcal{E}| = N\left|\frac{\Delta\Phi}{\Delta t}\right| = (1)\frac{B\,\Delta A}{\Delta t} = \frac{B(L\,\Delta x)}{\Delta t} = BLv$$

as before. Now proceed as in Method 1 or as follows:

mechanical power supplied to circuit = rate at which electrical work is done on charges

$$Fv = \frac{(\Delta q)|\mathcal{E}|}{\Delta t} = I|\mathcal{E}| = \frac{|\mathcal{E}|^2}{R}$$

Substitute for $|\mathcal{E}|$ and solve for F, to find

$$F = \frac{B^2 L^2 v}{R} = \frac{(0.15\ \text{T})^2 (0.50\ \text{m})^2 (2\ \text{m/s})}{3\ \Omega} = 3.75 \times 10^{-3}\ \text{N}$$

32.11. The rod shown in Fig. 32-5 rotates about point C as pivot with the constant frequency 5 rev/s. Find the potential difference between its two ends, 80 cm apart, because of the magnetic field, $B = 0.3$ T.

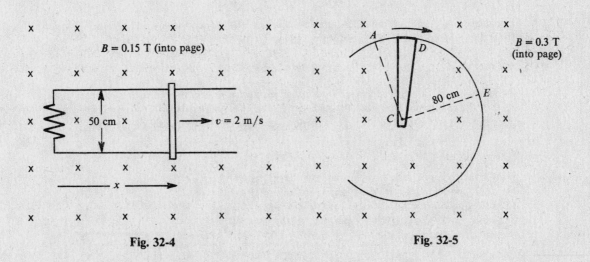

Fig. 32-4   Fig. 32-5

Consider a fictitious loop $CADC$. As time goes on, its area and the flux through it will increase. The induced emf in this loop will equal the potential difference we seek.

$$|\mathcal{E}| = N\left|\frac{\Delta\Phi}{\Delta t}\right| = (1)\frac{B\,\Delta A}{\Delta t}$$

It takes one-fifth second for the area to change from zero to that of a full circle, πr^2. Therefore,

$$|\mathcal{E}| = B\frac{\Delta A}{\Delta t} = B\frac{\pi r^2}{0.20\ \text{s}} = (0.3\ \text{T})\frac{\pi (0.8\ \text{m})^2}{0.20\ \text{s}} = 3.0\ \text{V}$$

Supplementary Problems

32.12. A flux of 9×10^{-4} Wb is produced in the iron core of a solenoid. When the core is removed, a flux (in air) of 5×10^{-7} Wb is produced in the same solenoid by the same current. What is the relative permeability of the iron?   *Ans.* 1800

32.13. In Fig. 32-6 there is a $+x$-directed magnetic field of 0.2 T. Find the magnetic flux through each face of the box shown.
Ans. Zero through bottom and rear and front sides; through top, 1 mWb; through left side, 1.8 mWb; through right side, 0.8 mWb.

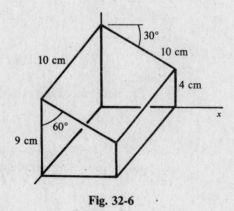

Fig. 32-6

32.14. A solenoid 60 cm long has 5000 turns on it and is wound on an iron rod of 0.75 cm radius. Find the flux through the solenoid when the current in it is 3 A. The relative permeability of the iron is 300.   *Ans.* 1.67 mWb

32.15. A room has its walls aligned accurately with respect to north, south, east, and west. The north wall has an area of 15 m², the east wall has an area of 12 m², and the floor's area is 35 m². At the site the earth's magnetic field has a value of 0.60 G and is directed 50° below the horizontal and 7° east of north. Find the fluxes through the north wall, the east wall, and the floor.
Ans. 5.7×10^{-4} Wb, 0.56×10^{-4} Wb, 16×10^{-4} Wb

32.16. For the solenoid of Problem 32.14, the flux through it is reduced to a value of 1 mWb in a time of 0.05 s. Find the induced emf in the solenoid.   *Ans.* 67 V

32.17. A flat coil with radius 8 mm has 50 loops of wire on it. It is placed in a magnetic field, $B = 0.30$ T, so that the maximum flux goes through it. Later, it is rotated in 0.02 s to a position such that no flux goes through it. Find the average emf induced between the terminals of the coil.   *Ans.* 0.15 V

32.18. How much charge will flow through a 200 Ω galvanometer connected to a 400 Ω circular coil of 1000 turns wound on a wooden stick 2 cm in diameter, if a magnetic field $B = 0.0113$ T parallel to the axis of the stick is decreased suddenly to zero?
Ans. 5.9 μC

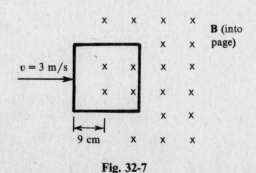

Fig. 32-7

32.19. The square coil shown in Fig. 32-7 is 20 cm on a side and has 15 loops of wire on it. It is moving to the right at 3 m/s. Find the induced emf (magnitude and direction) in it (*a*) at the instant shown, (*b*) when the coil is all in the field region. The magnetic field is 0.4 T into the page.
Ans. (*a*) 3.6 V counterclockwise; (*b*) zero

32.20. The magnet in Fig. 32-8 rotates as shown on a pivot through its center. At the instant shown, in what direction is the induced current flowing (*a*) in resistor *AB*, (*b*) in resistor *CD*?
      Ans. (*a*) *B* to *A*;   (*b*) *C* to *D*

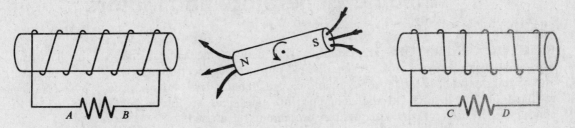

Fig. 32-8

32.21. A train is moving directly south with a speed of 10 m/s. If the downward vertical component of the earth's magnetic field is 0.54 G, compute the magnitude and direction of the emf induced in a rail car axle 1.2 m long.   *Ans.* 6.5×10^{-4} V from west to east

32.22. A copper disk of 10 cm radius is rotating at 20 rev/s about its axis and with its plane perpendicular to a uniform field with $B = 0.6$ T.   What is the potential difference between the center and rim of the disk? (*Hint*: There is some similarity with Problem 32.11.)   *Ans.* 0.38 V

Chapter 33

Electric Generators and Motors

ELECTRIC GENERATORS are machines that convert mechanical energy into electrical energy. They consist fundamentally of a magnetic field produced by electromagnets or by permanent magnets, and of an armature having an iron core and carrying conductors on its surface. When the armature is revolved the conductors cut the magnetic flux lines and an alternating emf is induced in each conductor. If a direct current is desired, a commutator is used.

In most types of direct-current generators the field is stationary and the armature rotates, while in alternating-current generators the armature usually is stationary and the field rotates. In either case, the emf is generated by the *relative* motion of the armature conductors and the magnetic field.

ELECTRIC MOTORS convert electrical energy into mechanical energy. The armature of a motor is closely similar to that of a generator. When a motor is running, the armature conductors cut magnetic flux lines; hence they develop an induced emf, called the *back emf* or *counter emf*, which opposes the impressed voltage.

The armature of a *shunt motor* is connected directly across the supply line. Since the back emf of the armature opposes the line voltage, the net potential difference which causes current through the armature is

$$\text{net p.d. across armature} = (\text{line voltage}) - (\text{back emf})$$

and
$$\text{armature current} = \frac{(\text{line voltage}) - (\text{back emf})}{\text{armature resistance}}$$

The mechanical power P developed within the armature of a motor is

$$P = (\text{armature current})(\text{back emf})$$

The useful mechanical power delivered by the motor is slightly less, due to friction, windage and iron losses.

Solved Problems

ELECTRIC GENERATORS

33.1. A two-pole generator has a drum armature with a cylindrical core 20 cm long and 10 cm in diameter, and 200 armature conductors arranged in two parallel paths. The flux density in the air gap is 0.5 Wb/m². If the armature is driven at 30 rps, what is the magnitude of the average emf generated?

The flux in the air gap is $\Phi = BA = (0.5 \text{ Wb/m}^2)(0.2 \text{ m} \times 0.1 \text{ m}) = 0.01 \text{ Wb}$. Each conductor cuts 0.01 Wb twice per revolution. Then in 1 s, or 30 revolutions, each conductor cuts $2(30)(0.01) = 0.6$ Wb. As there are 200 conductors in two parallel paths, the number of conductors in series in each path is $200/2 = 100$. (The emf's in series are additive.)

$$|\mathcal{E}| = \text{flux lines cut per s by 100 conductors in series} = (100)(0.60 \text{ Wb/s}) = 60 \text{ V}$$

236

33.2.   A 6-pole generator with fixed-field excitation develops an emf of 100 V when operating at 1500 rpm.   At what speed must it rotate to develop 120 V?

The flux cut per second, and hence the emf, varies directly as the speed.

$$\text{speed} = \frac{120}{100}\,(1500 \text{ rpm}) = 1800 \text{ rpm}$$

33.3.   A shunt generator has armature resistance 0.08 Ω and develops a total induced emf of 120 V when driven at its rated speed.   What is the terminal voltage V if the armature current is 50 A?

$$V = (\text{emf}) - (IR \text{ drop in armature}) = 120 \text{ V} - (50 \text{ A})(0.08 \text{ }\Omega) = 116 \text{ V}$$

33.4.   A shunt generator has armature resistance 0.06 Ω and shunt-field resistance 100 Ω. What power is developed in the armature when it delivers 40 kW at 250 V to an external circuit?   See Fig. 33-1.

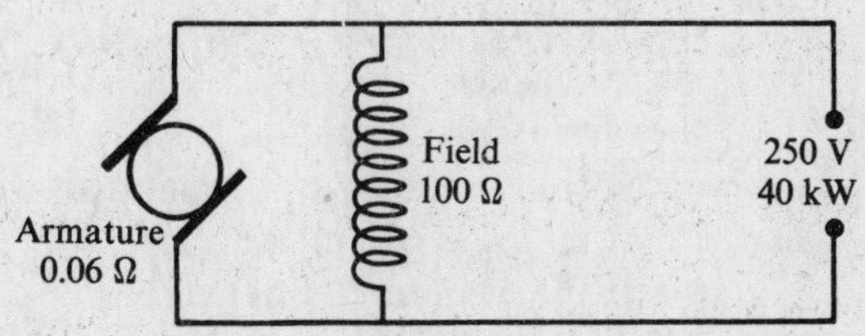

Fig. 33-1

$$\text{current in external circuit} = I_x = \frac{I_x V}{V}$$
$$= \frac{40\,000 \text{ W}}{250 \text{ V}} = 160 \text{ A}$$

$$\text{field current} = I_f = \frac{V_f}{R_f} = \frac{250 \text{ V}}{100 \text{ }\Omega} = 2.5 \text{ A}$$

$$\text{armature current} = I_a = I_x + I_f = 162.5 \text{ A}$$

total induced emf $= |\mathscr{E}| = (250 \text{ V} + I_a R_a \text{ drop in armature}) = 250 \text{ V} + (162.5 \text{ A})(0.06 \text{ }\Omega) = 260 \text{ V}$

power developed by armature $= I_a |\mathscr{E}| = (162.5 \text{ A})(260 \text{ V}) = 42.2 \text{ kW}$

Another Method

power loss in armature $= I_a^2 R_a = (162.5 \text{ A})^2 (0.06 \text{ }\Omega) = 1.6 \text{ kW}$

power loss in field $= I_f^2 R_f = (2.5 \text{ A})^2 (100 \text{ }\Omega) = 0.6 \text{ kW}$

power developed $=$ (power delivered) $+$ (power loss in armature) $+$ (power loss in field)
$$= 40 \text{ kW} + 1.6 \text{ kW} + 0.6 \text{ kW} = 42.2 \text{ kW}$$

33.5.   At what speed must an 8-pole ac generator rotate to produce an emf having a frequency of 60 Hz?

For each pair of poles there is produced one cycle (360 electrical degrees) per revolution.   Hence 8 poles give 4 cycles per revolution.

$$\text{required angular speed} = \frac{60 \text{ cycles/s}}{4 \text{ cycles/rev}} = 15 \text{ rev/s} = 900 \text{ rpm}$$

ELECTRIC MOTORS

33.6.   The armature of a 4-pole shunt motor carries 320 surface conductors, arranged in four parallel paths, which cut a flux of 0.025 Wb per pole.   (a) Determine the back emf developed by the armature when rotating at 900 rpm.   (b) What is the impressed voltage V if the armature resistance R_a is 0.1 Ω and the armature current I_a is 40 A?

(a) Each conductor cuts 0.025 Wb four times per revolution. Then in 1 s, or 15 revolutions, each conductor cuts 4(15)(0.025) = 1.5 Wb. Since the 320 conductors are in 4 parallel paths, there are 320/4 = 80   conductors in series in each path.

back emf = $|\mathcal{E}_b|$ = flux lines cut per s by 80 conductors in series = (80)(1.5 Wb/s) = 120 V

(b) $V = |\mathcal{E}_b| + I_a R_a = 120 \text{ V} + (40 \text{ A})(0.1 \text{ }\Omega) = 124 \text{ V}$

33.7. The shunt motor shown in Fig. 33-2 has armature resistance 0.05 Ω and is connected to 120 V mains. (a) What is the armature current at the starting instant, i.e. before the armature develops any back emf? (b) What starting rheostat resistance R, in series with the armature, will limit the starting current to 60 A? (c) With no starting resistance, what back emf is generated when the armature current is 20 A? (d) If this machine were running as a generator, what would be the total induced emf developed by the armature when the armature is delivering 20 A at 120 V to the shunt field and external circuit?

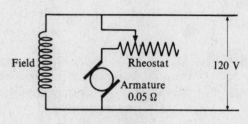

Fig. 33-2

(a)   armature current = $\dfrac{\text{impressed voltage}}{\text{armature resistance}} = \dfrac{120 \text{ V}}{0.05 \text{ }\Omega} = 2400 \text{ A}$

(b)   armature current = $\dfrac{\text{impressed voltage}}{0.05 \text{ }\Omega + R}$ or $60 \text{ A} = \dfrac{120 \text{ V}}{0.05 \text{ }\Omega + R}$

from which   R = 1.95 Ω.

(c)   back emf = (impressed voltage) − (voltage drop in armature resistance)

= 120 V − (20 A)(0.05 Ω) = 119 V

(d)   induced emf = (terminal voltage) + (voltage drop in armature resistance)

= 120 V + (20 A)(0.05 Ω) = 121 V

33.8. The shunt motor shown in Fig. 33-3 has armature resistance 0.25 Ω and field resistance 150 Ω. It is connected across 120 V mains and is generating a back emf of 115 V. Compute: (a) the armature current I_a, the field current I_f, and the total current I_t taken by the motor; (b) the total power taken by the motor; (c) the power lost in heat in the armature and field circuits; (d) the electrical efficiency of this machine (when only heat losses in the armature and field are considered).

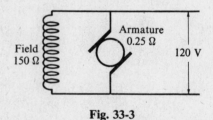

Fig. 33-3

(a) $I_a = \dfrac{\text{(impressed voltage)} - \text{(back emf)}}{\text{armature resistance}} = \dfrac{(120 - 115) \text{ V}}{0.25 \text{ }\Omega} = 20 \text{ A}$

$I_f = \dfrac{\text{impressed voltage}}{\text{field resistance}} = \dfrac{120 \text{ V}}{150 \text{ }\Omega} = 0.80 \text{ A}$ $I_t = I_a + I_f = 20.80 \text{ A}$

(b)   power input = (120 V)(20.80 A) = 2496 W

(c) $I_a^2 R_a$ loss in armature = $(20 \text{ A})^2 (0.25 \text{ }\Omega) = 100 \text{ W}$

$I_f^2 R_f$ loss in field = $(0.80 \text{ A})^2 (150 \text{ }\Omega) = 96 \text{ W}$

(d)   power output = (power input) − (power losses) = 2496 − (100 + 96) = 2300 W

= (armature current)(back emf) = (20 A)(115 V) = 2300 W

$$\text{efficiency} = \frac{\text{power output}}{\text{power input}} = \frac{2300 \text{ W}}{2496 \text{ W}} = 0.921 = 92.1\%$$

33.9. A motor has back emf 110 V and armature current 90 A when running at 1500 rpm. Determine the power and the torque developed within the armature.

power = (armature current)(back emf) = (90 A)(110 V) = 9900 W

$$\text{torque} = \frac{\text{power}}{\text{angular speed}} = \frac{9900 \text{ W}}{(2\pi \times 25) \text{ rad/s}} = 63.0 \text{ N} \cdot \text{m}$$

33.10. A motor armature develops a torque of 100 N·m when it draws 40 A from the line. Determine the torque developed if the armature current is increased to 70 A and the magnetic field strength is reduced to 80% of its initial value.

The torque developed by the armature of a given motor is proportional to the armature current and to the field strength (see Chapter 31).

$$\text{torque} = (100 \text{ N} \cdot \text{m}) \left(\frac{70}{40} \right) (0.80) = 140 \text{ N} \cdot \text{m}$$

Supplementary Problems

ELECTRIC GENERATORS

33.11. Determine the separate effects on the induced emf of a generator if (a) the flux per pole is doubled, (b) the speed of the armature is doubled.   *Ans.* (a) doubled; (b) doubled

33.12. The armature of a two-pole shunt generator has 400 conductors arranged in two parallel paths and cutting 6 mWb per pole at 1800 rpm. Compute the average emf generated.   *Ans.* $|\mathcal{E}| = 72$ V

33.13. A 4-pole dc generator has 240 armature conductors arranged in 4 parallel paths and cutting 20 mWb per pole at 1500 rpm. What is the average emf generated?   *Ans.* $|\mathcal{E}| = 120$ V

33.14. The emf induced in the armature of a shunt generator is 596 V. The armature resistance is 0.1 Ω. (a) Compute the terminal voltage when the armature current is 460 A. (b) The field resistance is 110 Ω. Determine the field current, and the current and power delivered to the external circuit.
Ans. (a) 550 V; (b) 5 A, 455 A, 250 kW

33.15. A dynamo delivers 30 A at 120 V to an external circuit when operating at 1200 rpm. What torque is required to drive the generator at this speed if the total power losses are 400 W?   *Ans.* 31.8 N·m

33.16. A 75 kW, 230 V shunt generator has a generated emf of 243.5 V. If the field current is 12.5 A at rated output, what is the armature resistance?   *Ans.* 0.04 Ω

33.17. What is the voltage frequency generated by a 12-pole alternator when driven at 250 rpm?
Ans. 25 Hz

ELECTRIC MOTORS

33.18. A bipolar shunt generator has an armature with 500 surface conductors, arranged in two parallel paths, which cut a flux of 8 mWb per pole. Compute the back emf it develops when run as a motor at 1500 rpm.   *Ans.*   100 V

33.19. The active length of each armature conductor of a motor is 30 cm and the flux density directly under the poles is 1 Wb/m^2. A current of 15 A is in each conductor. Determine the force acting on each conductor when directly under the poles.   *Ans.*   4.5 N

33.20. A shunt motor with armature resistance 0.08 Ω is connected to 120 V mains. With 50 A in the armature, compute the back emf and the mechanical power developed within the armature.
Ans.   116 V, 5.8 kW

33.21. A shunt motor is connected to a 110 V line. When the armature generates a back emf of 104 V the armature current is 15 A. Compute the armature resistance.   *Ans.*   0.4 Ω

33.22. A shunt dynamo has an armature resistance of 0.12 Ω. (*a*) If it is connected across 220 V mains and is running as a motor, what is the induced (back) emf when the armature current is 50 A? (*b*) If this machine is running as a generator, what is the induced emf when the armature is delivering 50 A at 220 V to the shunt field and external circuit?   *Ans.*   (*a*) 214 V;   (*b*) 226 V

33.23. A shunt motor has a speed of 900 rpm when connected to 120 V mains and delivering 12 hp. The total losses are 1048 W. Compute the power input, the line current and the motor torque.
Ans.   10 kW, 83.3 A,   70 lb·ft = 93 N·m

33.24. A shunt motor has armature resistance 0.20 Ω and field resistance 150 Ω, and draws 30 A when connected to a 120 V supply line. Determine the field current, the armature current, the back emf, the mechanical power developed within the armature, and the electrical efficiency of the machine.
Ans.   0.80 A, 29.2 A, 114.2 V, 3.33 kW, 92.5%

33.25. A shunt motor develops 80 N·m torque when the flux density in the air gap is 1 Wb/m^2 and the armature current is 15 A. What is the torque when the flux density is 1.3 Wb/m^2 and the armature current is 18 A?   *Ans.*   125 N·m

33.26. A shunt motor has a field resistance of 200 Ω, an armature resistance of 0.5 Ω, and is connected to 120 V mains. The motor draws a current of 4.6 A when running at full speed. What current will be drawn by the motor if the speed is reduced to 90% of full speed by application of a load?   *Ans.*   28.2 A

Chapter 34

Inductance; R-C and R-L Time Constants

SELF-INDUCTANCE: A coil can induce an emf in itself. If the current in a coil changes, the flux through the coil due to the current also changes. As a result, the changing current in a coil induces an emf in that same coil.

Because an induced emf $\mathcal{E}$ is proportional to $\Delta\Phi/\Delta t$ and because $\Delta\Phi$ is proportional to Δi, where i is the current that causes the flux,

$$\mathcal{E} = -(\text{constant})\,\frac{\Delta i}{\Delta t}$$

Here i is the current through the same coil in which $\mathcal{E}$ is induced. (We shall denote a time-varying current by i instead of by I.) The negative sign indicates that the self-induced emf, $\mathcal{E}$, is a back emf and opposes the change in current.

The proportionality constant depends upon the geometry of the coil. We represent it by L and call it the *self-inductance* of the coil. Then

$$\mathcal{E} = -L\,\frac{\Delta i}{\Delta t}$$

For $\mathcal{E}$ in V, i in A, and t in s, L is in *henries* (H).

MUTUAL INDUCTANCE: When the flux from one coil threads through another coil, an emf can be induced in either one by the other. The coil that contains the power source is called the *primary coil*. The other coil, in which an emf is induced by the changing current in the primary, is called the *secondary coil*. The induced secondary emf, $\mathcal{E}_s$, is proportional to the time rate of change of the primary current, $\Delta i_p/\Delta t$:

$$\mathcal{E}_s = M\,\frac{\Delta i_p}{\Delta t}$$

where M is a constant called the *mutual inductance* of the two-coil system.

ENERGY STORED IN AN INDUCTOR: Because of its self-induced back emf, work must be done to increase the current through an inductor from zero to I. The energy furnished to the coil in the process is stored in the coil and can be recovered as the coil's current is decreased once again to zero. If a current I is flowing in an inductor of self-inductance L, then the energy stored in the inductor is

$$\text{stored energy} = \tfrac{1}{2}LI^2$$

For L in henries and I in amperes, the energy is in J.

R-C TIME CONSTANT: Consider the circuit shown in Fig. 34-1(a). The capacitor is initially uncharged. If the switch is now closed, the current i in the circuit and the charge q on the capacitor vary as shown in Fig. 34-1(b). Writing the loop rule for this circuit gives, calling the p.d. across the capacitor v_c,

$$-iR - v_c + \mathcal{E} = 0 \qquad \text{or} \qquad i = \frac{\mathcal{E} - v_c}{R}$$

At the first instant after the switch is closed, $v_c = 0$ and $i = \mathcal{E}/R$. As time goes on, v_c increases and i decreases. The time, in seconds, taken for the current to drop to $1/2.718$ or 0.368 of its initial value is RC, which is called the *time constant* of the R-C circuit.

241

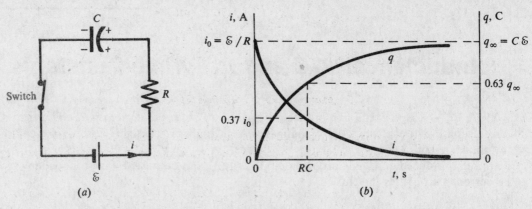

Fig. 34-1

Also shown in Fig. 34-1(b) is the variation of q, the charge on the capacitor, with time. At t = RC, q has attained 0.632 of its final value.

When a charged capacitor C with initial charge q_0 is discharged through a resistor R, its discharge current follows the same curve as for charging. The charge q on the capacitor follows a curve similar to that for the discharge current. At time RC, $i = 0.368\,i_0$ and $q = 0.368\,q_0$ during discharge.

R-L TIME CONSTANT: Consider the circuit in Fig. 34-2(a). The symbol —⎛⎞⎛⎞⎛⎞— represents a coil of self-inductance L henries. When the switch in the circuit is first closed, the current in the circuit rises as shown in Fig. 34-2(b). The current does not jump to its final value because the changing flux through the coil induces a back emf in the coil, which opposes the rising current. After L/R seconds, the current has risen to 0.632 of its final value i_∞. This time, t = L/R, is called the *time constant* of the R-L circuit. After a long time, the current is changing so slowly that the back emf in the inductor, $L(\Delta i/\Delta t)$, is negligible. Then $i = i_\infty = \mathcal{E}/R$.

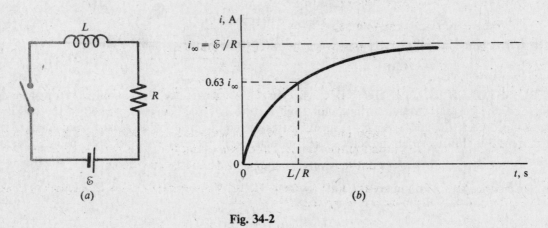

Fig. 34-2

Solved Problems

34.1. A steady current of 2 A in a coil of 400 turns causes a flux of 10^{-4} Wb to link (pass through) the loops of the coil. Compute (a) the average back emf induced in the coil if the current is stopped in 0.08 s, (b) the inductance of the coil, and (c) the energy stored in the coil.

(a)
$$|\mathcal{E}| = N\left|\frac{\Delta\Phi}{\Delta t}\right| = 400\frac{(10^{-4}-0)\text{ Wb}}{0.08\text{ s}} = 0.5\text{ V}$$

(b)
$$|\mathcal{E}| = L\left|\frac{\Delta i}{\Delta t}\right| \quad\text{or}\quad L = \left|\frac{\mathcal{E}\,\Delta t}{\Delta i}\right| = \frac{(0.5\text{ V})(0.08\text{ s})}{(2-0)\text{ A}} = 0.02\text{ H}$$

(c)
$$\text{energy} = \tfrac{1}{2}LI^2 = \tfrac{1}{2}(0.02\text{ H})(2\text{ A})^2 = 0.04\text{ J}$$

34.2. A long air core solenoid has a cross-sectional area A and N loops of wire on its length, d. (*a*) Find its self-inductance. (*b*) What is its inductance if the core material has permeability μ?

(*a*) We can write
$$|\mathcal{E}| = N\left|\frac{\Delta\Phi}{\Delta t}\right| \quad\text{and}\quad |\mathcal{E}| = L\left|\frac{\Delta i}{\Delta t}\right|$$

Upon equating these two expressions for $|\mathcal{E}|$,
$$L = N\left|\frac{\Delta\Phi}{\Delta i}\right|$$

If the current changes from zero to I, then the flux changes from zero to Φ. Therefore, $\Delta i = I$ and $\Delta\Phi = \Phi$ in this case. The self-inductance, assumed constant for all cases, is then
$$L = N\frac{\Phi}{I} = N\frac{BA}{I}$$

But, for an air core solenoid, $B = \mu_0 nI = \mu_0(N/d)I$. Substitution gives $L = \mu_0 N^2 A/d$.

(*b*) If the material of the core has permeability μ instead of μ_0, then B, and therefore L, will be increased by the factor μ/μ_0. In that case, $L = \mu N^2 A/d$. An iron core solenoid has a much higher self-inductance than an air core solenoid has.

34.3. A solenoid 30 cm long is made by winding 2000 loops of wire on an iron rod whose cross section is 1.5 cm². If the relative permeability of the iron is 600, what is the self-inductance of the solenoid? What average emf is induced in the solenoid as the current in it is decreased from 0.6 A to 0.1 A in a time of 0.03 s?

From Problem 34.2(*b*), with $k_m = \mu/\mu_0$,
$$L = \frac{k_m\mu_0 N^2 A}{d} = \frac{(600)(4\pi\times10^{-7}\text{ T}\cdot\text{m/A})(2000)^2(1.5\times10^{-4}\text{ m}^2)}{0.30\text{ m}} = 1.51\text{ H}$$

Then
$$|\mathcal{E}| = L\left|\frac{\Delta i}{\Delta t}\right| = (1.51\text{ H})\frac{0.5\text{ A}}{0.03\text{ s}} = 25\text{ V}$$

34.4. A coil of resistance 15 Ω and inductance 0.6 H is connected to a steady 120 V power source. At what rate will the current in the coil rise (*a*) at the instant the coil is connected to the power source, and (*b*) at the instant the current reaches 80% of its maximum value?

The effective driving voltage in the circuit is the 120 V power supply minus the induced back emf, $L(\Delta i/\Delta t)$. This equals the p.d. in the resistance of the coil:
$$120\text{ V} - L\frac{\Delta i}{\Delta t} = iR$$

(This same equation can be obtained by writing the loop equation for the circuit of Fig. 34-2(*a*). In doing so, remember that the inductance acts as a back emf of value $L\,\Delta i/\Delta t$.)

(*a*) At the first instant, i is essentially zero. Then
$$\frac{\Delta i}{\Delta t} = \frac{120\text{ V}}{L} = \frac{120\text{ V}}{0.6\text{ H}} = 200\text{ A/s}$$

(*b*) The current reaches a maximum value of $(120\text{ V})/R$ when the current finally stops changing (i.e. when $\Delta i/\Delta t = 0$). We are interested in the case when
$$i = (0.8)\left(\frac{120\text{ V}}{R}\right)$$

Then, substitution in the loop equation gives

$$120 \text{ V} - L\frac{\Delta i}{\Delta t} = (0.8)\left(\frac{120 \text{ V}}{R}\right)R$$

from which

$$\frac{\Delta i}{\Delta t} = \frac{(0.2)(120 \text{ V})}{L} = \frac{(0.2)(120 \text{ V})}{0.6 \text{ H}} = 40 \text{ A/s}$$

34.5. When the current in a certain coil is changing at a rate of 3 A/s, it is found that an emf of 7 mV is induced in a nearby coil. What is the mutual inductance of the combination?

$$\mathcal{E}_s = M\frac{\Delta i_p}{\Delta t} \qquad \text{or} \qquad M = \mathcal{E}_s\frac{\Delta t}{\Delta i_p} = (7 \times 10^{-3} \text{ V})\frac{1 \text{ s}}{3 \text{ A}} = 2.33 \text{ mH}$$

34.6. Two coils are wound on the same iron rod so that the flux generated by one passes through the other also. The primary coil has N_p loops on it and, when a current of 2 A flows through it, the flux in it is 2.5×10^{-4} Wb. Determine the mutual inductance of the two coils if the secondary coil has N_s loops on it.

$$|\mathcal{E}_s| = N_s\left|\frac{\Delta \Phi_s}{\Delta t}\right| \qquad \text{and} \qquad |\mathcal{E}_s| = M\left|\frac{\Delta i_p}{\Delta t}\right|$$

give

$$M = N_s\left|\frac{\Delta \Phi_s}{\Delta i_p}\right| = N_s\frac{(2.5 \times 10^{-4} - 0) \text{ Wb}}{(2 - 0) \text{ A}} = (1.25 \times 10^{-4}N_s) \text{ H}$$

34.7. A 2000-loop solenoid is wound uniformly on a long iron rod with length d and cross section A. The relative permeability of the iron is k_m. On top of this is wound a 50-loop coil which is used as a secondary. Find the mutual inductance of the system.

The flux through the solenoid is

$$\Phi = BA = (k_m\mu_0 n I_p)A = k_m\mu_0 I_p A(2000/d)$$

This same flux goes through the secondary. We have

$$|\mathcal{E}_s| = N_s\left|\frac{\Delta \Phi}{\Delta t}\right| \qquad \text{and} \qquad |\mathcal{E}_s| = M\left|\frac{\Delta i_p}{\Delta t}\right|$$

from which

$$M = N_s\left|\frac{\Delta \Phi}{\Delta i_p}\right| = N_s\frac{\Phi - 0}{I_p - 0} = (50)\frac{k_m\mu_0 I_p A(2000/d)}{I_p} = \frac{10^5 k_m\mu_0 A}{d}$$

34.8. A certain series circuit consists of a 12 V battery, a switch, a 1 MΩ resistor, and a 2 μF capacitor, initially uncharged. If the switch is now closed, find (*a*) initial current in the circuit, (*b*) time taken for the current to drop to 0.37 of its initial value, (*c*) charge on the capacitor then, (*d*) final charge on the capacitor.

(*a*) The loop rule applied to the circuit of Fig. 34-1(*a*) at any instant gives

$$12 \text{ V} - iR - v_c = 0$$

where v_c is the p.d. across the capacitor. At the first instant, q is essentially zero and so $v_c = 0$. Then

$$12 \text{ V} - iR - 0 = 0 \qquad \text{or} \qquad i = \frac{12 \text{ V}}{10^6 \text{ }\Omega} = 12 \text{ }\mu\text{A}$$

(*b*) The current drops to 0.37 of its initial value when

$$t = RC = (10^6 \text{ }\Omega)(2 \times 10^{-6} \text{ F}) = 2 \text{ s}$$

(c) At $t = 2$ s the charge on the capacitor has increased to 0.63 of its final value. (See (d) below.)

(d) The final value for the charge occurs when $i = 0$ and $v_c = 12$ V. Therefore,

$$q_{final} = Cv_c = (2 \times 10^{-6} \text{ F})(12 \text{ V}) = 24 \ \mu\text{C}$$

34.9. A 5 μF capacitor is charged to a potential of 20 kV. After being disconnected from the power source, it is connected across a 7 MΩ resistor to discharge. What is the initial discharge current and how long will it take for the capacitor voltage to decrease to 37% of the 20 kV?

The loop equation for the discharging capacitor is

$$v_c - iR = 0$$

where v_c is the p.d. across the capacitor. At the first instant, $v_c = 20$ kV, so

$$i = \frac{v_c}{R} = \frac{20 \times 10^3 \text{ V}}{7 \times 10^6 \ \Omega} = 2.86 \text{ mA}$$

The potential across the capacitor, as well as the charge on it, will decrease to 0.37 of its original value in one time constant. The required time is

$$RC = (7 \times 10^6 \ \Omega)(5 \times 10^{-6} \text{ F}) = 35 \text{ s}$$

34.10. A coil has an inductance of 1.5 H and a resistance of 0.6 Ω. If the coil is suddenly connected across a 12 V battery, find the time required for the current to rise to 0.63 of its final value. What will be the final current through the coil?

The time required is the time constant of the circuit.

$$\text{time constant} = \frac{L}{R} = \frac{1.5 \text{ H}}{0.6 \ \Omega} = 2.5 \text{ s}$$

At long times, the current will be steady and so no back emf will exist in the coil. Under those conditions,

$$I = \frac{\mathcal{E}}{R} = \frac{12 \text{ V}}{0.6 \ \Omega} = 20 \text{ A}$$

Supplementary Problems

34.11. An emf of 8 V is induced in a coil when the current in it changes at the rate of 32 A/s. Compute the inductance of the coil.   *Ans.*   0.25 H

34.12. A steady current of 2.5 A creates a flux of 1.4×10^{-4} Wb in a coil of 500 turns. What is the inductance of the coil?   *Ans.*   28 mH

34.13. The mutual inductance between the primary and secondary of a transformer is 0.3 H. Compute the induced emf in the secondary when the primary current changes at the rate of 4 A/s.   *Ans.*   1.2 V

34.14. A coil of inductance 0.2 H and 1.0 Ω resistance is connected to a 90 V source. At what rate will the current in the coil grow (a) at the instant the coil is connected to the source? (b) at the instant the current reaches two-thirds of its maximum value?   *Ans.*   (a) 450 A/s; (b) 150 A/s

34.15. Two neighboring coils, A and B, have 300 and 600 turns respectively. A current of 1.5 A in A causes 1.2×10^{-4} Wb to pass through A and 0.9×10^{-4} Wb through B. Determine (*a*) the self-inductance of A, (*b*) the mutual inductance of A and B, and (*c*) the average induced emf in B when the current in A is interrupted in 0.2 s.   *Ans.* (*a*) 24 mH; (*b*) 36 mH; (*c*) 0.27 V

34.16. A coil of 0.48 H carries a current of 5 A. Compute the energy stored in it.   *Ans.* 6 J

34.17. The iron core of a solenoid has a length of 40 cm, a cross section of 5.0 cm^2, and is wound with 10 turns of wire per cm of length. Compute the inductance of the solenoid, assuming the relative permeability of the iron to be constant at 500.   *Ans.* 126 mH

34.18. Show that (*a*) 1 N/A^2 = 1 T·m/A = 1 Wb/A·m = 1 H/m, (*b*) 1 C^2/N·m^2 = 1 F/m.

34.19. A series circuit consisting of an uncharged 2 μF capacitor and a 10 MΩ resistor is connected across a 100 V power source. What are the current in the circuit and the charge on the capacitor (*a*) after one time constant, (*b*) when the capacitor has acquired 90% of its final charge?
Ans. (*a*) 3.7 μA, 126 μC; (*b*) 1 μA, 180 μC

34.20. A charged capacitor is connected across a 10^4 Ω resistor and allowed to discharge. The potential difference across the capacitor drops to 0.37 of its original value after a time of 7 s. What is the capacitance of the capacitor?   *Ans.* 700 μF

34.21. When a long, iron core solenoid is connected across a 6 V battery, the current rises to 0.63 of its maximum value after a time of 0.75 s. The experiment is then repeated with the iron core removed. Now the time required to reach 0.63 of the maximum is 0.0025 s. Calculate (*a*) the relative permeability of the iron and (*b*) L for the air core solenoid if the maximum current is 0.5 A.
Ans. (*a*) 300; (*b*) 0.03 H

Chapter 35

Alternating Currents

THE EMF GENERATED BY A ROTATING COIL in a magnetic field has a graph similar to the one shown in Fig. 35-1. It is called an *ac voltage*. If the coil rotates with frequency f revolutions per second, then the emf has frequency f hertz (cycles per second). The instantaneous voltage v generated has the form

$$v = v_0 \sin \omega t = v_0 \sin 2\pi ft$$

where v_0 is the amplitude (maximum value) of the voltage in volts, $\omega = 2\pi f$ is the angular velocity in rad/s, and f is the frequency in hertz. The frequency f of the voltage is related to its period T by

$$T = \frac{1}{f}$$

where T is in seconds.

Rotating coils are not the only source of ac voltages. Electronic devices for generating ac voltages are very common. Alternating voltages produce alternating currents. The alternating current has a graph much like that for the voltage shown in Fig. 35-1. Its instantaneous value is i and its amplitude is i_0. Often the current and voltage do not reach a maximum at the same time, even though they both have the same frequency.

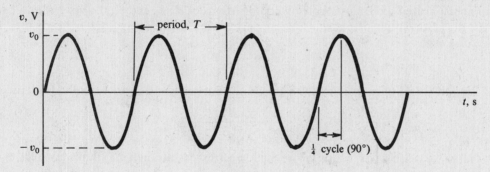

Fig. 35-1

METERS for use in ac circuits read the *effective*, or *root mean square* (rms), values of the current and voltage. These values are always positive and are related to the amplitudes of the instantaneous sinusoidal values through

$$V = V_{\text{rms}} = \frac{v_0}{\sqrt{2}} = 0.707\, v_0$$

$$I = I_{\text{rms}} = \frac{i_0}{\sqrt{2}} = 0.707\, i_0$$

It is customary to represent meter readings by capital letters (V, I), while instantaneous values are represented by small letters (v, i).

HEAT GENERATED OR POWER LOST by an effective current I in a resistor R is given by I^2R.

FORMS OF OHM'S LAW: Suppose that a sinusoidal current of frequency f with effective value I (I is the value read by an ordinary ac ammeter) flows through a pure resistor R, or a pure inductor L, or a pure capacitor C. Then an ac voltmeter placed across the element in question will read an rms voltage V as follows:

$$\text{pure resistor:} \qquad V = IR$$
$$\text{pure inductor :} \qquad V = IX_L$$

where $X_L = 2\pi fL$ is called the *inductive reactance*. Its units are ohms when L is in henries and f is in hertz.

$$\text{pure capacitor :} \qquad V = IX_C$$

where $X_C = 1/2\pi fC$ is called the *capacitive reactance*. Its units are ohms when C is in farads.

PHASE: When an ac voltage is applied to a pure resistance, the voltage across the resistance and the current through it attain their maximum values at the same instant and their zero values at the same instant; the voltage and current are said to be *in phase*.

When an ac voltage is applied to a pure inductance, the voltage across the inductance reaches its maximum value $\frac{1}{4}$ cycle ahead of the current, i.e. when the current is zero. The back emf of the inductance causes the current through the inductance to lag behind the voltage by $\frac{1}{4}$ cycle (or 90°), and the two are 90° *out of phase*.

When an ac voltage is applied to a pure capacitor, the voltage across it lags 90° behind the current flowing through it. Current must flow before the voltage across (and charge on) the capacitor can build up.

In more complicated situations involving combinations of R, L, and C, the voltage and current are usually (but not always) out of phase. The angle by which the voltage lags or leads the current is called the *phase angle*.

THE IMPEDANCE (Z) of a series circuit containing resistance, inductance, and capacitance is given by

$$Z = \sqrt{R^2 + (X_L - X_C)^2}$$

with Z in ohms. If a voltage V is applied to such a series circuit, then an Ohm's law relates V to the current I through it:

$$V = IZ$$

The phase angle ϕ between V and I is given by

$$\tan\phi = \frac{X_L - X_C}{R} \qquad \text{or} \qquad \cos\phi = \frac{R}{Z}$$

VECTOR REPRESENTATIONS in a series R-L-C circuit are possible because the above expression for the impedance can be associated with the Pythagorean theorem for a right triangle. As shown in Fig. 35-2(a), Z is the hypotenuse of the right triangle, while R and $(X_L - X_C)$ are its two legs. The angle labeled ϕ is the phase angle between the current and the voltage.

A similar relation applies to the voltages across the elements in the series circuit. As shown in Fig. 35-2(b), it is

$$V^2 = V_R^2 + (V_L - V_C)^2$$

Unlike the instantaneous voltages, the ac voltmeter readings are never negative. As a result, the voltmeter reading across a series circuit is not equal to the sum of the individual voltage readings across its elements. Instead, the above relation must be used.

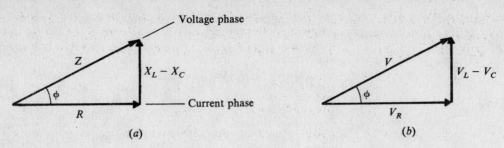

Fig. 35-2

RESONANCE occurs in a series R-L-C circuit when $X_L = X_C$. Under this condition Z is minimum, so that I is maximum for a given value of V. Equating X_L to X_C, we find for the *resonant* (or *natural*) *frequency* of the circuit

$$f_0 = \frac{1}{2\pi\sqrt{LC}}$$

POWER: Suppose that an ac voltage V is impressed across an impedance of any type. It gives rise to a current I through the impedance, and the phase angle between V and I is ϕ. The power loss in the impedance is given by

$$\text{power} = VI\cos\phi$$

The quantity $\cos\phi$ is called the *power factor*. It is unity for a pure resistor; but it is zero for a pure inductor or capacitor (no power loss occurs in a pure inductor or capacitor).

A TRANSFORMER is a device to raise or lower the voltage in an ac circuit. It consists of a primary and a secondary coil wound on the same iron core. An alternating current in one coil creates a continuously changing magnetic flux through the core. This change of flux induces an alternating emf in the other coil. The efficiency of a transformer is usually very high. Thus, neglecting losses,

$$\text{power in primary} = \text{power in secondary}$$

$$V_1 I_1 = V_2 I_2$$

The voltage ratio is the ratio of the numbers of turns; the current ratio is the inverse ratio of the numbers of turns.

$$\frac{V_1}{V_2} = \frac{N_1}{N_2} \qquad \text{and} \qquad \frac{I_1}{I_2} = \frac{N_2}{N_1}$$

Solved Problems

35.1. A sinusoidal, 60 cycle, ac voltage is read to be 120 V by an ordinary voltmeter. (*a*) What is the maximum value the voltage takes on during a cycle? (*b*) What is the equation for the voltage?

(*a*) $$V = \frac{v_0}{\sqrt{2}} \qquad \text{or} \qquad v_0 = \sqrt{2}\, V = \sqrt{2}\,(120\text{ V}) = 170\text{ V}$$

(*b*) $$v = v_0 \sin 2\pi f t = (170\text{ V}) \sin 120\pi t$$

where t is in s.

35.2. A voltage $v = (60 \text{ V}) \sin 120\pi t$ is applied across a 20 Ω resistor. What will an ac ammeter in series with the resistor read?

The rms voltage across the resistor is

$$V = 0.707\, v_0 = (0.707)(60 \text{ V}) = 42.4 \text{ V}$$

Then

$$I = \frac{V}{R} = \frac{42.4 \text{ V}}{20 \text{ Ω}} = 2.12 \text{ A}$$

35.3. A 120 V ac voltage source is connected across a 2 μF capacitor. Find the current to the capacitor if the frequency of the source is (a) 60 Hz, (b) 60 kHz. (c) What is the power loss in the capacitor?

(a)

$$X_C = \frac{1}{2\pi f C} = \frac{1}{2\pi (60 \text{ s}^{-1})(2 \times 10^{-6} \text{ F})} = 1330 \text{ Ω}$$

$$I = \frac{V}{X_C} = \frac{120 \text{ V}}{1330 \text{ Ω}} = 0.090 \text{ A}$$

(b) Now $X_C = 1.33$ Ω, so $I = 90$ A. Notice that the impeding effect of a capacitor varies inversely with the frequency.

(c)

$$\text{power} = VI \cos \phi = VI \cos 90° = 0$$

35.4. A 120 V ac voltage source is connected across a pure 0.70 H inductor. Find the current through it if the frequency of the source is (a) 60 Hz, (b) 60 kHz. (c) What is the power loss in it?

(a)

$$X_L = 2\pi f L = 2\pi (60 \text{ s}^{-1})(0.7 \text{ H}) = 264 \text{ Ω}$$

$$I = \frac{V}{X_L} = \frac{120 \text{ V}}{264 \text{ Ω}} = 0.455 \text{ A}$$

(b) Now $X_L = 264 \times 10^3$ Ω, so $I = 0.455 \times 10^{-3}$ A. Notice that the impeding effect of an inductor varies directly with the frequency.

(c)

$$\text{power} = VI \cos \phi = VI \cos 90° = 0$$

35.5. A coil having inductance 0.14 H and resistance 12 Ω is connected across a 110 V, 25 Hz line. Compute (a) the current in the coil, (b) the phase angle between the current and supply voltage, (c) the power factor, (d) the power loss in the coil.

(a)

$$X_L = 2\pi f L = 2\pi (25)(0.14) = 22.0 \text{ Ω}$$

$$Z = \sqrt{R^2 + (X_L - X_C)^2} = \sqrt{(12)^2 + (22 - 0)^2} = 25.1 \text{ Ω}$$

$$I = \frac{V}{Z} = \frac{110 \text{ V}}{25.1 \text{ Ω}} = 4.4 \text{ A}$$

(b)

$$\tan \phi = \frac{X_L - X_C}{R} = \frac{22 - 0}{12} = 1.83 \quad \text{or} \quad \phi = 61.3°$$

The voltage leads the current by 61.3°.

(c)

$$\text{power factor} = \cos \phi = \cos 61.3° = 0.48$$

(d)

$$\text{power loss} = VI \cos \phi = (110 \text{ V})(4.4 \text{ A})(0.48) = 230 \text{ W}$$

Or, since power loss occurs only because of the resistance of the coil,

$$\text{power loss} = I^2 R = (4.4 \text{ A})^2 (12 \text{ Ω}) = 230 \text{ W}$$

35.6. A capacitor is in series with a resistance of 30 Ω and is connected to a 220 V ac line. The reactance of the capacitor is 40 Ω. Determine (a) the current in the circuit, (b) the phase angle between the current and the supply voltage, and (c) the power loss in the circuit.

(a)
$$Z = \sqrt{R^2 + (X_L - X_C)^2} = \sqrt{(30)^2 + (0 - 40)^2} = 50 \; \Omega$$

$$I = \frac{V}{Z} = \frac{220 \text{ V}}{50 \; \Omega} = 4.4 \text{ A}$$

(b)
$$\tan \phi = \frac{X_L - X_C}{R} = \frac{0 - 40}{30} = -1.33 \qquad \text{or} \qquad \phi = -53°$$

The negative sign tells us that the voltage *lags* the current by 53°. The angle φ in Fig. 35-2 would lie below the horizontal axis.

(c) **Method 1**
$$\text{power} = VI \cos \phi = (220)(4.4) \cos(-53°) = (220)(4.4) \cos 53° = 580 \text{ W}$$

Method 2
The power loss occurs only in the resistor, not in the pure capacitor.
$$\text{power} = I^2 R = (4.4 \text{ A})^2 (30 \; \Omega) = 580 \text{ W}$$

35.7. A series circuit consisting of a 100 Ω noninductive resistor, a coil of 0.10 H inductance and negligible resistance, and a 20 μF capacitor, is connected across a 110 V, 60 Hz power source. Find (a) the current, (b) the power loss, (c) the phase angle between the current and the source voltage, (d) voltmeter readings across the three elements.

(a) For the entire circuit, $Z = \sqrt{R^2 + (X_L - X_C)^2}$, with

$$R = 100 \; \Omega$$
$$X_L = 2\pi f L = 2\pi(60 \text{ s}^{-1})(0.10 \text{ H}) = 37.7 \; \Omega$$
$$X_C = \frac{1}{2\pi f C} = \frac{1}{2\pi(60 \text{ s}^{-1})(20 \times 10^{-6} \text{ F})} = 132.7 \; \Omega$$

from which

$$Z = \sqrt{(100)^2 + (38 - 133)^2} = 138 \; \Omega \qquad \text{and} \qquad I = \frac{V}{Z} = \frac{110 \text{ V}}{138 \; \Omega} = 0.79 \text{ A}$$

(b) The power loss all occurs in the resistor.
$$\text{power} = I^2 R = (0.79 \text{ A})^2 (100 \; \Omega) = 63 \text{ W}$$

(c)
$$\tan \phi = \frac{X_L - X_C}{R} = \frac{-95 \; \Omega}{100 \; \Omega} = -0.95 \qquad \text{or} \qquad \phi = -43.5°$$

The voltage lags the current.

(d)
$$V_R = IR = (0.79 \text{ A})(100 \; \Omega) = 79 \text{ V}$$
$$V_C = IX_C = (0.79 \text{ A})(132.7 \; \Omega) = 105 \text{ V}$$
$$V_L = IX_L = (0.79 \text{ A})(37.7 \; \Omega) = 30 \text{ V}$$

Notice that $V_C + V_L + V_R$ does not equal the source voltage. From Fig. 35-2(b), the correct relationship is

$$V = \sqrt{V_R^2 + (V_L - V_C)^2} = \sqrt{(79)^2 + (-75)^2} = 109 \text{ V}$$

which checks within the limits of rounding-off errors.

35.8. A 5 Ω resistance is in a series circuit with a 0.2 H pure inductance and a 4×10^{-8} F pure capacitance. The combination is placed across a 30 V, 1780 Hz power supply. Find (a) the current in the circuit, (b) the phase angle between source voltage and current, (c) the power loss in the circuit, and (d) the voltmeter reading across each element of the circuit.

(a)
$$X_L = 2\pi fL = 2\pi(1780\ \text{s}^{-1})(0.2\ \text{H}) = 2240\ \Omega$$

$$X_C = \frac{1}{2\pi fC} = \frac{1}{2\pi(1780\ \text{s}^{-1})(4 \times 10^{-8}\ \text{F})} = 2240\ \Omega$$

$$Z = \sqrt{R^2 + (X_L - X_C)^2} = R = 5\ \Omega$$
$$I = \frac{V}{Z} = \frac{30\ \text{V}}{5\ \Omega} = 6\ \text{A}$$

(b)
$$\tan\phi = \frac{X_L - X_C}{R} = 0 \qquad \text{or} \qquad \phi = 0°$$

(c)
$$\text{power} = VI\cos\phi = (30\ \text{V})(6\ \text{A})(1) = 180\ \text{W}$$

or
$$\text{power} = I^2R = (6\ \text{A})^2(5\ \Omega) = 180\ \text{W}$$

(d)
$$V_R = IR = (6\ \text{A})(5\ \Omega) = 30\ \text{V}$$
$$V_C = IX_C = (6\ \text{A})(2240\ \Omega) = 13.44\ \text{kV}$$
$$V_L = IX_L = (6\ \text{A})(2240\ \Omega) = 13.44\ \text{kV}$$

This circuit is in resonance because $X_C = X_L$. Notice how very large the voltages across the inductor and capacitor can become, even though the source voltage is low.

35.9. As shown in Fig. 35-3, a series circuit connected across a 200 V, 60 cycle line consists of a capacitor of capacitive reactance 30 Ω, a noninductive resistor of 44 Ω, and a coil of inductive reactance 90 Ω and resistance 36 Ω. Determine (a) the current in the circuit, (b) the potential difference across each unit, (c) the power factor of the circuit, and (d) the power absorbed by the circuit.

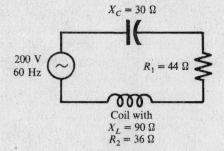

Fig. 35-3

(a)
$$Z = \sqrt{(R_1 + R_2)^2 + (X_L - X_C)^2}$$
$$= \sqrt{(44 + 36)^2 + (90 - 30)^2} = 100\ \Omega$$
$$I = \frac{V}{Z} = \frac{200\ \text{V}}{100\ \Omega} = 2\ \text{A}$$

(b)
$$\text{p.d. across capacitor} = IX_C = (2\ \text{A})(30\ \Omega) = 60\ \text{V}$$
$$\text{p.d. across resistor} = IR_1 = (2\ \text{A})(44\ \Omega) = 88\ \text{V}$$
$$\text{impedance of coil} = \sqrt{R_2^2 + X_L^2} = \sqrt{(36)^2 + (90)^2} = 97\ \Omega$$
$$\text{p.d. across coil} = (2\ \text{A})(97\ \Omega) = 194\ \text{V}$$

(c)
$$\text{power factor} = \cos\phi = \frac{R}{Z} = \frac{80}{100} = 0.80$$

(d)
$$\text{power used} = VI\cos\phi = (200\ \text{V})(2\ \text{A})(0.80) = 320\ \text{W}$$
$$= I^2R = (2\ \text{A})^2(80\ \Omega) = 320\ \text{W}$$

35.10. Calculate the resonant frequency of a circuit of negligible resistance containing an inductance of 40 mH and a capacitance of 600 pF.

$$f_0 = \frac{1}{2\pi\sqrt{LC}} = \frac{1}{2\pi\sqrt{(40 \times 10^{-3}\ \text{H})(600 \times 10^{-12}\ \text{F})}} = 32.5\ \text{kHz}$$

35.11. A step-up transformer is used on a 120 V line to furnish 1800 V. The primary has 100 turns. How many turns are on the secondary?

$$\frac{V_1}{V_2} = \frac{N_1}{N_2} \qquad \text{or} \qquad \frac{120 \text{ V}}{1800 \text{ V}} = \frac{100 \text{ turns}}{N_2}$$

from which $N_2 = 1500$ turns.

35.12. A transformer used on a 120 V line delivers 2 A at 900 V. What current is drawn from the line? Assume 100% efficiency.

$$\text{power in primary} = \text{power in secondary}$$
$$I_1(120 \text{ V}) = (2 \text{ A})(900 \text{ V})$$
$$I_1 = 15 \text{ A}$$

35.13. A step-down transformer operates on a 2.5 kV line and supplies a load with 80 A. The ratio of the primary winding to the secondary winding is 20 : 1. Assuming 100% efficiency, determine the secondary voltage, V_2, the primary current, I_1, and the power output, P_2.

$$V_2 = \left(\frac{1}{20}\right)V_1 = 125 \text{ V} \qquad I_1 = \left(\frac{1}{20}\right)I_2 = 4 \text{ A} \qquad P_2 = V_2 I_2 = 10 \text{ kW}$$

The last expression is correct only if it is assumed that the load is resistive, so that the power factor is unity.

Supplementary Problems

35.14. A voltmeter reads 80 V when connected across the terminals of a sinusoidal power source with $f = 1000$ Hz. Write the equation for the instantaneous voltage provided by the source.
Ans. $v = (113 \text{ V}) \sin 2000\pi t$ for t in seconds

35.15. An ac current in a 10 Ω resistance produces heat at the rate of 360 W. Determine the effective values of the current and voltage. *Ans.* 6 A, 60 V

35.16. A 40 Ω resistor is connected across a 15 V, variable frequency, electronic oscillator. Find the current through the resistor when the frequency is (a) 100 Hz and (b) 100 kHz.
Ans. (a) 0.375 A; (b) 0.375 A

35.17. Solve Problem 35.16 if the 40 Ω resistor is replaced by a 2 mH inductor.
Ans. (a) 11.9 A; (b) 11.9 mA

35.18. Solve Problem 35.16 if the 40 Ω resistor is replaced by a 0.3 μF capacitor.
Ans. (a) 2.83 mA; (b) 2.83 A

35.19. A coil has resistance 20 Ω and inductance 0.35 H. Compute its reactance and its impedance to an alternating current of 25 cycles/s. *Ans.* 55.0 Ω, 58.5 Ω

35.20. A current of 30 mA is taken by a 4 μF capacitor connected across an alternating current line having a frequency of 500 Hz. Compute the reactance of the capacitor and the voltage across the capacitor.
Ans. 80 Ω, 2.4 V

35.21. A coil has an inductance of 0.10 H and a resistance of 12 Ω. It is connected to a 110 V, 60 Hz line. Determine (a) the reactance of the coil, (b) the impedance of the coil, (c) the current through the coil, (d) the phase angle between current and supply voltage, (e) the power factor of the circuit, (f) the reading of a wattmeter connected in the circuit.
Ans.   (a) 37.7 Ω;   (b) 39.6 Ω;   (c) 2.78 A;   (d) voltage leads by 72.3°;   (e) 0.303;   (f) 92.6 W

35.22. A 10 μF capacitor is in series with a 40 Ω resistance and the combination is connected to a 110 V, 60 Hz line. Calculate (a) the capacitive reactance, (b) the impedance of the circuit, (c) the current in the circuit, (d) the phase angle between current and supply voltage, (e) the power factor for the circuit.
Ans.   (a) 266 Ω;   (b) 269 Ω;   (c) 0.409 A;   (d) voltage leads by 81.4°;   (d) 0.149

35.23. A circuit having a resistance, an inductance, and a capacitance in series is connected to a 110 V ac line. For it, $R = 9.0$ Ω, $X_L = 28$ Ω, and $X_C = 16$ Ω. Compute (a) the impedance of the circuit, (b) the current, (c) the phase angle between the current and the supply voltage, (d) the power factor of the circuit.   Ans.   (a) 15 Ω;   (b) 7.33 A;   (c) voltage leads by 53.1°;   (d) 0.60

35.24. An experimenter has a coil of inductance 3 mH and wishes to construct a circuit whose resonant frequency is 1 MHz. What should be the value of the capacitor used?   Ans.   8.44 pF

35.25. A circuit has a resistance 11 Ω, a coil of inductive reactance 120 Ω, and a capacitor with 120 Ω reactance, all connected in series with a 110 V, 60 Hz power source. What is the potential difference across each circuit element?   Ans. $V_R = 110$ V, $V_L = V_C = 1200$ V

35.26. A 120 V, 60 Hz power source is connected across an 800 Ω noninductive resistance and an unknown capacitance in series. The voltage drop across the resistor is 102 V. (a) What is the voltage drop across the capacitor? (b) What is the reactance of the capacitor?   Ans.   (a) 63 V;   (b) 496 Ω

35.27. A coil of negligible resistance is connected in series with a 90 Ω resistor across a 120 V, 60 Hz line. A voltmeter reads 36 V across the resistance. Find the voltage across the coil and the inductance of the coil.   Ans.   114.5 V, 0.76 H

35.28. A step-down transformer is used on a 2200 V line to deliver 110 V. How many turns are on the primary winding if the secondary has 25 turns?   Ans.   500

35.29. A step-down transformer is used on a 1650 V line to deliver 45 A at 110 V. What current is drawn from the line? Assume 100% efficiency.   Ans.   3 A

35.30. A step-up transformer operates on a 110 V line and supplies a load with 2 A. The ratio of the primary and secondary windings is 1 : 25. Determine the secondary voltage, the primary current, and the power output. Assume a resistive load and 100% efficiency.   Ans.   2750 V, 50 A, 5.5 kW

Chapter 36

Illumination and Photometry

LUMINOUS INTENSITY (I) of a light source is a measure of the source intensity as seen by the eye. Because the eye is less sensitive to blue light than to green light, for example, a blue light source must radiate more power in watts than must a green source if the two are to have the same luminous intensity.

A source of any color composition is said to have a luminous intensity I of 1 *candela* (1 cd) if its visual brightness is the same as that of a certain standard source maintained at a white-hot temperature of 2046 K. Most sources have different luminous intensities when viewed from different directions and so the value of I for a source may vary with the angle at which it is viewed. An older unit for I is the *candle* or *candlepower*:

$$1 \text{ candle} = 1 \text{ candlepower} = 0.981 \text{ cd}$$

LUMINOUS FLUX (ΔF) through a small area ΔA is a measure of the light (as perceived by the eye) that passes through the area ΔA. Consider a spherical surface of radius r centered on a point source of light (Fig. 36-1). Let the luminous intensity of the source in the direction of the surface element ΔA be I. Then, by definition, the light flux through ΔA is

$$\Delta F = I \, \frac{\Delta A}{r^2} = I \, \Delta \omega$$

where $\Delta\omega = (\Delta A)/r^2$ is the solid angle subtended at the center of the sphere by the surface element ΔA. Solid angles are measured in dimensionless units called *steradians* (sr). With I in candelas and $\Delta\omega$ in steradians, the flux ΔF is in *lumens* (lm). Thus

$$1 \text{ lm} = 1 \text{ cd} \cdot \text{sr} \qquad \text{or} \qquad 1 \text{ cd} = 1 \text{ lm/sr}$$

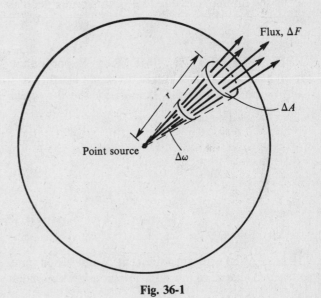

Fig. 36-1

TOTAL FLUX (F) can be found by summing the contributions ΔF over all directions. For an *isotropic source* of light, the intensity I is the same in every direction and so

$$\text{total flux } F = \sum_{\text{sphere}} \Delta F = I \sum_{\text{sphere}} \Delta \omega = 4\pi I \qquad \text{(isotropic source)}$$

For a nonisotropic source, we define the *mean spherical luminous intensity*, $\bar{I}$, by the relation

$$\text{total flux } F = 4\pi\bar{I}$$

In other words, $\bar{I}$ is the luminous intensity of the isotropic source that would emit the same total flux as the actual source does; $\bar{I}$ is the average of the luminous intensities of the actual source taken over all directions of observation.

255

ILLUMINANCE (E) of a surface (also called *illumination*) is a measure of how much light falls on unit area of the surface.   If a flux ΔF strikes a tiny area ΔA, then the illuminance at that point is given by

$$E = \frac{\Delta F}{\Delta A}$$

For the uniform illumination of a finite area A by a flux F, the illuminance is $E = F/A$.   The units of E are lm/m^2 (also called *lux*, lx) or lm/ft^2 (also called the *foot-candle*).

At a given point in a light beam, maximum illumination of a surface occurs if the flux is perpendicular to the surface, because maximum flux then goes through the area. If the area is inclined to the flux so that the angle between the flux and the normal to the area is θ, then a fraction $\cos\theta$ of the maximum possible flux goes through the area.   Therefore

$$E = E_{max}\cos\theta$$

ILLUMINATION BY AN ISOTROPIC POINT SOURCE: The luminous flux coming from an isotropic point source is independent of direction and is intercepted normally by the surface of a sphere of radius r concentric to the source.   The illuminance at each point of the surface is then

$$E = \frac{F}{A} = \frac{4\pi I}{4\pi r^2} = \frac{I}{r^2}$$

Thus the illuminance due to a point source decreases inversely as the square of the distance from the source.

PRINCIPLE OF PHOTOMETRY: Suppose that two point sources, I_1 and I_2, at distances r_1 and r_2 from a screen produce fluxes perpendicular to (or at the same angle to) the screen.   If the distances r_1 and r_2 are such that the illuminations at the screen are the same for the two sources, then $E_1 = E_2$ and

$$\frac{I_1}{I_2} = \left(\frac{r_1}{r_2}\right)^2$$

Solved Problems

36.1. A 60 watt incandescent lamp has a mean spherical luminous intensity of 66.5 candelas. Determine the total luminous flux F radiated by the lamp and the *luminous efficiency* of the lamp.

$$F = 4\pi\bar{I} = (4\pi\ \text{sr})(66.5\ \text{cd}) = 836\ \text{lm}$$

$$\text{luminous efficiency} \equiv \frac{\text{output in lumens}}{\text{input in watts}} = \frac{836\ \text{lm}}{60\ \text{W}} = 13.9\ \text{lm/W}$$

36.2. Compute the illumination E of a small surface at a distance of 120 cm from an isotropic point source of luminous intensity 72 cd, (*a*) if the surface is normal to the luminous flux and (*b*) if the normal to the surface makes an angle of 30° with the light rays.

(*a*)
$$E = \frac{I}{r^2} = \frac{72\ \text{lm/sr}}{(1.2\ \text{m})^2} = 50\ \text{lm/m}^2$$

As in the case of other angle measures, steradians is not a unit in the usual sense and does not carry through in equations.

(*b*)
$$E = E_{max}\cos\theta = \frac{I}{r^2}\cos\theta = (50\ \text{lm/m}^2)(\cos 30°) = 43\ \text{lm/m}^2$$

36.3. A light meter reads the illumination received from the sun as 10^5 lm/m^2. If the distance from the earth to the sun is 1.5×10^{11} m, compute the luminous intensity of the sun.

$$I = Er^2 = (10^5 \text{ lm/m}^2)(1.5 \times 10^{11} \text{ m})^2 = 2.25 \times 10^{27} \text{ lm/sr} = 2.25 \times 10^{27} \text{ cd}$$

36.4. In Fig. 36-2, the isotropic point source pro-vides an illumination of 1000 lux (or lm/m^2) at point A. (*a*) What is the luminous intensity of the source? (*b*) What is the illumination at point B?

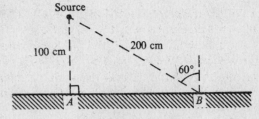

For an isotropic source, $E = (I/r^2) \cos \theta$.

(*a*) $$E_A = \frac{I}{r_A^2} \cos 0°$$

Fig. 36-2

or $I = E_A r_A{}^2 = (1000 \text{ lm/m}^2)(1 \text{ m})^2 = 1000 \text{ lm/sr}$

(*b*) The angle between the normal at B and the incident flux is 60°.

$$E_B = \frac{I}{r_B^2} \cos 60° = \frac{1000 \text{ lm/sr}}{(2 \text{ m})^2} \cos 60° = 125 \text{ lm/m}^2$$

36.5. A certain spotlight has a mean spherical luminous intensity of 1000 cd. The total light from the spotlight is concentrated by a reflector and lens onto a 200 cm^2 area on a wall. Find the illumination of the area, supposing it to be uniform.

$$\text{total flux from spotlight} = 4\pi I = (4\pi \text{ sr})(1000 \text{ cd}) = 12\,600 \text{ lm}$$

$$E = \frac{F}{A} = \frac{12\,600 \text{ lm}}{200 \times 10^{-4} \text{ m}^2} = 6.3 \times 10^5 \text{ lm/m}^2 = 6.3 \times 10^5 \text{ lx}$$

36.6. Light from a projection lantern provides an illumination of $12\,000$ lm/m^2 on a wall perpen-dicular to the beam and at a distance of 5 m from the source. What intensity must an isotropic source have to give this same illumination at a distance of 5 m?

For an isotropic source and normal incidence, $E = I/r^2$. Therefore,

$$I = r^2 E = (5 \text{ m})^2 (12\,000 \text{ lm/m}^2) = 3 \times 10^5 \text{ cd}$$

36.7. A long, straight, fluorescent lamp gives an illumination of E_1 at a radial distance of r_1. Find E_2 at r_2. Assume the length of the bulb to be much greater than r_1 and r_2, so that the effects of the ends of the bulb can be ignored.

Consider two short cylindrical shells, of length L and radii r_1 and r_2, coaxial with the lamp. The flux through each is the same. Assume negligible flux through the ends. Then

$$A_1 E_1 = A_2 E_2 \qquad \text{or} \qquad 2\pi r_1 L E_1 = 2\pi r_2 L E_2$$

from which

$$E_2 = E_1 \frac{r_1}{r_2}$$

Notice that the illumination decreases as $1/r$, unlike the $1/r^2$ relation that applies to a point source.

36.8. An unknown lamp placed 90 cm from a photometer screen gives the same illumination as a standard lamp of 32 cd placed 60 cm from the screen. Compute the luminous intensity I_1 of the lamp under test.

$$\frac{I_1}{I_2} = \frac{r_1^2}{r_2^2} \qquad \text{or} \qquad I_1 = (32 \text{ cd}) \frac{(90 \text{ cm})^2}{(60 \text{ cm})^2} = 72 \text{ cd}$$

36.9. Two lamps of 20 cd and 40 cd are 10 m apart. At what two points on the straight line passing through the lamps will the illuminations produced by them be equal?

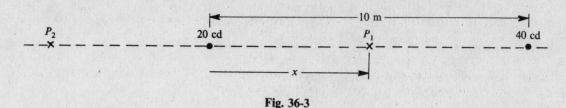

Fig. 36-3

Refer to Fig. 36-3.   Let

$$x = \text{distance of 20 cd lamp from point of equal illumination}$$
$$10 - x = \text{distance of 40 cd lamp from point of equal illumination}$$

At the point of equal illumination,

$$\frac{I_1}{I_2} = \frac{r_1^2}{r_2^2} \qquad \text{or} \qquad \frac{20 \text{ cd}}{40 \text{ cd}} = \frac{x^2}{(10 - x)^2}$$

Simplifying and solving,

$$x^2 + 20x - 100 = 0 \qquad \text{or} \qquad x = \frac{-20 \pm \sqrt{(20)^2 - 4(1)(-100)}}{2(1)}$$

There are two solutions, 4.14 m and −24.14 m.   The first solution represents a point P_1, between the lamps and 4.14 m from the 20 cd lamp.   The second point is P_2, 24.14 m from the 20 cd lamp and 34.14 m from the 40 cd lamp.

Supplementary Problems

36.10. An isotropic point source has an intensity of 200 cd.   (*a*) How much luminous flux is emitted by the source?   (*b*) How much flux strikes a 2 cm^2 area on a tabletop 80 cm directly below the source?   (*c*) What is the illuminance at that point on the tabletop?
Ans.   (*a*) 2510 lm;   (*b*) (2 cm^2/4πr^2)(2510) = 0.0625 lm;   (*c*) 313 lm/m^2

36.11. How many lumens pass through an area of 0.3 m^2 on a sphere of radius 4 m, if an 800 cd isotropic source is at its center?   What is E for the area?   *Ans.*   15 lm,   50 lm/m^2 = 50 lx

36.12. What is the luminous intensity of a 200 W tungsten lamp whose efficiency is 18 lm/W?
Ans.   3600/4π = 286 cd

36.13. At what distance from a screen will a 27 cd lamp provide the same illumination as a 75 cd lamp 15 ft from the screen?   *Ans.*   9 ft

36.14. Determine the illumination on a surface 7 ft distant from a 125 cd source (*a*) if the surface is normal to the light rays, (*b*) if the normal to the surface makes an angle of 15° with the rays.
Ans.   (*a*) 2.55 lm/ft^2;   (*b*) 2.46 lm/ft^2

36.15. Compute the illumination at the edge of a circular table of radius 1 m, if a 200 cd source is suspended 3 m above its center.   *Ans.*   20 lm/m^2

36.16. How much should a 60 W lamp be lowered to double the illumination on an object which is 60 cm directly under it?   *Ans.*   17.6 cm

36.17. A 40 W, 110 V lamp is rated 11.0 lm/W.   At what distance from the lamp is the maximum illumination 5 lm/m^2?   *Ans.*   2.65 m

36.18. Two lamps of 5 cd and 20 cd are 150 cm apart.   At what point between them will the same illumination be produced by each?   *Ans.*   50 cm from 5 cd lamp

36.19. A frosted light bulb, which may be taken as an isotropic source, gives an illuminance of 8 lm/m^2 at a distance of 5 m.   Compute (*a*) the total luminous flux from the bulb and (*b*) its luminous intensity. *Ans.*   (*a*) 800π lm;   (*b*) 200 cd

36.20. An air-cooled searchlight bulb has an efficiency of 25 lm/W.   The searchlight is powered by a generator of efficiency 80% to which power is supplied by a belt passing around a pulley of 16 cm diameter.   If the difference in tension of the two portions of the belt is 98 N, calculate the angular speed of the generator required to give 40 lm/m^2 of illumination on the ground.   The reflector system of the searchlight concentrates all the light into a circle on the ground 50 m in diameter.   *Ans.* 80 rev/s

Chapter 37

Reflection of Light

LAWS OF REFLECTION: The *angle of incidence* is the angle between the incident ray and the normal to the reflecting surface at the point of incidence. The *angle of reflection* is the angle between the reflected ray and the normal to the surface. These angles are shown in Fig. 37-1.

In *specular* (or *mirror*) reflection: (1) The incident ray, reflected ray, and normal to the surface lie in the same plane. (2) The angle of incidence equals the angle of reflection.

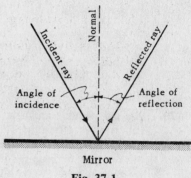

Fig. 37-1

PLANE MIRRORS form images that are erect, of the same size as the object, and as far behind the reflecting surface as the object is in front of it. The images are *virtual*, i.e. the image cannot be captured on a screen because the light does not converge on the image position.

SPHERICAL MIRRORS: The *principal focus* of a spherical mirror, such as the ones shown in Fig. 37-2, is the point *F* where rays parallel to and very close to the principal axis *XX* of the mirror are focused. This focus is real for a concave mirror and virtual for a convex mirror. It is located on the principal axis *XX* and midway between the center of curvature *C* and the mirror.

Concave mirrors form real and inverted images of objects located beyond the principal focus. If the object is between the principal focus and the mirror, the image is virtual, erect, and enlarged.

Convex mirrors produce only virtual, erect, and diminished images of objects placed in front of them.

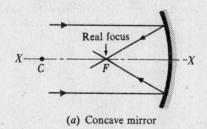

(a) Concave mirror   (b) Convex mirror

Fig. 37-2

MIRROR EQUATION for both concave and convex spherical mirrors:

$$\frac{1}{p} + \frac{1}{q} = \frac{2}{R} = \frac{1}{f}$$

where p = object distance from the mirror
q = image distance from the mirror
R = radius of curvature of the mirror
f = focal length of the mirror = $R/2$

p is positive when the object is in front of mirror

q is positive when the image is real, i.e. in front of mirror

q is negative when the image is virtual, i.e. behind mirror

R and f are positive for a concave mirror and negative for a convex mirror

SIZE OF IMAGE formed by a spherical mirror:

$$\text{linear magnification} = \frac{\text{length of image}}{\text{length of object}} = \frac{\text{image distance from mirror}}{\text{object distance from mirror}} = \left| \frac{q}{p} \right|$$

Solved Problems

37.1. Two plane mirrors make an angle of 30° with each other. Locate graphically four images of a luminous point A placed between the two mirrors. See Fig. 37-3.

From A draw normals AA' and AB' to mirrors OY and OX respectively, making $\overline{AL} = \overline{LA'}$ and $\overline{AM} = \overline{MB'}$. Then A' and B' are images of A.

Next, from A' and B' draw normals to OX and OY respectively, making $\overline{A'N} = \overline{NA''}$ and $\overline{B'P} = \overline{PB''}$. Then A'' is the image of A' in OX and B'' is the image of B' in OY.

The four images of A are A', B', A'', B''. Additional images also exist; for example, images of A'' and B''.

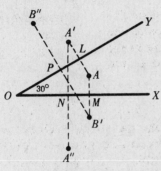

Fig. 37-3

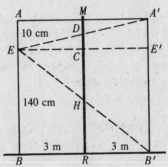

Fig. 37-4

37.2. A boy is 1.50 m tall and can just see his image in a vertical plane mirror 3 m away. His eyes are 1.40 m from the floor level. Determine the vertical dimension and elevation of the mirror.

In Fig. 37-4, let AB represent the boy. His eyes are at E. Then $A'B'$ is the image of AB in mirror MR, and DH represents the shortest mirror necessary for the eye to view the image $A'B'$.

Triangles DEC and $DA'M$ are congruent and so

$$\overline{CD} = \overline{DM} = 5 \text{ cm}$$

Triangles HRB' and HCE are congruent and so

$$\overline{RH} = \overline{HC} = 70 \text{ cm}$$

The dimension of the mirror is $\overline{HC} + \overline{CD} = 75$ cm and its elevation is $\overline{RH} = 70$ cm.

37.3. As shown in Fig. 37-5, a light ray IO is incident on a small plane mirror which is attached to a galvanometer coil. The mirror reflects this ray upon a straight scale SC which is 1 m distant and parallel to the undeflected mirror MM. When the current through the instrument has a certain value, the mirror turns through an angle of 8° and assumes the position $M'M'$. Across what distance on the scale will the spot of light move?

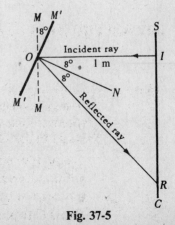

When the mirror turns through 8°, the normal to it also turns through 8° and the incident ray makes an angle of 8° with the normal NO to the deflected mirror $M'M'$. Because the incident ray IO and the reflected ray OR make equal angles with the normal, angle IOR is twice the angle through which the mirror has turned, or 16°.

$$\overline{IR} = \overline{IO} \tan 16° = (1 \text{ m})(0.287) = 28.7 \text{ cm}$$

Fig. 37-5

37.4. The concave spherical mirror shown in Fig. 37-6 has a radius of curvature 4 m. A real object *OO'*, 5 cm high, is placed 3 m in front of the mirror. By (a) construction and (b) computation, determine the position and height of the image *II'*.

C = center of curvature, 4 m from mirror   F = principal focus, 2 m from mirror

(a) Two of the following three convenient rays from *O* will locate the image.

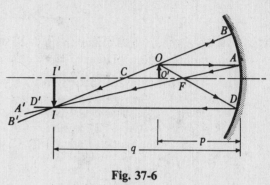

     1. Draw the ray *OA* parallel to the principal axis. This ray, like all parallel rays, is reflected through the principal focus *F* in the direction *AFA'*.

     2. Draw the ray *OB* in the same direction as if it passed through the center of curvature *C*. This ray is normal to the mirror and is reflected back on itself in the direction *BCB'*.

     3. A ray *OFD* passes through the principal focus *F* and, like all rays passing through *F*, is reflected parallel to the principal axis in the direction *DD'*.

Fig. 37-6

The intersection *I* of any two of these reflected rays is the image of *O*. Thus *II'* represents the position and size of the image of *OO'*. The image is real, inverted, magnified, and at a greater distance from the mirror than the object. (*Note*: If the object were at *II'*, the image would be at *OO'* (real, inverted, smaller).)

(b) By the mirror equation,

$$\frac{1}{p} + \frac{1}{q} = \frac{2}{r} \qquad \text{or} \qquad \frac{1}{3} + \frac{1}{q} = \frac{2}{4} \qquad \text{or} \qquad q = 6 \text{ m}$$

The image is real (since *q* is positive) and 6 m from the mirror.

$$\frac{\text{height of image}}{\text{height of object}} = \left| \frac{q}{p} \right| = \frac{6 \text{ m}}{3 \text{ m}} = 2 \qquad \text{or} \qquad \text{height of image} = 2(5 \text{ cm}) = 10 \text{ cm}$$

37.5. An object *OO'* is 25 cm from a concave spherical mirror of radius 80 cm (Fig. 37-7). Determine the position and relative size of its image *II'*, (a) by construction and (b) by use of the mirror equation.

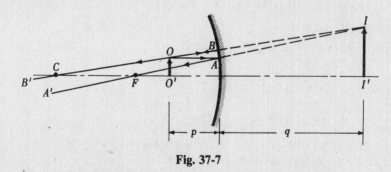

Fig. 37-7

(a) Choose two convenient rays coming from *O*.

     1. A ray *OA*, parallel to the principal axis, is reflected through the principal focus *F*, 40 cm from the mirror.

     2. A ray *OB*, in the line of the radius *COB*, is normal to the mirror and is reflected back on itself through the center of curvature *C*.

The reflected rays, *AA'* and *BB'*, do not meet, but appear to originate from a point *I* behind the mirror. Thus *II'* represents the relative position and size of the image of *OO'*. The image is virtual (behind the mirror), erect and magnified.

(b) $$\frac{1}{p} + \frac{1}{q} = \frac{2}{r} \qquad \text{or} \qquad \frac{1}{25} + \frac{1}{q} = \frac{2}{80} \qquad \text{or} \qquad q = -66.7 \text{ cm}$$

The image is virtual (since q is negative) and 66.7 cm behind the mirror.

$$\text{linear magnification} = \frac{\text{size of image}}{\text{size of object}} = \left| \frac{q}{p} \right| = \frac{66.7 \text{ cm}}{25 \text{ cm}} = 2.67 \text{ times}$$

37.6. As shown in Fig. 37-8, an object 6 cm high is located 30 cm in front of a convex spherical mirror of radius 40 cm.   Determine the position and height of its image, (a) by construction and (b) by use of the mirror equation.

(a)   Choose two convenient rays coming from O.
1. A ray OA, parallel to the principal axis, is reflected in the direction AA' as if it passed through the principal focus F.
2. A ray OB, directed toward the center of curvature C, is normal to the mirror and is reflected back on itself.

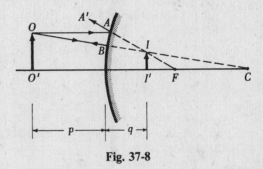

The reflected rays, AA' and BO, never meet but appear to originate from a point I behind the mirror. Then II' represents the size and position of the image of OO'.

All images formed by convex mirrors are virtual, erect, and reduced in size provided the object is in front of the mirror (i.e. a real object).

Fig. 37-8

(b) $$\frac{1}{p} + \frac{1}{q} = \frac{2}{r} \qquad \text{or} \qquad \frac{1}{30} + \frac{1}{q} = -\frac{2}{40} \qquad \text{or} \qquad q = -12 \text{ cm}$$

The image is virtual (q is negative) and 12 cm behind the mirror.

$$\frac{\text{height of image}}{\text{height of object}} = \left| \frac{q}{p} \right| = \frac{12 \text{ cm}}{30 \text{ cm}} = 0.4 \qquad \text{or} \qquad \text{height of image} = (0.4)(6 \text{ cm}) = 2.4 \text{ cm}$$

37.7. Where should an object be placed with reference to a concave spherical mirror of radius 180 cm in order to form a real image having half its linear dimensions?

The magnification is to be 1/2; hence $q = p/2$.

$$\frac{1}{p} + \frac{1}{q} = \frac{2}{r} \qquad \text{or} \qquad \frac{1}{p} + \frac{2}{p} = \frac{2}{180} \qquad \text{or} \qquad p = 270 \text{ cm from mirror}$$

37.8. How far must a man stand in front of a concave spherical mirror of radius 120 cm in order to see an erect image of his face four times its natural size?

The erect image must be virtual; hence q is negative, and $q = -4p$.

$$\frac{1}{p} + \frac{1}{q} = \frac{2}{r} \qquad \text{or} \qquad \frac{1}{p} - \frac{1}{4p} = \frac{2}{120} \qquad \text{or} \qquad p = 45 \text{ cm from mirror}$$

37.9. What kind of spherical mirror must be used, and what must be its radius, in order to give an erect image 1/5 as large as an object placed 15 cm in front of it?

An erect image produced by a spherical mirror is virtual; hence $q = -p/5 = -15/5 = -3$ cm. As the virtual image is smaller than the object, a convex mirror is required.

$$\frac{1}{p} + \frac{1}{q} = \frac{2}{r} \qquad \text{or} \qquad \frac{1}{15} - \frac{1}{3} = \frac{2}{r} \qquad \text{or} \qquad r = -7.5 \text{ cm (convex mirror)}$$

37.10. The diameter of the sun subtends an angle of approximately 32 minutes (32') at any point on the earth. Determine the position and diameter of the solar image formed by a concave spherical mirror of radius 400 cm. Refer to Fig. 37-9.

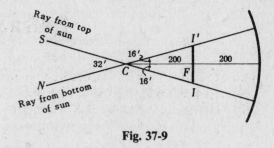

Fig. 37-9

Since the sun is very distant, p is very large and $1/p$ is practically zero.

$$\frac{1}{p} + \frac{1}{q} = \frac{2}{r} \qquad \text{or} \qquad 0 + \frac{1}{q} = \frac{2}{400}$$

Solving, $q = 200$ cm. The image is at the principal focus F, 200 cm from the mirror.

The diameter of the sun and its image II' subtend equal angles at the center of curvature C of the mirror.

$$\overline{I'I} = 2(\overline{I'F}) = 2(\overline{CF} \tan 16') = 2(200 \text{ cm})(0.00465) = 1.86 \text{ cm}$$

Supplementary Problems

37.11. Two plane mirrors are parallel to each other and spaced 20 cm apart. A luminous point is placed between them and 5 cm from one mirror. Determine the distance from each mirror of the three nearest images in each. *Ans.* 5, 35, 45 cm; 15, 25, 55 cm

37.12. A ray of light makes an angle of 25° with the normal to a plane mirror. If the mirror is turned through 6°, making the angle of incidence 31°, through what angle is the reflected ray rotated? *Ans.* 12°

37.13. Describe the image of a candle flame located 40 cm from a concave spherical mirror of radius 64 cm. *Ans.* real, inverted, 160 cm before mirror, magnified 4 times

37.14. Describe the image of an object positioned 20 cm from a concave spherical mirror of radius 60 cm. *Ans.* virtual, erect, 60 cm behind mirror, magnified 3 times

37.15. How far should an object be from a concave spherical mirror of radius 36 cm to form a real image 1/9 its size? *Ans.* 180 cm

37.16. An object 7 cm high is placed 15 cm from a convex spherical mirror of radius 45 cm. Describe its image. *Ans.* virtual, erect, 9 cm behind mirror, 4.2 cm high

37.17. What is the focal length of a convex spherical mirror which produces an image 1/6 the size of an object located 12 cm from the mirror? *Ans.* −2.4 cm

37.18. It is desired to cast the image of a lamp, magnified 5 times, upon a wall 12 ft distant from the lamp. What kind of spherical mirror is required and what is its position?
Ans. concave, radius 5 ft, 3 ft from lamp

37.19. Compute the position and diameter of the image of the moon in a polished sphere of diameter 20 cm. The diameter of the moon is 3500 km and its distance from the earth is 384 000 km, approximately.
Ans. 5 cm inside sphere, 0.46 mm

Chapter 38

Refraction of Light

THE SPEED OF LIGHT varies from material to material. Light travels fastest in vacuum, where its speed is $c = 2.998 \times 10^8$ m/s. Its speed in air is $c/1.0003$. In water its speed is $c/1.33$ and, in ordinary glass, it is about $c/1.5$.

INDEX OF REFRACTION (n or μ): The absolute index of refraction of a material is defined by

$$n = \frac{\text{speed of light in vacuum}}{\text{speed of light in material}} = \frac{c}{v}$$

where v is the speed of light in the material.

Sometimes use is made of the *relative index of refraction* of one material, A, to another material, B.

$$\text{relative index} = \frac{n_A}{n_B}$$

where n_A and n_B are the absolute refractive indices of the two materials.

SNELL'S LAW: Consider a ray of light passing from medium A to medium B, as shown in Fig. 38-1. If $n_B > n_A$, the ray will be refracted (bent) towards the normal NN as it enters medium B. If $n_B < n_A$, the ray will deflect away from the normal as it enters medium B. In either case, the incident and refracted rays and the normal will all lie in the same plane.

If we define the *angle of incidence*, i, and the *angle of refraction*, r, as shown in the figure, then *Snell's law* states that

$$n_A \sin i = n_B \sin r$$

Or, if we call $n_A = n_1$, $i = \theta_1$, $n_B = n_2$, and $r = \theta_2$, we have

$$n_1 \sin \theta_1 = n_2 \sin \theta_2$$

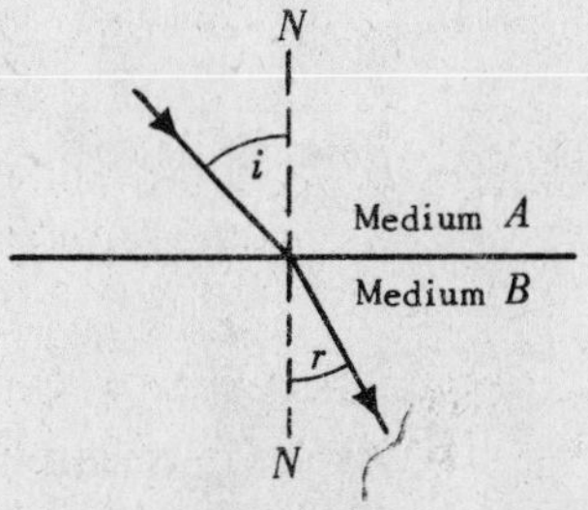

and this form of the equation shows that it makes no difference which material we call material 1 and which we call material 2. As a result, a ray of light deflects along the same path when its direction is reversed.

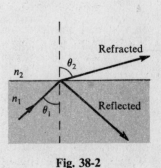

Fig. 38-1

CRITICAL ANGLE FOR TOTAL INTERNAL REFLECTION:
Suppose that a ray of light passes from a material of higher index of refraction to one of lower index, as shown in Fig. 38-2. Part of the incident ray is refracted and part is reflected at the interface. Because θ_2 must be larger than θ_1, it is possible to make θ_1 large enough so that $\theta_2 = 90°$. This value for θ_1 is called the *critical angle*. For θ_1 larger than this, no refracted ray can exist; all of the light is reflected.

The condition for total internal reflection is that θ_1 exceed the critical angle θ_c given by

$$n_1 \sin \theta_c = n_2 \sin 90° \qquad \text{or} \qquad \sin \theta_c = \frac{n_2}{n_1}$$

Fig. 38-2

Because the sine of an angle can never be larger than unity, this relation confirms that total internal reflection can occur only if $n_1 > n_2$.

REFRACTION BY A PRISM: The refractive index n of the material of a triangular prism can be measured by determining the angle of minimum deviation, D_{min}, for the prism in vacuum (or air, to good approximation).   (The angle of deviation, D, is shown in Fig. 38-3.) One has

$$n = \frac{\sin \frac{1}{2}(A + D_{min})}{\sin \frac{1}{2}A}$$

where A is the apex (or *refracting*) angle of the prism.

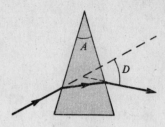

Fig. 38-3

Solved Problems

38.1.   The speed of light in water is $(3/4)c$.   What is the effect on the frequency and wavelength of light in passing from vacuum (or air, to good approximation) into water?   Compute the refractive index of water.

The same number of wave peaks leave the air each second as enter into the water.   Hence the frequency is the same in the two materials.

Because   wavelength = (speed)/(frequency),   the wavelength in water is 3/4 that in air.   To find the refractive index,

$$n = \frac{\text{speed in vacuum}}{\text{speed in water}} = \frac{c}{(3/4)c} = \frac{4}{3} = 1.33$$

38.2.   A glass plate is 0.6 cm thick and has a refractive index of 1.55.   How long does it take for a pulse of light to pass through the plate?

$$t = \frac{x}{v} = \frac{0.006 \text{ m}}{(3 \times 10^8/1.55) \text{ m/s}} = 3.1 \times 10^{-11} \text{ s}$$

38.3.   As is shown in Fig. 38-4, a ray of light in air strikes a glass plate ($n = 1.50$) at an incidence angle of 50°.   Determine the angles of the reflected and refracted rays.

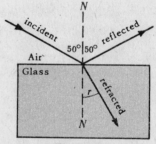

The law of reflection applies to the reflected ray.   Therefore the angle of reflection is 50°, as shown.

For the refracted ray, $n_1 \sin \theta_1 = n_2 \sin \theta_2$ becomes

$$\sin r = \frac{n_1}{n_2} \sin i = \frac{1.0}{1.5} \sin 50° = 0.51$$

from which $r = 30.7°$.

Fig. 38-4

38.4.   The refractive index of diamond is 2.42.   What is the critical angle for light passing from diamond to air?

We use $n_1 \sin \theta_1 = n_2 \sin \theta_2$ to give

$$(2.42) \sin \theta_c = (1) \sin 90°$$

from which $\sin \theta_c = 0.413$ and $\theta_c = 24.4°$.

38.5.   What is the critical angle for light passing from glass ($n = 1.54$) to water ($n = 1.33$)?

$$n_1 \sin \theta_1 = n_2 \sin \theta_2 \qquad \text{becomes} \qquad n_1 \sin \theta_c = n_2 \sin 90°$$

from which $$\sin \theta_c = \frac{n_2}{n_1} = \frac{1.33}{1.54} = 0.864 \qquad \text{or} \qquad \theta_c = 59.7°$$

38.6.   A layer of oil ($n = 1.45$) floats on water ($n = 1.33$).   A ray of light shines onto the oil with an incidence angle of $40°$.   Find the angle the ray makes in the water.

At the air-oil interface, Snell's law gives

$$n_{\text{air}} \sin 40° = n_{\text{oil}} \sin \theta_{\text{oil}}$$

At the oil-water interface, we have (using the equality of alternate angles)

$$n_{\text{oil}} \sin \theta_{\text{oil}} = n_{\text{water}} \sin \theta_{\text{water}}$$

Thus, $n_{\text{air}} \sin 40° = n_{\text{water}} \sin \theta_{\text{water}}$;   the overall refraction is just as though the oil layer were absent. Solving,

$$\sin \theta_{\text{water}} = \frac{n_{\text{air}} \sin 40°}{n_{\text{water}}} = \frac{(1)(0.643)}{1.33} \qquad \text{or} \qquad \theta_{\text{water}} = 28.9°$$

38.7.   As shown in Fig. 38-5, a small luminous body, at the bottom of a pool of water ($n = 4/3$) 1 m deep, emits rays upward in all directions.   A circular area of light is formed at the surface of the water.   Determine the radius R of the circle.

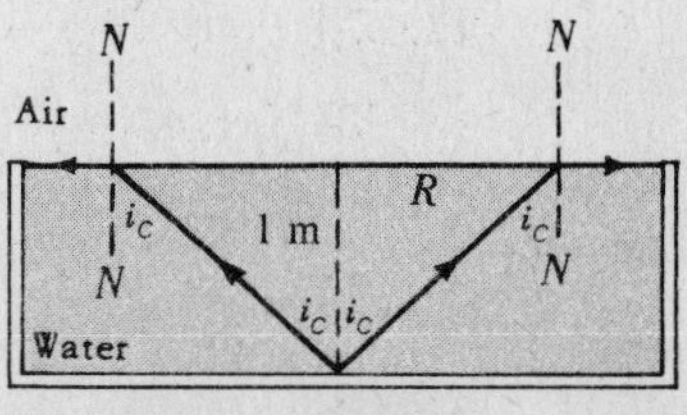

The circular area is formed by rays refracted into the air. Total internal reflection, and hence no refraction, occurs when the angle of incidence in water is greater than the critical angle, i_c.   We have

$$\sin i_c = \frac{n_a}{n_w} = \frac{1}{4/3} \qquad \text{or} \qquad i_c = 48.6°$$

From the figure,

$$R = (1 \text{ m}) \tan i_c = (1 \text{ m})(1.13) = 1.13 \text{ m}$$

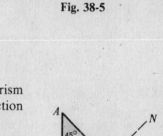

Fig. 38-5

38.8.   What is the minimum value of the refractive index for a $45°$ prism which is used to turn a beam of light by total internal reflection through a right angle?   See Fig. 38-6.

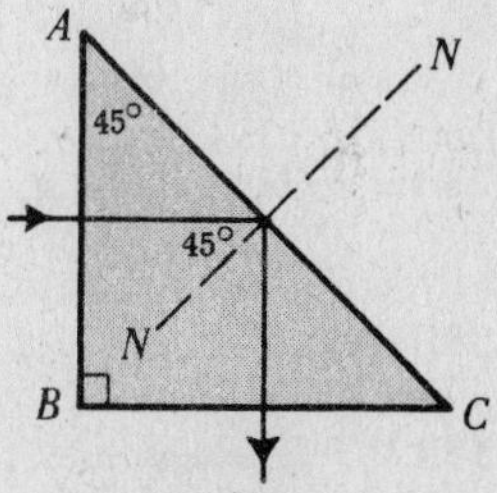

The ray enters the prism without deviation, since it strikes side AB normally.   It then makes an incidence angle of $45°$ with normal NN to side AC.   The critical angle of the prism must be smaller than $45°$ if the ray is to be totally reflected at side AC and thus turned through $90°$. From $n_1 \sin \theta_c = n_2 \sin 90°$,

$$\text{minimum } n = \frac{1}{\sin 45°} = \frac{1}{1/\sqrt{2}} = \sqrt{2} = 1.414$$

Fig. 38-6

38.9.   A prism whose refracting angle is $60°$ causes a minimum deviation of $48°$ in a monochromatic beam of light.   Calculate the refractive index of the prism for that wavelength.

$$n = \frac{\sin \frac{1}{2}(A + D_{\text{min}})}{\sin \frac{1}{2}A} = \frac{\sin \frac{1}{2}(60° + 48°)}{\sin \frac{1}{2}(60°)} = \frac{0.809}{0.500} = 1.62$$

38.10.   The glass prism shown in Fig. 38-7 has an index of refraction of 1.55.   Find the angle of deviation D for the case shown.

No deflection occurs at the entering surface, because the incidence angle is zero.   At the second surface, $i = 30°$ (because its sides are mutually perpendicular to the sides of the apex angle).   Then, Snell's law becomes

$$n_1 \sin i = n_2 \sin r \qquad \text{or} \qquad \sin r = \frac{1.55}{1} \sin 30°$$

from which $r = 50.8°$.   But $D = r - i$ and so $D = 20.8°$.

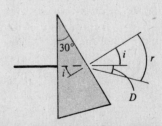

Fig. 38-7

38.11. As in Fig. 38-8, an object is at a depth d beneath the surface of a transparent material of refractive index n. As viewed from a point directly above, how deep does the object appear to be?

The two rays shown emerging into the air appear to come from point B. Therefore the apparent depth is CB. We have

$$\frac{b}{CB} = \tan r \qquad \text{and} \qquad \frac{b}{CA} = \tan i$$

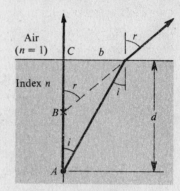

If the object is viewed from nearly straight above, then angles r and i will be very small. For small angles, the sine and tangent are nearly equal. Therefore,

$$\frac{\overline{CB}}{\overline{CA}} = \frac{\tan i}{\tan r} \approx \frac{\sin i}{\sin r}$$

But $n \sin i = (1) \sin r$, from which

$$\frac{\sin i}{\sin r} = \frac{1}{n}$$

Hence   apparent depth $\overline{CB} = \dfrac{\text{actual depth } \overline{CA}}{n}$

Fig. 38-8

Supplementary Problems

38.12. The speed of light in a certain glass is 1.91×10^8 m/s. What is the refractive index of the glass?
Ans. 1.57

38.13. What is the frequency of light which has a wavelength in air of 546 nm? What is its frequency in water ($n = 1.33$)? Its speed in water? Its wavelength in water?
Ans. 550 THz, 550 THz, 2.26×10^8 m/s, 410 nm

38.14. A beam of light strikes the surface of water at an incidence angle of 60°. Determine the directions of the reflected and refracted rays. For water, $n = 1.33$.
Ans. 60° reflected into air, 40.6° refracted into water

38.15. The critical angle for light passing from rock salt into air is 40.5°. Calculate the index of refraction of rock salt. *Ans.* 1.54

38.16. What is the critical angle when light passes from glass ($n = 1.50$) into air? *Ans.* 41.8°

38.17. The absolute indices of refraction of diamond and crown glass are 5/2 and 3/2 respectively. Compute (*a*) the refractive index of diamond relative to crown glass and (*b*) the critical angle between diamond and crown glass. *Ans.* (*a*) 5/3; (*b*) 37°

38.18. A pool of water ($n = 4/3$) is 60 cm deep. Find its apparent depth when viewed vertically through air.
Ans. 45 cm

38.19. A layer of benzene ($n = 1.50$) 6 cm deep floats on water ($n = 1.33$) 4 cm deep. Determine the apparent distance of the bottom of the vessel below the surface of the benzene when viewed vertically from air.
Ans. 7 cm

38.20. A mirror is made of plate glass ($n = 3/2$) 1 cm thick and silvered on the back. A man is 50 cm from the front face of the mirror. If he looks perpendicularly into it, at what distance behind the front face of the mirror will his image appear to be? *Ans.* 51.3 cm

38.21. A straight rod is partially immersed in water ($n = 1.33$). Its submerged portion appears to be inclined 45° with the surface when viewed vertically from air. What is the actual inclination of the rod?
Ans. arctan 1.33 = 53°

38.22. A prism whose refracting angle is 46° causes a minimum deviation of 32° in a monochromatic beam of light. Compute the refractive index of the prism for that wavelength of light. *Ans.* 1.61

Chapter 39

Thin Lenses

TYPES OF LENSES: As indicated in Fig. 39-1, *converging*, or *positive*, lenses are thicker at the center than at the rim and will converge a beam of parallel light to a real focus. *Diverging*, or *negative*, lenses are thinner at the center than at the rim and will diverge a beam of parallel light from a virtual focus.

The *principal focus* of a thin lens with spherical surfaces is the point F where rays parallel to and near the principal axis XX are brought to a focus; this focus is real for a converging lens and virtual for a diverging lens. The *focal length* is the distance of the principal focus from the lens.

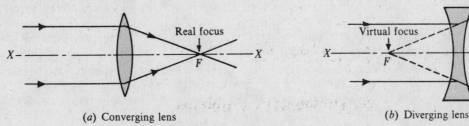

(a) Converging lens   (b) Diverging lens

Fig. 39-1

OBJECT AND IMAGE RELATION for converging and diverging lenses:

$$\frac{1}{p} + \frac{1}{q} = \frac{1}{f}$$

where p is the object distance from lens, q is the image distance from lens, f is the focal length of lens. The lens is assumed to be thin and the light rays *paraxial* (close to the principal axis).

p is positive for a real object, and negative for a virtual object (see Chapter 40)
q is positive for a real image, and negative for a virtual image
f is positive for a converging lens, and negative for a diverging lens

$$\text{linear magnification} = \frac{\text{size of image}}{\text{size of object}} = \frac{\text{image distance from lens}}{\text{object distance from lens}} = \left| \frac{q}{p} \right|$$

Converging lenses form real and inverted images of objects located outside of the principal focus. When the object is between the principal focus and the lens, the image is virtual (on the same side of the lens as the object), erect and enlarged.

Diverging lenses produce only virtual, erect and smaller images of real objects.

LENSMAKER'S EQUATION:

$$\frac{1}{f} = (n - 1)\left(\frac{1}{r_1} + \frac{1}{r_2} \right)$$

where n is the refractive index of the lens material; r_1 and r_2 are the radii of curvature of the two lens surfaces. This equation holds for all types of thin lenses. The radii of curvature are considered to be positive for convex surfaces and negative for concave surfaces.

If a lens with refractive index n_1 is immersed in a material with index n_2, then n in the lensmaker's equation is to be replaced by n_1/n_2.

LENS POWER in *diopters* (m^{-1}) is equal to $1/f$, where f is the focal length expressed in meters.

269

LENSES IN CONTACT: When two thin lenses having focal lengths f_1 and f_2 are in contact, the focal length f of the combination is given by

$$\frac{1}{f} = \frac{1}{f_1} + \frac{1}{f_2}$$

Solved Problems

39.1. An object OO', 4 cm high, is 20 cm in front of a thin convex lens of focal length $+12$ cm. Determine the position and height of its image II', (a) by construction and (b) by computation.

(a) The following two convenient rays from O will locate the image (see Fig. 39-2).
1. A ray OP, parallel to the principal axis, must after refraction pass through the principal focus F.
2. A ray passing through the optical center C of a thin lens is not appreciably deviated. Hence ray OCI may be drawn as a straight line.

The intersection I of these two rays is the image of O. Thus II' represents the position and size of the image of OO'. The image is real, inverted, enlarged, and at a greater distance from the lens than the object. (If the object were at II', the image, at OO', would be real, inverted, smaller.)

(b) $\qquad \frac{1}{p} + \frac{1}{q} = \frac{1}{f} \qquad$ or $\qquad \frac{1}{20} + \frac{1}{q} = \frac{1}{12} \qquad$ or $\qquad q = 30$ cm

The image is real (since q is positive) and 30 cm behind the lens.

$$\frac{\text{height of image}}{\text{height of object}} = \left| \frac{q}{p} \right| = \frac{30 \text{ cm}}{20 \text{ cm}} = 1.5 \qquad \text{or} \qquad \text{height of image} = (1.5)(4 \text{ cm}) = 6 \text{ cm}$$

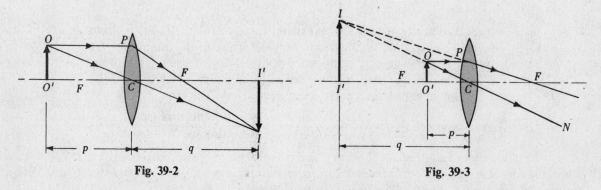

Fig. 39-2   Fig. 39-3

39.2. An object OO' is 5 cm in front of a convex lens of focal length $+7.5$ cm. Determine the position and magnification of its image II', (a) by construction and (b) by computation.

(a) Choose two convenient rays from O, as in Fig. 39-3.
1. A ray OP, parallel to the principal axis, is refracted so as to pass through the principal focus F.
2. A ray OCN, through the optical center of the lens, is drawn as a straight line.
   These two rays do not meet, but appear to originate from a point I. Thus II' represents the position and size of the image of OO'.
   When the object is between F and C, the image is virtual, erect, and enlarged.

(b) $\qquad \frac{1}{p} + \frac{1}{q} = \frac{1}{f} \qquad$ or $\qquad \frac{1}{5} + \frac{1}{q} = \frac{1}{7.5} \qquad$ or $\qquad q = -15$ cm

Since q is negative, the image is virtual (on the same side of the lens as the object) and 15 cm in front of the lens.

$$\text{linear magnification} = \frac{\text{size of image}}{\text{size of object}} = \left| \frac{q}{p} \right| = \frac{15 \text{ cm}}{5 \text{ cm}} = 3 \text{ diameters}$$

39.3. An object OO', 9 cm high, is 27 cm in front of a concave lens of focal length -18 cm. Determine the position and height of its image II', (a) by construction and (b) by computation.

(a) Choose the two convenient rays from O shown in Fig. 39-4.
   1. A ray OP, parallel to the principal axis, is refracted outward in the direction D as if it passed through the principal focus F.
   2. A ray through the optical center of the lens is drawn as a straight line OC.

   Then II' is the image of OO'. Images formed by concave or divergent lenses are virtual, erect, and smaller.

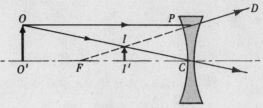

Fig. 39-4

(b) $$\frac{1}{p} + \frac{1}{q} = \frac{1}{f} \qquad \text{or} \qquad \frac{1}{27} + \frac{1}{q} = -\frac{1}{18} \qquad \text{or} \qquad q = -10.8 \text{ cm}$$

Since q is negative, the image is virtual and 10.8 cm in front of the lens.

$$\text{linear magnification} = \left| \frac{q}{p} \right| = \frac{10.8 \text{ cm}}{27 \text{ cm}} = 0.40 \qquad \text{or} \qquad \text{height of image} = (0.40)(9 \text{ cm}) = 3.6 \text{ cm}$$

39.4. Compute the position and focal length of the converging lens which will project the image of a lamp, magnified 4 diameters, upon a screen 10 m from the lamp.

From $p + q = 10$ and $q = 4p$, we find $p = 2$ and $q = 8$.

$$\frac{1}{f} = \frac{1}{p} + \frac{1}{q} = \frac{1}{2} + \frac{1}{8} = \frac{5}{8} \qquad \text{or} \qquad f = \frac{8}{5} = +1.6 \text{ m}$$

39.5. In what two positions will a converging lens of focal length $+7.5$ cm form images of a luminous object on a screen located 40 cm from the object?

Given $p + q = 40$ and $f = +7.5$,

$$\frac{1}{p} + \frac{1}{40 - p} = \frac{1}{7.5} \qquad \text{or} \qquad p^2 - 40p + 300 = 0 \qquad \text{or} \qquad (p - 10)(p - 30) = 0$$

from which $p = 10$ and $p = 30$. The two lens positions are 10 cm and 30 cm from the object.

39.6. A lens has a convex surface of radius 20 cm and a concave surface of radius 40 cm, and is made of glass of refractive index 1.54. Compute the focal length of the lens, and state whether it is a converging or a diverging lens.

$$\frac{1}{f} = (n - 1)\left(\frac{1}{r_1} + \frac{1}{r_2} \right) = (1.54 - 1)\left(\frac{1}{20} - \frac{1}{40} \right) = \frac{0.54}{40} \qquad \text{or} \qquad f = +74.1 \text{ cm}$$

Since f turns out to be positive, the lens is converging.

39.7. A double convex lens has faces of radii 18 and 20 cm. When an object is 24 cm from the lens, a real image is formed 32 cm from the lens. Determine (a) the focal length of the lens and (b) the refractive index of the lens material.

(a) $$\frac{1}{f} = \frac{1}{p} + \frac{1}{q} = \frac{1}{24} + \frac{1}{32} = \frac{7}{96} \qquad \text{or} \qquad f = \frac{96}{7} = +13.7 \text{ cm}$$

(b) $\dfrac{1}{f} = (n - 1)\left(\dfrac{1}{r_1} + \dfrac{1}{r_2} \right) \qquad$ or $\qquad \dfrac{1}{13.7} = (n - 1)\left(\dfrac{1}{18} + \dfrac{1}{20} \right) \qquad$ or $\qquad n = 1.69$

39.8. A glass lens ($n = 1.50$) has a focal length of $+10$ cm in air. Compute its focal length in water ($n = 1.33$).

Using
$$\frac{1}{f} = \left(\frac{n_1}{n_2} - 1 \right)\left(\frac{1}{r_1} + \frac{1}{r_2} \right)$$

for air :
$$\frac{1}{10} = (1.50 - 1)\left(\frac{1}{r_1} + \frac{1}{r_2} \right)$$

for water :
$$\frac{1}{f} = \left(\frac{1.50}{1.33} - 1 \right)\left(\frac{1}{r_1} + \frac{1}{r_2} \right)$$

Divide one equation by the other to obtain $f = 5/0.128 = 39$ cm.

39.9. A double convex lens has faces with 20 cm radii for each. The index of refraction of the glass is 1.50. Compute the focal length of this lens (a) in air and (b) when immersed in carbon disulfide ($n = 1.63$).

Using
$$\frac{1}{f} = \left(\frac{n_1}{n_2} - 1 \right)\left(\frac{1}{r_1} + \frac{1}{r_2} \right)$$

(a)
$$\frac{1}{f} = (1.50 - 1)\left(\frac{1}{20} + \frac{1}{20} \right) \qquad \text{or} \qquad f = +20 \text{ cm}$$

(b)
$$\frac{1}{f} = \left(\frac{1.50}{1.63} - 1 \right)\left(\frac{1}{20} + \frac{1}{20} \right) \qquad \text{or} \qquad f = -125 \text{ cm}$$

Here, the focal length is negative and so the lens is diverging.

39.10. Two thin lenses, of focal lengths $+9$ and -6 cm, are placed in contact. Calculate the focal length of the combination.

$$\frac{1}{f} = \frac{1}{f_1} + \frac{1}{f_2} = \frac{1}{9} - \frac{1}{6} = \frac{-1}{18} \qquad \text{or} \qquad f = -18 \text{ cm (diverging)}$$

39.11. A certain achromatic lens is formed from two thin lenses in contact, having powers of $+10$ and -6 diopters. Determine the power and focal length of the combination.

Since reciprocal focal lengths add,

$$\text{power} = +10 - 6 = +4 \text{ diopters} \qquad \text{and} \qquad \text{focal length} = \frac{1}{\text{power}} = \frac{1}{+4 \text{ m}^{-1}} = +25 \text{ cm}$$

39.12. A lens of focal length f projects upon a screen the image of a luminous object magnified M times. Show that the lens distance from the screen is $f(M + 1)$.

The image is real, since it can be shown on a screen, and so $q > 0$. We then have

$$M = \frac{q}{p} = q\left(\frac{1}{p} \right) = q\left(\frac{1}{f} - \frac{1}{q} \right) = \frac{q}{f} - 1 \qquad \text{or} \qquad q = f(M + 1)$$

Supplementary Problems

39.13. Draw diagrams to locate qualitatively the position, nature and size of image formed by a converging lens of focal length f for the following object distances: (a) infinity, (b) greater than $2f$, (c) equal to $2f$, (d) between $2f$ and f, (e) equal to f, (f) less than f.

39.14. Determine the nature, position and linear magnification of the image formed by a thin converging lens of focal length $+100$ cm when the object distance from the lens is (a) 150 cm, (b) 75 cm.
Ans. (a) real, inverted, 300 cm beyond lens, 2 : 1; (b) virtual, erect, 300 cm in front of lens, 4 : 1

39.15. In what two positions of the object will its image be enlarged 8 times by a lens of focal length $+4$ cm?
Ans. 4.5 cm from lens (image is real and inverted), 3.5 cm from lens (image is virtual and erect)

39.16. What is the nature and focal length of the lens that will form a real image having one-third the dimensions of an object located 9 cm from the lens?   *Ans.* converging, $+2.25$ cm

39.17. Describe fully the image of an object which is 10 cm high and 28 cm from a diverging lens of focal length -7 cm.   *Ans.* virtual, erect, smaller, 5.6 cm in front of lens, 2 cm high

39.18. Compute the focal length of the lens which will give an erect image 10 cm from the lens when the object distance from the lens is (a) 200 cm, (b) very great.   *Ans.* (a) -10.5 cm; (b) -10 cm

39.19. A luminous object and a screen are 12.5 m apart. What are the position and focal length of the lens which will throw upon the screen an image of the object magnified 24 diameters?
Ans. 0.50 m from object, $+0.48$ m

39.20. A plano-concave lens has a spherical surface of radius 12 cm, and its focal length is -22.2 cm. Compute the refractive index of the lens material.   *Ans.* 1.54

39.21. A convexo-concave lens has faces of radii 3 and 4 cm respectively, and is made of glass of refractive index 1.6. Determine (a) its focal length and (b) the linear magnification of the image when the object is 28 cm from the lens.   *Ans.* (a) $+20$ cm; (b) 2.5 : 1

39.22. A double convex glass lens has faces of radii 8 cm each. Compute its focal length in air and when immersed in water. Refractive index of the glass, 1.50; of water, 1.33.   *Ans.* $+8$ cm, $+32$ cm

39.23. Two thin lenses, of focal lengths $+12$ and -30 cm, are in contact. Compute the focal length and power of the combination.   *Ans.* $+20$ cm, $+5$ diopters

Chapter 40

Optical Instruments

COMBINATION OF THIN LENSES: When a lens is used in combination with another lens to form an image: (1) compute the position of the image produced by the first lens alone, disregarding the second lens; (2) then consider this image as the object for the second lens and locate its image as produced by the second lens alone. This latter image is the final image required.

If the image formed by the first lens alone is computed to be in back of the second lens, then that image is a virtual object for the second lens and its distance from the second lens is considered negative.

Solved Problems

40.1. A certain nearsighted person cannot see distinctly objects beyond 80 cm from the eye. What is the power in diopters of the spectacle lenses which will enable him to see distant objects clearly?

The image must be on the same side of the lens as the distant object (hence the image is virtual and $q = -80$ cm), and nearer to the lens than the object (hence diverging or negative lenses are indicated). As the object is at a great distance, p is very large and $1/p$ is practically zero.

$$\frac{1}{p} + \frac{1}{q} = \frac{1}{f} \qquad \text{or} \qquad 0 - \frac{1}{80} = \frac{1}{f} \qquad \text{or} \qquad f = -80 \text{ cm (diverging)}$$

$$\text{power in diopters} = \frac{1}{f \text{ in meters}} = \frac{1}{-0.80 \text{ m}} = -1.25 \text{ diopters}$$

40.2. A certain farsighted person cannot see clearly objects closer to the eye than 75 cm. Determine the power of the spectacle lenses which will enable her to read type at a distance of 25 cm.

The image must be on the same side of the lens as the type (hence the image is virtual and $q = -75$ cm), and farther from the lens than the type (hence converging or positive lenses are prescribed).

$$\frac{1}{f} = \frac{1}{25} - \frac{1}{75} \qquad \text{or} \qquad f = +37.5 \text{ cm}$$

$$\text{power} = \frac{1}{+0.375 \text{ m}} = +2.67 \text{ diopters}$$

40.3. A projection lens of focal length 30 cm throws a picture of a slide 2 cm $\times$ 3 cm upon a screen 10 m from the lens. Compute the dimensions of the image.

$$\frac{1}{p} = \frac{1}{f} - \frac{1}{q} = \frac{1}{0.30} - \frac{1}{10} = 3.23 \text{ m}^{-1}$$

and so

$$\text{linear magnification of image} = \left| \frac{q}{p} \right| = \frac{10 \text{ m}}{(1/3.23) \text{ m}} = 32$$

The length and width of the slide are each magnified 32 times.

$$\text{size of image} = (32 \times 2 \text{ cm}) \times (32 \times 3 \text{ cm}) = 64 \text{ cm} \times 96 \text{ cm}$$

40.4. A camera gives a clear image of a distant landscape when the lens is 8 cm from the film. What adjustment is required to get a good photograph of a map placed 72 cm from the lens?

When the camera is focused for distant objects (for parallel rays), the distance between lens and film is the focal length of the lens, 8 cm.   For an object 72 cm distant:

$$\frac{1}{q} = \frac{1}{f} - \frac{1}{p} = \frac{1}{8} - \frac{1}{72} \qquad \text{or} \qquad q = 9 \text{ cm}$$

The lens should be moved away from the film a distance of $(9 - 8)$ cm $= 1$ cm.

40.5. With a given illumination and film, the correct exposure for a camera lens set at $f/12$ is $(1/5)$ s.   What is the proper exposure time with the lens working at $f/4$?

A setting of $f/12$ means that the diameter of the opening, or stop, of the lens is $1/12$ of the focal length; $f/4$ means that it is $1/4$ of the focal length.

The amount of light passing through the opening is proportional to its area, and therefore to the square of its diameter.   As the diameter of the stop at $f/4$ is three times that at $f/12$, $3^2 = 9$ times as much light will pass through the lens at $f/4$, and the correct exposure at $f/4$ is

$$(1/9)(\text{exposure time at } f/12) = (1/45) \text{ s}$$

40.6. An engraver who has normal eyesight uses a converging lens of focal length 8 cm which he holds very close to his eye.   At what distance from the work should the lens be placed, and what is the magnifying power of the lens?

When a converging lens is used as a magnifying glass, the object is between the lens and the principal focal plane.   The virtual, erect, and enlarged image forms at the distance of distinct vision, 25 cm from the eye.

$$\frac{1}{p} + \frac{1}{q} = \frac{1}{f} \qquad \text{or} \qquad \frac{1}{p} + \frac{1}{-25} = \frac{1}{8} \qquad \text{or} \qquad p = \frac{200}{33} = 6.06 \text{ cm}$$

$$\text{magnifying power} = \left| \frac{q}{p} \right| = \frac{25}{6.06} = 4.13 \text{ diameters}$$

40.7. In the compound microscope shown in Fig. 40-1, the objective and eyepiece have focal lengths of $+0.8$ and $+2.5$ cm respectively.   The real image $A'B'$ formed by the objective is 16 cm from the objective.   Determine the total magnification if the eye is held close to the eyepiece and views the virtual image $A''B''$ at a distance of 25 cm.

Let p_o = object distance from objective
$\quad q_o$ = real-image distance from objective

$$\frac{1}{p_o} = \frac{1}{f_o} - \frac{1}{q_o} = \frac{1}{0.8} - \frac{1}{16} = \frac{19}{16} \text{ cm}^{-1}$$

and so linear magnification by the objective is

$$\left| \frac{q_o}{p_o} \right| = (16 \text{ cm})\left(\frac{19}{16} \text{ cm}^{-1} \right) = 19$$

The magnifying power of the eyepiece is

$$\left| \frac{q_e}{p_e} \right| = \left| q_e \left(\frac{1}{f_e} - \frac{1}{q_e} \right) \right| = \left| \frac{q_e}{f_e} - 1 \right|$$

$$= \left| \frac{-25}{+2.5} - 1 \right| = 11$$

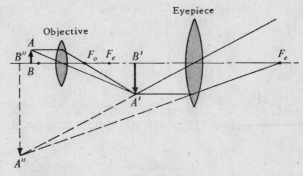

Fig. 40-1

Therefore, the magnifying power of the instrument is $19 \times 11 = 209$ diameters.

Alternately, under the conditions stated, the magnifying power of the eyepiece is

$$\frac{25}{f_e} + 1 = \frac{25}{2.5} + 1 = 11$$

40.8. Two positive lenses, having focal lengths of $+2$ and $+5$ cm, are 14 cm apart as shown in Fig. 40-2. An object AB is placed 3 cm in front of the $+2$ lens. Determine the position and magnification of the final image $A''B''$ formed by this combination of lenses.

To locate image $A'B'$ by the $+2$ lens alone:

$$\frac{1}{q} = \frac{1}{f} - \frac{1}{p} = \frac{1}{2} - \frac{1}{3} = \frac{1}{6} \qquad \text{or} \qquad q = 6 \text{ cm}$$

The image $A'B'$ is real, inverted, and 6 cm beyond the $+2$ lens.

To locate the final image $A''B''$: The image $A'B'$ is $(14 - 6)$ cm $= 8$ cm in front of the $+5$ lens and is taken as a real object for the $+5$ lens.

$$\frac{1}{q'} = \frac{1}{5} - \frac{1}{8} \qquad \text{or} \qquad q' = 13.3 \text{ cm}$$

$A''B''$ is real, erect, and 13.3 cm from the $+5$ lens.

$$\text{total linear magnification} = \frac{\overline{A''B''}}{\overline{AB}} = \frac{\overline{A'B'}}{\overline{AB}} \times \frac{\overline{A''B''}}{\overline{A'B'}} = \frac{6}{3} \times \frac{13.3}{8} = 3.33$$

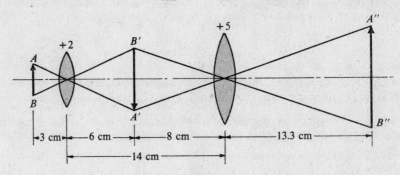

Fig. 40-2

40.9. The telephoto lens shown in Fig. 40-3 consists of a converging lens of focal length $+6$ cm placed 4 cm in front of a diverging lens of focal length -2.5 cm. (*a*) Locate the image of a very distant object. (*b*) Compare the size of the image formed by this lens combination with the size of the image that could be produced by the positive lens alone.

(*a*) If the negative lens were not employed, the image AB would be formed at the principal focal plane of the $+6$ lens, 6 cm distant from the $+6$ lens. The negative lens decreases the convergence of the rays refracted by the positive lens and causes them to focus at $A'B'$ instead of at AB.

The image AB (that would have been formed by the $+6$ lens alone) is $6 - 4 = 2$ cm beyond the -2.5 lens and is taken as the (virtual) object for the -2.5 lens. Then $p = -2$ cm (negative because AB is virtual).

$$\frac{1}{q} = \frac{1}{f} - \frac{1}{p} = -\frac{1}{2.5} + \frac{1}{2} = \frac{1}{10} \qquad \text{or} \qquad q = +10 \text{ cm}$$

The final image $A'B'$ is real and 10 cm beyond the negative lens.

(*b*) $\qquad$ linear magnification by negative lens $= \dfrac{\overline{A'B'}}{\overline{AB}} = \left| \dfrac{q}{p} \right| = \dfrac{10 \text{ cm}}{2 \text{ cm}} = 5$

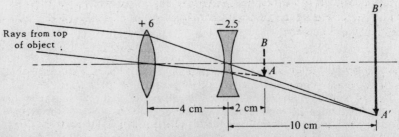

Fig. 40-3

40.10. Compute the magnifying power of a telescope, having objective and eyepiece lenses of focal lengths $+60$ and $+3$ cm respectively, when focused for parallel rays.

$$\text{magnifying power} = \frac{\text{focal length of objective}}{\text{focal length of eyepiece}} = \frac{60 \text{ cm}}{3 \text{ cm}} = 20 \text{ diameters}$$

40.11. As shown in Fig. 40-4, an object is placed 40 cm in front of a converging lens that has $f = +8$ cm. A plane mirror is 30 cm beyond the lens. Find the positions of all images formed by this system.

For the lens,

$$\frac{1}{q} = \frac{1}{f} - \frac{1}{p} = \frac{1}{8} - \frac{1}{40} = \frac{4}{40} \qquad \text{or} \qquad q = 10 \text{ cm}$$

This is image $A'B'$ in the figure. It is real and inverted.

$A'B'$ acts as an object for the plane mirror, 20 cm away. The mirror causes a virtual image, CD, at 20 cm behind the mirror.

Light reflected by the mirror appears to come from the image at CD. The lens uses CD as an object and gives an image of it to the left of the lens. The distance from the lens to this latter image, q', is given by

$$\frac{1}{q'} = \frac{1}{f} - \frac{1}{p'} = \frac{1}{8} - \frac{1}{50} = 0.105 \qquad \text{or} \qquad q' = 9.5 \text{ cm}$$

The real images are therefore located 10 cm to the right of the lens, and 9.5 cm to the left of the lens. This latter image is upright. A virtual inverted image is found 20 cm behind the mirror.

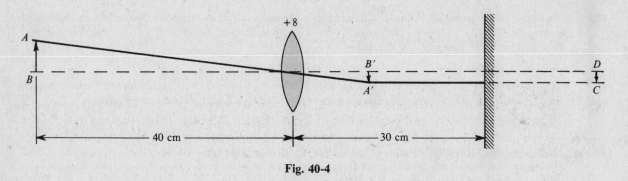

Fig. 40-4

Supplementary Problems

40.12. A certain farsighted person cannot see clearly objects closer to the eye than 60 cm. Determine the focal length and power of the spectacle lenses which will enable her to read a book at a distance of 25 cm. *Ans.* $+42.9$ cm, $+2.33$ diopters

40.13. A certain nearsighted person cannot see clearly objects beyond 50 cm from the eye. Determine the focal length and power of the glasses which will enable him to see distant objects clearly. *Ans.* -50 cm, -2 diopters

40.14. A projection lens is employed to produce pictures 2.4 m $\times$ 3.2 m from slides 3 cm $\times$ 4 cm on a screen 25 m from the lens. Compute its focal length. *Ans.* 30.9 cm

40.15. A camera gives a life-size picture of a flower when the lens is 20 cm from the film. What is the distance between lens and film when snapping a flock of birds high overhead? *Ans.* 10 cm

40.16. What is the maximum stop rating of a camera lens having a focal length of +10 cm and a diameter of 2 cm? If the correct exposure at $f/6$ is $(1/90)$ s, what exposure is needed when the diaphragm is changed to $f/9$?   *Ans.* $f/5$, $(1/40)$ s

40.17. What is the magnifying power of a lens of focal length +2 cm when used as a magnifying glass (or simple microscope)? The lens is held close to the eye, and the virtual image forms at the distance of distinct vision, 25 cm from the eye.   *Ans.* 13.5 ×

40.18. When the object distance from a converging lens is 5 cm, a real image is formed 20 cm from the lens. What magnification is produced by this lens when it is used as a magnifying glass, the distance of most distinct vision being 25 cm?   *Ans.* 7.3 ×

40.19. In a compound microscope, the focal lengths of the objective and eyepiece are +0.5 and +2 cm respectively. The instrument is focused on an object 0.52 cm from the objective lens. Compute the magnifying power of the microscope if the virtual image is viewed by the eye at a distance of 25 cm.   *Ans.* 338 ×

40.20. A refracting astronomical telescope has a magnifying power of 150 × when adjusted for minimum eyestrain using an eyepiece of focal length +1.2 cm. (*a*) Determine the focal length of the objective lens. (*b*) How far apart must the two lenses be in order to project a real image of a distant object on a screen 12 cm from the eyepiece?   *Ans.* (*a*) +180 cm; (*b*) 181.33 cm

40.21. The large telescope at Mt. Palomar has a concave objective mirror of diameter 200 inches and radius of curvature 150 ft. What is the magnifying power of the instrument when used with an eyepiece of focal length 0.5 inch?   *Ans.* 1800 diameters

40.22. An astronomical telescope, having an objective lens of focal length +80 cm, is focused on the moon. How much must the eyepiece be drawn out in order to focus on an object 40 meters distant?
Ans. 1.63 cm

40.23. A lens combination consists of two lenses, having focal lengths of +4 and +8 cm, which are spaced 16 cm apart. Locate and describe the final image of an object placed 12 cm in front of the +4 lens.
Ans. 40 cm beyond +8 lens, real, erect

40.24. Two lenses, having focal lengths of +6 and −10 cm, are spaced 1.5 cm apart. Locate and describe the image of an object 30 cm in front of the +6 lens.
Ans. 15 cm beyond negative lens, real, inverted, 5/8 as large as the object

40.25. A telephoto lens consists of a positive lens of focal length +3.5 cm placed 2 cm in front of a negative lens of focal length −1.8 cm. (*a*) Locate the image of a very distant object. (*b*) Determine the focal length of the single lens that would form as large an image of a distant object as is formed by this lens combination.   *Ans.* (*a*) real image 9 cm in back of negative lens; (*b*) +21 cm

40.26. An opera glass has an objective lens of focal length +3.60 cm and a negative eyepiece of focal length −1.20 cm. How far apart must the two lenses be in order for the viewer to see a distant object at 25 cm from the eye?   *Ans.* 2.34 cm

40.27. Repeat Problem 40.11 if the distance between the plane mirror and the lens is 8 cm.
Ans. at 6 cm (real) and 24 cm (virtual) to the right of lens

40.28. Solve Problem 40.11 if the plane mirror is replaced by a concave mirror with a 20 cm radius of curvature.
Ans. at 10 cm (real, inverted), 10 cm (real, upright), −40 cm (real, inverted) to the right of the lens

Chapter 41

Dispersion of Light

DISPERSION OF LIGHT: White light is composed of a large number of waves of different frequencies. Each component frequency (or corresponding wavelength) produces a characteristic color sensation. The wavelengths (in air) comprising the visible spectrum range from about 400 nm (violet) to 700 nm (red). In order of increasing wavelength, the color sensations are violet, blue, green, yellow, and red. The following units are often used to measure wavelengths:

$$1 \text{ micrometer } (\mu m) = 10^{-6} \text{ m}$$

$$1 \text{ nanometer (nm)} = 10^{-9} \text{ m}$$

$$1 \text{ angstrom } (\mathring{A}) = 10^{-10} \text{ m}$$

The first two of these were formerly called the "micron" and the "millimicron."

The refractive index of transparent optical materials increases with decreasing wavelength. It is greater for blue light than for red light. Hence, when a beam of white light passes through a refracting material, each wavelength is refracted differently. The white light is thereby separated (or *dispersed*) into its different component colors. A colored spectrum can thus be obtained.

CHROMATIC ABERRATION is a lens defect whereby a lens acts much like a prism and refracts each color differently. As a result, images formed by the lens have colored edges.

Solved Problems

The light reaching the earth from the sun consists of a continuous spectrum of color except for several distinct wavelengths (called the *Fraunhofer lines*) that are absorbed by the gases in space between the earth and sun. These lines are designated by letters such as C, D, and F. We shall use these convenient wavelengths in the problems that follow.

41.1. What angular dispersion between the C (red) and F (blue) spectral lines is produced by a flint-glass prism whose refracting angle is 12°? Assume the prism is being used under conditions of minimum deviation. For flint glass, $n_C = 1.644$ and $n_F = 1.664$.

From Chapter 38, the index of refraction n is related to the prism refracting angle A and minimum deviation angle D_{min} (see Fig. 41-1) by

$$n = \frac{\sin \frac{1}{2}(A + D_{min})}{\sin \frac{1}{2}A}$$

But if an angle θ is small, then $\sin \theta \approx \theta$ provided θ is in radians. The above equation then becomes

$$n = \frac{A + D_{min}}{A} \qquad \text{or} \qquad D_{min} = A(n - 1)$$

In this particular case, only the ratio of angles occurs and so the angles can be measured in either degrees or radians, so long as we use the same angular measure throughout. We can now write

dispersion between C and F lines = (deviation of F line) − (deviation of C line)

$$= D_{min, F} - D_{min, C} = A(n_F - 1) - A(n_C - 1)$$

$$= A(n_F - n_C) = (12°)(1.664 - 1.644) = 0.24°$$

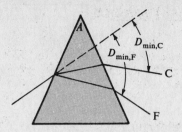
Fig. 41-1

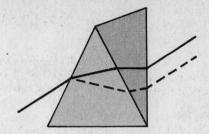
Fig. 41-2

41.2. By joining prisms of two different glasses (Fig. 41-2), it is possible to deviate two wavelengths by the same amount. We then say that the prism is *achromatic* for these two wavelengths. The same principle is used to make achromatic lenses. It is required to fit a 10° crown-glass prism with a flint-glass prism so as to achromatize the wavelength interval between the C and F spectral lines. (a) What must be the angle of the flint-glass prism? (b) What is the deviation produced by the prism combination, as computed for the D line? (The D line is located in the yellow region of the spectrum and is taken as the mean ray between the C and F lines.) Indices of the crown glass for the C, D, F lines are $n_C = 1.514$, $n_D = 1.517$, $n_F = 1.523$; indices of the flint glass for C, D, F lines are $n'_C = 1.644$, $n'_D = 1.650$, $n'_F = 1.664$.

(a) As in Problem 41.1,

$$\text{dispersion by crown-glass prism} = A(n_F - n_C)$$
$$\text{dispersion by flint-glass prism} = A'(n'_F - n'_C)$$

Achromatism obtains when the dispersion produced by the crown-glass prism equals (and so annuls) that produced by the flint-glass prism. Then

$$A(n_F - n_C) = A'(n'_F - n'_C)$$
$$(10°)(1.523 - 1.514) = A'(1.664 - 1.644)$$
$$A' = 4.5°$$

(b) resultant deviation of D line by prism combination

= (deviation of D line by crown-glass prism) − (deviation of D line by flint-glass prism)

$$= A(n_D - 1) - A'(n'_D - 1) = (10°)(1.517 - 1) - (4.5°)(1.650 - 1) = 2.24°$$

41.3. It is desired to fit a 12° crown-glass prism with a flint-glass prism so that the combination will produce dispersion, but without deviation of the mean D line. The indices of the crown and flint glass for the D line are respectively 1.520 and 1.650. What should be the angle of the flint-glass prism?

$$\text{deviation of D line by crown-glass prism} = \text{deviation of D line by flint-glass prism}$$
$$A(n_D - 1) = A'(n'_D - 1)$$
$$(12°)(1.520 - 1) = A'(1.650 - 1)$$
$$A' = 9.6°$$

41.4. Compute the *dispersive power* of a light flint glass in the region between the C and F spectral lines. The D line is taken as the mean ray between lines C and F. Indices of the glass for C, D, F lines: $n_C = 1.571$, $n_D = 1.575$, $n_F = 1.585$.

dispersive power of the flint glass between the Fraunhofer lines C and F

$$= \frac{\text{dispersion by small-angle prism between the C and F lines}}{\text{deviation of the mean D line}}$$
$$= \frac{A(n_F - n_C)}{A(n_D - 1)} = \frac{n_F - n_C}{n_D - 1} = \frac{1.585 - 1.571}{1.575 - 1.000} = 0.0243$$

Supplementary Problems

41.5. A beam of white light is incident upon water at an angle of 60°. The indices of water for the Fraunhofer lines A (red region) and H (violet region) are 1.330 and 1.344 respectively. Calculate the angular separation of the A and H lines in water.   *Ans.* 0.51°

41.6. A parallel beam of white light strikes a double convex lens having faces of radii +32 and +48 cm. The refractive indices of the glass for the A (red region) and H (violet region) spectral lines are respectively 1.578 and 1.614. Determine the distance between the focal points of the red and violet radiations. *Ans.* 1.95 cm

41.7. The blue F line of the hydrogen spectrum has a wavelength of 486.1 nm. (*a*) Express this length in meters, in micrometers, and in angstroms. (*b*) What is the frequency of this radiation? The speed of light in air is 3.00×10^8 m/s.   *Ans.* (*a*) 4.861×10^{-7} m, 0.4861 μm, 4861 Å; (*b*) 617 THz

41.8. A 10° prism is made of the heaviest flint glass, having refractive indices of 1.879 and 1.919 for the C and F lines respectively. Compute the dispersion and the length of the spectrum between the C and F lines produced by the prism on a screen 3 m distant and normal to the average direction of the emerging light. *Ans.* 0.40° or 24′, 21 mm

41.9. An achromatic lens having a focal length of +20 cm is to be constructed by combining a crown-glass lens with a flint-glass lens. What must be the focal lengths (f_1 and f_2) of the component lenses if the dispersive powers (δ_1 and δ_2) of the crown and flint glass used are 0.0158 and 0.0324 respectively? (*Hint:* Achromatism obtains when $(\delta_1/f_1) + (\delta_2/f_2) = 0$.) *Ans.* +10.2 cm (crown), −20.9 cm (flint)

Chapter 42

Interference and Diffraction of Light

COHERENT WAVES have the same form, the same frequency and a fixed phase difference (i.e. the amount by which the peaks of one wave lead or lag those of the other wave is fixed in time).

INTERFERENCE effects occur when two or more coherent waves are combined. If two coherent waves of the same amplitude are superposed, *destructive interference* (annulment, darkness) occurs at those points where the waves are 180° out of phase. *Constructive interference* (reinforcement, brightness) occurs where they are in phase.

DIFFRACTION refers to the bending or spreading of waves around the edges of apertures and opaque obstacles. This deviation of light from a straight-line path gives rise to interference patterns that blur the edges of shadows. It also places a limit on the size of the details that can be observed and limits the precision of even the best measurements.

SINGLE-SLIT DIFFRACTION: When parallel rays of light of wavelength λ are incident normally upon a slit of width s, a diffraction pattern is observed beyond the slit. Complete darkness is observed at angles θ_n to the straight-through beam, where

$$n\lambda = s \sin \theta_n$$

Here, $n = 1, 2, 3, \ldots$ is the *order number* of the diffraction dark band.

LIMIT OF RESOLUTION of two objects due to diffraction: If two objects are viewed in an optical instrument, the diffraction patterns caused by the aperture of the instrument limit our ability to distinguish the objects from each other. For distinguishability, the angle θ subtended at the aperture by the objects must be larger than a critical value θ_c, given by

$$\sin \theta_c = (1.22) \frac{\lambda}{D}$$

where D is the diameter of the aperture.

As a rough rule of thumb, it is impossible to observe detail that is smaller than the wavelength of the radiation used for the observation.

DIFFRACTION GRATING EQUATION: When waves of wavelength λ are incident normally upon a grating with grating spacing d, brightness is observed beyond the grating at angles θ_n to the normal, where

$$n\lambda = d \sin \theta_n$$

The *order number* n of the diffracted image is $0, 1, 2, 3, \ldots$.

This same relation applies to the major maxima in the interference patterns of even two and three slits. In these cases, however, the maxima are not nearly so sharply defined as for a grating consisting of hundreds of slits. The pattern may become quite complex if the slits are wide enough so that the single-slit diffraction pattern from each slit shows several minima.

DIFFRACTION OF X-RAYS of wavelength λ by reflection from a crystal is described by the *Bragg equation*. Strong reflections are observed at grazing angles ϕ_n (where ϕ is the angle between the face of the crystal and the reflected beam) given by

$$n\lambda = 2d \sin \phi_n$$

where d is the distance between reflecting planes in the crystal and $n = 1, 2, 3, \ldots$ is the *order* of reflection.

EQUIVALENT OPTICAL PATH LENGTH: In the same time that it takes a beam of light to travel a distance d in a material of index of refraction n, the beam would travel a distance nd in air or vacuum. For this reason, nd is defined as the *equivalent optical path length*.

Solved Problems

42.1. Monochromatic light from a point source illuminates two narrow, parallel slits. The centers of the two slits are $d = 0.8$ mm apart, as shown in Fig. 42-1. An interference pattern forms on the screen, 50 cm away. In the pattern, the bright and dark fringes are evenly spaced. The distance y shown is 0.304 mm. Compute the wavelength λ of the light.

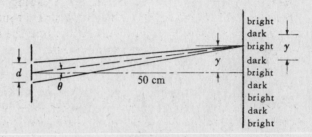

Fig. 42-1

Using the grating equation for this case of two slits, the angle θ shown in the figure is given by

$$n\lambda = d \sin \theta_n$$

In the present case we are concerned with the first-order maximum and so $n = 1$. The equation becomes

$$\lambda = d \sin \theta = (0.08 \text{ cm}) \frac{y}{\sqrt{(50)^2 + y^2}} \approx (0.08 \text{ cm}) \frac{y}{50 \text{ cm}}$$

$$= \frac{(0.08 \text{ cm})(0.0304 \text{ cm})}{50 \text{ cm}} = 4.86 \times 10^{-5} \text{ cm} = 486 \text{ nm}$$

42.2. When one leg of a Michelson interferometer is lengthened slightly, 150 dark fringes sweep through the field of view. If the light used has λ = 480 nm, how far was the mirror in that leg moved?

Darkness is observed when the light beams from the two legs are 180° out of phase. As the length of one leg is increased by $\frac{1}{2}\lambda$, the path length (down and back) increases by λ and the field of view changes from dark to bright to dark. When 150 fringes pass, the leg is lengthened by an amount

$$(150)(\tfrac{1}{2}\lambda) = (150)(240 \text{ nm}) = 36\,000 \text{ nm} = 0.036 \text{ mm}$$

42.3. As shown in Fig. 42-2, two flat glass plates touch at one edge and are separated at the other edge by a spacer. Using vertical viewing and light with λ = 589 nm, five dark fringes (D) are obtained from edge to edge. What is the thickness of the spacer?

The pattern is caused by interference between a beam reflected from the upper surface of the air wedge and a beam reflected from the lower surface of the wedge. The two reflections are of different nature in that reflection at the upper surface takes place at the boundary of a medium (air) of lower refractive index, while reflection at the lower surface occurs at the boundary of a medium (glass) of higher refractive index. In such cases, the act of reflection by itself involves a phase displacement of 180° between the two reflected beams. This explains the presence of a dark fringe at the left-hand edge.

As we move from a dark fringe to the next dark fringe, the beam that traverses the wedge must be held back by a path-length difference of λ. Because the beam travels twice through the wedge (down and back up), the wedge thickness changes by only $\frac{1}{2}\lambda$ as we move from fringe to fringe.

$$\text{spacer thickness} = 4(\tfrac{1}{2}\lambda) = 2(589 \text{ nm}) = 1178 \text{ nm}$$

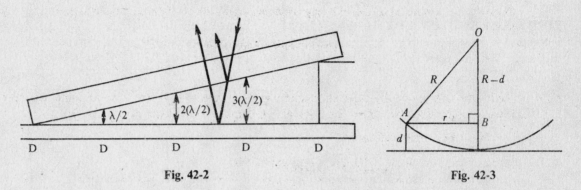

Fig. 42-2   Fig. 42-3

42.4. A convex lens is placed on a plane glass surface as shown in Fig. 42-3. It is illuminated from above with red light of wavelength 6700 Å. The interference pattern, caused by the light reflected from the convex and plane surfaces of the air film, consists of a dark spot at the point of contact surrounded by bright and dark rings. The radius of the 20th dark ring is 11 mm. Compute the radius of curvature R of the lens.

As in Problem 42.3, the gap thickness changes by $\frac{1}{2}\lambda$ as we move from one fringe to the next fringe of like type. Hence, its thickness d at the 20th dark ring is

$$d = 20(\tfrac{1}{2} \times 6.7 \times 10^{-7} \text{ m}) = 6.7 \times 10^{-6} \text{ m}$$

Let us now refer to Fig. 42-3. Let O be the center of curvature of the lens and r be the radius of the 20th dark ring. Then, in right triangle ABO,

$$R^2 = r^2 + (R - d)^2 = r^2 + R^2 - 2Rd + d^2$$

which gives $2Rd = r^2 + d^2$. Because d^2 is negligibly small compared with r^2,

$$R = \frac{r^2}{2d} = \frac{(11 \times 10^{-3} \text{ m})^2}{2(6.7 \times 10^{-6} \text{ m})} = 9.03 \text{ m}$$

42.5. What is the least thickness of a soap film which will appear black when viewed with sodium light ($\lambda = 589.3$ nm) reflected perpendicular to the film? The refractive index for soap solution is $n = 1.38$.

The situation is shown in Fig. 42-4. Ray b has an extra equivalent path length of $2nd = 2.76\,d$. In addition, there is a retardation of 180°, or $\frac{1}{2}\lambda$, between the beams because of the reflection process, as described in Problem 42.3.

Cancellation (and darkness) occurs if the retardation between the two beams is $\frac{1}{2}\lambda$, or $\frac{3}{2}\lambda$, or $\frac{5}{2}\lambda$, and so on. Therefore, for darkness,

$$2.76\,d + \tfrac{1}{2}\lambda = \tfrac{1}{2}\lambda, \qquad 2.76d + \tfrac{1}{2}\lambda = \tfrac{3}{2}\lambda, \qquad \dots$$

The first equation gives $d = 0$. The second one gives

$$d = \frac{\lambda}{2.76} = \frac{589.3 \text{ nm}}{2.76} = 214 \text{ nm}$$

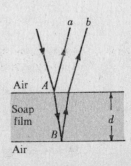

Fig. 42-4

as the thinnest possible film other than zero.

42.6. A single slit of width 0.1 mm is illuminated by parallel light of wavelength 6000 Å, and diffraction bands are observed on a screen 40 cm from the slit. How far is the third dark band from the central bright band? Refer to Fig. 42-5.

For a single slit, the dark bands are located from the equation $n\lambda = s \sin \theta_n$.

$$\sin \theta_3 = \frac{3\lambda}{s} = \frac{3(6 \times 10^{-7} \text{ m})}{10^{-4} \text{ m}} = 0.018$$

Since θ is a very small angle, $\tan \theta$ is approximately equal to $\sin \theta$. Hence the required distance y on the screen is

$$y = (40 \text{ cm}) \tan \theta = (40 \text{ cm})(0.018) = 0.72 \text{ cm}$$

42.7. Red light falls normally on a diffraction grating ruled 4000 lines/cm, and the second-order image is diffracted 34° from the normal. Compute the wavelength of the light.

From the grating equation,

$$\lambda = \frac{d \sin \theta_2}{2} = \frac{\left(\frac{1}{4000} \text{ cm}\right)(0.559)}{2} = 6.99 \times 10^{-5} \text{ cm} = 699 \text{ nm}$$

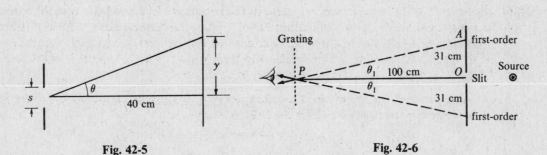

Fig. 42-5   **Fig. 42-6**

42.8. Figure 42-6 shows a laboratory setup for grating experiments. The diffraction grating has 5000 lines/cm and is 1 m from the slit, which is illuminated with sodium light. On either side of the slit, and parallel to the grating, is a meterstick. The eye, placed close to the grating, sees virtual images of the slit as if projected along the lengths of the metersticks. Determine the wavelength of the light if each first-order image is 31 cm from the slit.

$$\sin \theta_1 = \frac{31}{\sqrt{(100)^2 + (31)^2}} = 0.295$$

$$\lambda = \frac{d \sin \theta_1}{1} = \frac{(0.0002 \text{ cm})(0.295)}{1} = 59 \times 10^{-6} \text{ cm} = 590 \text{ nm}$$

42.9. Green light of wavelength 5400 Å is diffracted by a grating ruled 2000 lines/cm. (a) Compute the angular deviation of the third-order image. (b) Is a 10th-order image possible?

(a)
$$\sin \theta_3 = \frac{3\lambda}{d} = \frac{3(5.4 \times 10^{-5} \text{ cm})}{5 \times 10^{-4} \text{ cm}} = 0.324 \qquad \text{or} \qquad \theta = 18.9°$$

(b)
$$\sin \theta_{10} = \frac{10\lambda}{d} = \frac{10(5.4 \times 10^{-5} \text{ cm})}{5 \times 10^{-4} \text{ cm}} = 1.08 \text{ (impossible)}$$

Since the value of $\sin \theta$ cannot exceed 1, a 10th-order image is impossible.

42.10. Show that, in a spectrum of white light obtained with a grating, the red ($\lambda_r = 7000$ Å) of the second order overlaps the violet ($\lambda_v = 4000$ A) of the third order.

$$\text{For the red:} \quad \sin \theta_2 = \frac{2\lambda_r}{d} = \frac{2(7000)}{d} = \frac{14\,000}{d} \quad (d \text{ in Å})$$

$$\text{For the violet:} \quad \sin \theta_3 = \frac{3\lambda_v}{d} = \frac{3(4000)}{d} = \frac{12\,000}{d}$$

As $\sin \theta_2 > \sin \theta_3$, $\theta_2 > \theta_3$. Thus the angle of diffraction of red in the second order is greater than that of violet in the third order.

42.11. A parallel beam of X-rays is diffracted by a rock salt crystal. The first-order strong reflection is obtained when the glancing angle (the angle between the crystal face and the beam) is 6°50'. The distance between reflection planes in the crystal is 2.81 Å. What is the wavelength of the X-rays?

Note that the Bragg equation involves the glancing angle, not the angle of incidence.

$$\lambda = \frac{2d \sin \phi_1}{1} = \frac{(2)(2.81 \text{ Å})(0.119)}{1} = 0.67 \text{ Å}$$

42.12. Two light sources are 50 cm apart, as shown in Fig. 42-7. They are viewed by the eye at a distance L. The entrance opening (pupil) of the viewer's eye has a diameter of 3 mm. If the eye were perfect, the limiting factor for resolution of the two sources would be diffraction. In that limit, how large could L be and still have the sources seen as separate entities?

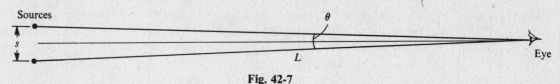

Fig. 42-7

In the limiting case, $\theta = \theta_c$, where $\sin \theta_c = (1.22)(\lambda/D)$. But, we see from the figure that $\sin \theta_c$ is nearly equal to s/L, because s is so much smaller than L. Substitution of this value gives

$$L = \frac{sD}{1.22\lambda} \approx \frac{(0.50 \text{ m})(3 \times 10^{-3} \text{ m})}{(1.22)(5 \times 10^{-7} \text{ m})} = 2500 \text{ m}$$

We have taken $\lambda = 500$ nm, about the middle of the visible range.

Supplementary Problems

42.13. Red light of wavelength 6438 Å, from a point source, passes through two parallel and narrow slits which are 1 mm apart. Determine the distance between the central bright fringe and the third dark interference fringe formed on a screen parallel to the plane of the slits and 1 m away. *Ans.* 1.61 mm

42.14. Two flat glass plates are pressed together at the top edge and separated at the bottom edge by a strip of tinfoil. The air wedge is examined in yellow sodium light (5893 Å) reflected normally from its two surfaces, and 42 dark interference fringes are observed. Compute the thickness of the tinfoil. *Ans.* 12.4 μm

42.15. A mixture of yellow light of wavelength 580 nm and blue light of wavelength 450 nm is incident normally on an air film 290 nm thick. What is the color of the reflected light? *Ans.* blue

42.16. A transparent film of glass of refractive index 1.50 is introduced normally in the path of one of the interfering beams of a Michelson interferometer which is illuminated with blue light of wavelength 486 nm. This causes 500 dark fringes to sweep across the field. Determine the thickness of the film. (*Hint*: Considering that the light passes twice through the glass film, the increased optical path of light reflected by one of the mirrors is $2(n-1)d$, where n = refractive index of the glass, 1 = refractive index of the air it replaces, and d = thickness of the film.)   *Ans.* 243 μm

42.17. A single slit of width 0.14 mm is illuminated by monochromatic light and diffraction bands are observed on a screen 2 m away. If the second dark band is 16 mm from the central bright band, what is the wavelength of the light?   *Ans.* 5600 Å

42.18. Green light of wavelength 5000 Å is incident normally on a grating, and the second-order image is diffracted 32° from the normal. How many lines/cm are marked on the grating?
Ans. 5300 lines/cm

42.19. A narrow beam of yellow light of wavelength 600 nm is incident normally on a diffraction grating ruled 2000 lines/cm, and images are formed on a screen parallel to the grating and 1 m distant. Compute the distance along the screen from the central bright line to the first-order lines.   *Ans.* 12.1 cm

42.20. Blue light of wavelength 470 nm is diffracted by a grating ruled 5000 lines/cm. (*a*) Compute the angular deviation of the second-order image. (*b*) What is the highest-order image theoretically possible with this wavelength and grating?   *Ans.* (*a*) 28°; (*b*) fourth

42.21. Determine the ratio of the wavelengths of two spectral lines if the second-order image of one line coincides with the third-order image of the other line, both lines being examined by means of the same grating.   *Ans.* 3 : 2

42.22. A spectrum of white light is obtained with a grating ruled 2500 lines/cm. Compute the angular separation between the violet (λ_v = 4000 Å) and red (λ_r = 7000 Å) in the (*a*) first order, (*b*) second order. (*c*) Does yellow (λ_y = 6000 Å) in the third order overlap the violet in the fourth order?
Ans. (*a*) 4°20′; (*b*) 8°57′; (*c*) yes

42.23. A spectrum of the sun's radiation in the infrared region is produced by a grating. What is the wavelength being studied, if the infrared line in the first order occurs at an angle of 25° with the normal, and the fourth-order image of the hydrogen line of wavelength 6563 Å occurs at 30°?
Ans. 22 200 Å

42.24. How far apart are the diffracting planes in a NaCl crystal for which X-rays of wavelength 1.54 Å make a glancing angle of 15°54′ in the first order?   *Ans.* 2.81 Å

Chapter 43

Relativity

THE SPECIAL THEORY OF RELATIVITY was proposed by A. Einstein and is concerned with bodies that are moving with constant velocity. Einstein's postulates were
(1) All motion is relative. Velocities of objects can only be given relative to some other object. It is impossible to determine an absolute velocity for an object.
(2) The accurately measured speed of light in free space, c, has the same value for all observers independent of the motion of the source or the motion of the observer.
These postulates lead to the following predictions.

VARIATION OF MASS: An object's mass measured when the object is at rest relative to the measurer is denoted by m_0 and is called the *rest mass* of the object. If the object is moving with speed v past an observer, the object has for that observer the mass

$$m = \frac{m_0}{\sqrt{1 - (v/c)^2}}$$

where $c = 2.998 \times 10^8$ m/s is the velocity of light in free space (vacuum). Notice that $m \to \infty$ as $v \to c$.

LIMITING SPEED: When $v = c$, the mass of an object becomes infinite. Hence, infinite forces would be required to accelerate the object to the speed of light. We conclude that no object can be accelerated to the speed of light, c, and so c is an upper limit for speed.

MASS–ENERGY CONVERSION: If an object's energy is changed by an amount ΔE, then its mass changes by an amount given by

$$\Delta E = (\Delta m)c^2$$

(This relation is often written as $E = mc^2$.) The relation is true for any type of energy change.
When an object is given translational kinetic energy, its mass, m, is made larger than its rest mass, m_0. The relation is

$$\text{translational KE} = (m - m_0)c^2$$

If the speed of the object is not too large, then this reduces to the usual expression

$$\text{translational KE} = \tfrac{1}{2}m_0 v^2 \qquad (v \ll c)$$

See Problem 43.7.

TIME DILATION: Two identical clocks sitting side by side tick out time in unison. However, if one clock is now accelerated to a high speed v and moves past the stationary clock and stationary observer, the moving clock appears to the stationary observer to tick out time too slowly. While the stationary clock ticks out a time t_s, the stationary observer will measure the moving clock to have ticked out a time $t_m < t_s$, where

$$t_m = t_s\sqrt{1 - (v/c)^2}$$

Because only relative motion is measurable, it is purely arbitrary which observer is taken to be "at rest" and which "moving". A clock that is moving relative to one observer may actually be the stationary clock for some other observer. Even so, an observer always notes that a clock moving relative to his stationary clock ticks out time more slowly than does his own stationary clock.

Suppose one times the duration of an event, the time Δt taken for an egg to cook, for example. The time Δt measured by a clock and observer at rest relative to the location of the event (the position of the egg in this case) is called the *proper time*. An observer moving with speed v past the event will measure the same event to take a time

$$\frac{\Delta t}{\sqrt{1 - (v/c)^2}}$$

according to his clock. The proper time for the duration of an event is smaller than the time measured by an observer moving past the location of the event.

SIMULTANEITY: Suppose that for an observer two events occur *at different locations* but at the same time. The events are simultaneous for this observer, but in general they are not simultaneous for a second observer moving relative to the first.

LENGTH CONTRACTION: An object has an x-component length L_0 when at rest relative to an observer (L_0 is called the *proper length*). If the object is now given an x-directed speed v, it will appear to the stationary observer to have been shortened in the x-direction (but not in the y- and z-directions). Its observed x-length will now be

$$L = L_0\sqrt{1 - (v/c)^2}$$

VELOCITY ADDITION FORMULA: Suppose that a spacecraft is moving in the x-direction with speed v relative to the earth. It shoots out a particle in the x-direction with speed u relative to the spacecraft. The speed of the particle relative to the earth is *not* $u + v$ (because this could be larger than c). Instead, the speed of the particle as measured by an observer on the earth is given by

$$\frac{v + u}{1 + (vu/c^2)}$$

Notice that even when $v = u = c$, this speed is only c. This agrees with the fact that the particle cannot be accelerated to a speed in excess of c.

Solved Problems

43.1. How fast must an object be moving if its mass is to be 1% larger than its rest mass?

We make use of $m = m_0/\sqrt{1 - (v/c)^2}$ to give

$$1 - \left(\frac{v}{c}\right)^2 = \left(\frac{m_0}{m}\right)^2 = \left(\frac{m_0}{1.01\, m_0}\right)^2 = 0.9803$$

Solving, $v = 0.14\, c = 4.2 \times 10^7$ m/s.

43.2. Compute the mass of an electron traveling at half the speed of light.

The rest mass of an electron is 9.1×10^{-31} kg.

$$m = \frac{m_0}{\sqrt{1 - (v/c)^2}} = \frac{m_0}{\sqrt{1 - (0.5)^2}} = \frac{m_0}{\sqrt{0.75}} = \frac{9.1 \times 10^{-31}\ \text{kg}}{0.866} = 1.05 \times 10^{-30}\ \text{kg}$$

43.3. If one gram of matter could be converted entirely into energy, what would be the value of the energy so produced, at 1 cent per kW·h?

We make use of $\Delta E = (\Delta m)c^2$.

$$\text{energy gained} = (\text{mass lost})c^2 = (10^{-3}\ \text{kg})(3 \times 10^8\ \text{m/s})^2 = 9 \times 10^{13}\ \text{J}$$

$$\text{value of energy} = (9 \times 10^{13}\ \text{J})\left(\frac{1\ \text{kW·h}}{3.6 \times 10^6\ \text{J}}\right)\left(\frac{\$0.01}{\text{kW·h}}\right) = \$250\,000$$

43.4. A 2 kg object is lifted from the floor to a tabletop 30 cm above the floor. By how much did the mass of the object increase because of its increased PE?

We use $\Delta E = (\Delta m)c^2$, with $\Delta E = mgh$. Therefore

$$\Delta m = \frac{\Delta E}{c^2} = \frac{mgh}{c^2} = \frac{(2\ \text{kg})(9.8\ \text{m/s}^2)(0.3\ \text{m})}{(3 \times 10^8\ \text{m/s})^2} = 6.5 \times 10^{-17}\ \text{kg}$$

43.5. An electron is accelerated from rest through a potential difference of 1.5 MV and thereby acquires 1.5 MeV of energy. Find its speed and mass.

We use $\Delta E = (\Delta m)c^2$, with

$$\Delta E = (1.5 \times 10^6\ \text{eV})(1.6 \times 10^{-19}\ \text{J/eV}) = 2.4 \times 10^{-13}\ \text{J}$$

Then

$$m - m_0 = \frac{\Delta E}{c^2} = \frac{2.4 \times 10^{-13}\ \text{J}}{(3 \times 10^8\ \text{m/s})^2} = 2.67 \times 10^{-30}\ \text{kg}$$

But $m_0 = 9.1 \times 10^{-31}\ \text{kg} = 0.91 \times 10^{-30}\ \text{kg}$ and so $m = 3.58 \times 10^{-30}\ \text{kg}$.

To find its speed, we use $m = m_0/\sqrt{1 - (v/c)^2}$. Then

$$1 - \left(\frac{v}{c}\right)^2 = \left(\frac{m_0}{m}\right)^2 = \left(\frac{0.91}{3.58}\right)^2 = 0.0646$$

from which

$$v = c\sqrt{1 - 0.0646} = 0.967\,c = 2.90 \times 10^8\ \text{m/s}$$

43.6. Determine the energy required to give an electron a speed 0.9 that of light, starting from rest.

$$\text{KE} = (m - m_0)c^2 = \left(\frac{m_0}{\sqrt{1 - (v/c)^2}} - m_0\right)c^2 = m_0 c^2\left(\frac{1}{\sqrt{1 - (v/c)^2}} - 1\right)$$

$$= (9.11 \times 10^{-31}\ \text{kg})(3 \times 10^8\ \text{m/s})^2\left(\frac{1}{\sqrt{1 - (0.9)^2}} - 1\right) = 1.061 \times 10^{-13}\ \text{J} = 0.663\ \text{MeV}$$

43.7. Show that $\text{KE} = (m - m_0)c^2$ reduces to $\text{KE} = \frac{1}{2}m_0 v^2$ when v is very much smaller than c.

$$\text{KE} = (m - m_0)c^2 = \left(\frac{m_0}{\sqrt{1 - (v/c)^2}} - m_0\right)c^2 = m_0 c^2\left[(1 - v^2/c^2)^{-1/2} - 1\right]$$

Let $b = -v^2/c^2$ and expand $(1 + b)^{-1/2}$ by the binomial theorem. Then

$$(1+b)^{-1/2} = 1 + (-1/2)b + \frac{(-1/2)(-3/2)}{2!}b^2 + \cdots = 1 + \frac{1}{2}\frac{v^2}{c^2} + \frac{3}{8}\frac{v^4}{c^4} + \cdots$$

and

$$KE = m_0 c^2\left[\left(1 + \frac{1}{2}\frac{v^2}{c^2} + \frac{3}{8}\frac{v^4}{c^4} + \cdots\right) - 1\right] = \frac{1}{2}m_0 v^2 + \frac{3}{8}m_0 v^2 \frac{v^2}{c^2} + \cdots$$

If v is very much smaller than c, the terms after $\frac{1}{2}m_0 v^2$ are negligibly small.

43.8. A beam of radioactive particles is measured as it shoots through the laboratory. It is found that, on the average, each particle "lives" for a time of 2×10^{-8} s; after that time, the particle changes to a new form. When at rest in the laboratory, the same particles "live" 0.75×10^{-8} s on the average. How fast are the particles in the beam moving?

Some sort of timing mechanism within the particle determines how long it "lives." This internal clock, which gives the proper lifetime, must obey the time dilation relation. We have

$$t_m = t_s\sqrt{1 - (v/c)^2} \qquad \text{or} \qquad 0.75 \times 10^{-8} = (2 \times 10^{-8})\sqrt{1 - (v/c)^2}$$

Note that t_0 is the time that the moving clock ticks out during the 2×10^{-8} s ticked out by the laboratory clock. Squaring each side of the equation and solving for v gives $v = 0.927 c = 2.78 \times 10^8$ m/s.

43.9. Two twins are 25 years old when one of them sets out on a journey through space at nearly constant speed. The twin in the spaceship measures time with an accurate watch. When he returns to the earth, he claims to be 31 years old, while the twin left on earth is then 43 years old. What was the speed of the spaceship?

The spaceship clock reads the trip to be only 6 years long while the earth clock reads it to be 18 years.

$$t_m = t_s\sqrt{1 - (v/c)^2} \qquad \text{becomes} \qquad 6 = 18\sqrt{1 - (v/c)^2}$$

from which

$$(v/c)^2 = 1 - 0.111 \qquad \text{or} \qquad v = 0.943 c = 2.83 \times 10^8 \text{ m/s}$$

43.10. A person in a spaceship holds a meterstick as the ship shoots past the earth with a speed v parallel to the earth's surface. What does the person in the ship notice as the stick is rotated from parallel to perpendicular to the ship's motion?

The stick behaves normally; it does not appear to change its length, because it has no translational motion relative to the observer in the spaceship. However, an observer on earth would measure the stick to be $(1 \text{ m})\sqrt{1 - (v/c)^2}$ long when parallel to the ship's motion, and 1 m long when perpendicular to the ship's motion.

43.11. As a rocket ship sweeps past the earth with speed v, it sends out a pulse of light ahead of it. How fast does the light pulse move according to people on the earth?

Method 1
c (by the second postulate of Special Relativity).

Method 2
According to the velocity addition formula, the observed speed will be (since $u = c$ in this case)

$$\frac{v + u}{1 + (uv/c^2)} = \frac{v + c}{1 + (v/c)} = \frac{(v + c)c}{c + v} = c$$

Supplementary Problems

43.12. At what speed must a particle move in order to double its mass?   *Ans.* 2.6×10^8 m/s

43.13. A particle is traveling at a speed v such that $v/c = 0.9900$.   Find m/m_0 for the particle.   *Ans.*   7.1

43.14. Compute the *rest energy* of an electron, i.e. the energy equivalent of its rest mass, 9.11×10^{-31} kg.   *Ans.*   0.512 MeV $= 8.2 \times 10^{-14}$ J

43.15. Determine the mass and speed of an electron having kinetic energy of 1×10^5 eV (1.6×10^{-14} J).   *Ans.* 1.09×10^{-30} kg, 1.64×10^8 m/s

43.16. A 2000 kg car is moving at 15 m/s. How much larger than its rest mass is its mass at this speed? (*Hint*: For x very small, $1/\sqrt{1-x} \approx 1 + \frac{1}{2}x$.)   *Ans.* 2.5×10^{-12} kg

43.17. A proton ($m_0 = 1.67 \times 10^{-27}$ kg) is accelerated to a kinetic energy of 200 MeV. What are its mass and its speed at this energy?   *Ans.* 2.03×10^{-27} kg, $0.57\,c = 1.70 \times 10^8$ m/s

43.18. A certain strain of bacteria doubles in number each 20 days. Two of these bacteria are placed on a spaceship and sent away from the earth for 1000 earth days. During this time, the speed of the ship was $0.9950\,c$. How many bacteria would be aboard when the ship lands on the earth?   *Ans.*   64

43.19. A certain light source sends out 2×10^{15} pulses each second. As a spaceship travels parallel to the earth's surface with a speed of $0.90\,c$, it uses this source to send pulses to the earth. The pulses are sent perpendicular to the path of the ship. How many pulses are recorded on earth each second? *Ans.* 8.72×10^{14} pulses/s

43.20. The insignia painted on the side of a spaceship is a circle with a line across it at 45° to the vertical. As the ship shoots past another ship in space, with a relative speed of $0.95\,c$, the second ship observes the insignia. What angle does the observed line make to the vertical? *Ans.* $\tan \theta = 0.31$ and $\theta = 17°$

Chapter 44

Quantum Physics and Wave Mechanics

QUANTA OF RADIATION: All electromagnetic waves, including light, have a dual nature. When traveling through space, they act like waves to give rise to interference and diffraction effects. But when electromagnetic radiation interacts with atoms and molecules, the beam acts like a stream of energy corpuscles called *photons* or *light quanta*.

The energy of each photon depends upon the frequency f (or wavelength λ) of the radiation in the beam. The interrelation is

$$\text{photon energy} = hf = \frac{hc}{\lambda}$$

where $h = 6.626 \times 10^{-34}$ J·s is a constant of nature called *Planck's constant*.

PHOTOELECTRIC EFFECT: When light is incident on a surface, under certain conditions electrons are ejected. Consider that a photon of energy hf collides with an electron at or near the surface of the material, that the photon transfers all its energy to the electron, and that the *work function*, the minimum work required to free an electron from the surface, is W_{min}. Then the maximum kinetic energy ($\frac{1}{2} m v_{max}^2$) of the ejected electron is given by *Einstein's photoelectric equation* as

$$\tfrac{1}{2} m v_{max}^2 = hf - W_{min}$$

The energy of the ejected electron may be found by determining what potential difference V must be applied to stop its motion; then $\frac{1}{2} m v^2 = Ve$.

For any surface, the light must be of short enough wavelength so that the photon energy hf is large enough to eject the electron. At the threshold, the photon's energy just equals the work function. For ordinary metals the threshold lies in the visible or ultraviolet. X-rays will eject photoelectrons readily; far-infrared or heat photons will never do so.

THE PHOTON HAS ZERO REST MASS: All of its mass is the result of its motion with speed c. Because $\Delta E = (\Delta m)c^2$, and because the energy of a photon is hf, we have, for a photon,

$$mc^2 = hf \qquad \text{or} \qquad m = \frac{hf}{c^2} = \frac{h}{c\lambda}$$

The momentum of a photon is $mc = h/\lambda$.

COMPTON EFFECT: A photon can collide with a particle of nonzero rest mass, such as an electron. When it does so, its energy and momentum may change due to the collision. The photon may also be deflected in the process. If a photon of original wavelength λ collides with a stationary particle of mass m and is deflected through an angle ϕ, then its wavelength is increased to λ', where

$$\lambda' = \lambda + \frac{h}{mc}(1 - \cos\phi)$$

The fractional change in wavelength is very small except for high-energy radiation such as X-rays or γ-rays.

DE BROGLIE WAVES: A particle of mass m moving with speed v has associated with it a *de Broglie wavelength*

$$\lambda = \frac{h}{mv}$$

A beam of particles can be diffracted and can undergo interference phenomena. These wavelike properties of particles can be computed by assuming the particles to behave like waves having the de Broglie wavelength.

Solved Problems

44.1. Show that the photons in a 1240 nm infrared light beam have energies of 1 eV.

$$\text{energy} = hf = \frac{hc}{\lambda} = \frac{(6.63 \times 10^{-34} \text{ J} \cdot \text{s})(3 \times 10^8 \text{ m/s})}{1240 \times 10^{-9} \text{ m}} = 1.60 \times 10^{-19} \text{ J} = 1 \text{ eV}$$

44.2. Compute the energy of a photon of blue light of wavelength 450 nm.

$$\text{energy} = \frac{hc}{\lambda} = \frac{(6.63 \times 10^{-34} \text{ J} \cdot \text{s})(3 \times 10^8 \text{ m/s})}{450 \times 10^{-9} \text{ m}} = 4.42 \times 10^{-19} \text{ J} = 2.76 \text{ eV}$$

44.3. In order to break a chemical bond in the molecules of human skin, causing sunburn, a photon energy of about 3.5 eV is required. To what wavelength does this correspond?

$$\lambda = \frac{hc}{\text{energy}} = \frac{(6.63 \times 10^{-34} \text{ J} \cdot \text{s})(3 \times 10^8 \text{ m/s})}{(3.5 \text{ eV})(1.6 \times 10^{-19} \text{ J/eV})} = 355 \text{ nm}$$

Ultraviolet light causes sunburn.

44.4. The work function of sodium metal is 2.3 eV. What is the longest-wavelength light that can cause photoelectron emission from sodium?

At threshold, the photon energy just equals the energy required to tear the electron loose from the metal, namely the work function, W_{min}.

$$W_{min} = \frac{hc}{\lambda}$$

$$(2.3 \text{ eV})\left(\frac{1.6 \times 10^{-19} \text{ J}}{1 \text{ eV}} \right) = \frac{(6.63 \times 10^{-34} \text{ J} \cdot \text{s})(3 \times 10^8 \text{ m/s})}{\lambda}$$

$$\lambda = 540 \text{ nm}$$

44.5. What potential difference must be applied to stop the fastest photoelectrons emitted by a nickel surface under the action of ultraviolet light of wavelength 2000 Å? The work function of nickel is 5.01 eV.

$$\text{energy of photon} = \frac{hc}{\lambda} = \frac{(6.63 \times 10^{-34} \text{ J} \cdot \text{s})(3 \times 10^8 \text{ m/s})}{2000 \times 10^{-10} \text{ m}} = 9.95 \times 10^{-19} \text{ J} = 6.21 \text{ eV}$$

Then, from the photoelectric equation, the energy of the fastest emitted electron is

$$6.21 \text{ eV} - 5.01 \text{ eV} = 1.20 \text{ eV}$$

Hence a negative retarding potential of 1.20 V is required.

44.6. Will photoelectrons be emitted by a copper surface, of work function 4.4 eV, when illuminated by visible light?

As in Problem 44.4,

$$\text{threshold } \lambda = \frac{hc}{W_{min}} = \frac{(6.63 \times 10^{-34} \text{ J} \cdot \text{s})(3 \times 10^8 \text{ m/s})}{4.4(1.60 \times 10^{-19}) \text{ J}} = 282 \text{ nm}$$

Hence visible light (400 nm to 700 nm) cannot eject photoelectrons from copper.

44.7. What wavelength must electromagnetic radiation have if a photon in the beam is to have the same momentum as an electron moving with a speed 2×10^5 m/s?

The requirement is that $(mv)_{electron} = (h/\lambda)_{photon}$. From this,

$$\lambda = \frac{h}{mv} = \frac{6.63 \times 10^{-34} \text{ J} \cdot \text{s}}{(9.1 \times 10^{-31} \text{ kg})(2 \times 10^5 \text{ m/s})} = 3.64 \text{ nm}$$

This wavelength is in the X-ray region.

44.8. Suppose that a 3.64 nm photon going in the $+x$-direction collides head-on with a 2×10^5 m/s electron moving in the $-x$-direction. If the collision is perfectly elastic, find the conditions after collision.

From the law of conservation of momentum,

$$\text{momentum before} = \text{momentum after}$$

$$\frac{h}{\lambda_0} - mv_0 = \frac{h}{\lambda} - mv$$

But, from Problem 44.7, $h/\lambda_0 = mv_0$ in this case. Hence, $h/\lambda = mv$. Also, for a perfectly elastic collision,

$$\text{KE before} = \text{KE after}$$

$$\frac{hc}{\lambda_0} + \frac{1}{2} mv_0^2 = \frac{hc}{\lambda} + \frac{1}{2} mv^2$$

Using the facts that $h/\lambda_0 = mv_0$ and $h/\lambda = mv$, we find

$$v_0\left(c + \frac{1}{2} v_0\right) = v\left(c + \frac{1}{2} v\right)$$

Therefore $v = v_0$ and the electron rebounds with its original speed. Because $h/\lambda = mv = mv_0$, the photon also rebounds and with its original wavelength.

44.9. A photon ($\lambda = 0.400$ nm) strikes an electron at rest and rebounds at an angle of 150° to its original direction. Find the speed and wavelength of the photon after the collision.

The speed of a photon is always the speed of light in vacuum, c. To obtain the wavelength after collision, we use the equation for the Compton effect:

$$\lambda' = \lambda + \frac{h}{mc}(1 - \cos\phi)$$

$$= 4 \times 10^{-10} \text{ m} + \frac{6.63 \times 10^{-34} \text{ J·s}}{(9.1 \times 10^{-31} \text{ kg})(3 \times 10^8 \text{ ms})}(1 - \cos 150°)$$

$$= 4 \times 10^{-10} \text{ m} + (2.43 \times 10^{-12} \text{ m})(1 + 0.866) = 0.4045 \text{ nm}$$

44.10. What is the de Broglie wavelength for a particle moving with speed 2×10^6 m/s if the particle is (a) an electron, (b) a proton, (c) a 0.2 kg ball?

We make use of the definition of the de Broglie wavelength:

$$\lambda = \frac{h}{mv} = \frac{6.63 \times 10^{-34} \text{ J·s}}{m(2 \times 10^6 \text{ m/s})} = \frac{3.3 \times 10^{-40} \text{ m·kg}}{m}$$

Substituting the required values for m, one finds that the wavelength is 3.6×10^{-10} m for the electron, 2×10^{-13} m for the proton, and 1.65×10^{-39} m for the 0.2 kg ball.

44.11. An electron falls from rest through a potential difference of 100 V. What is its de Broglie wavelength?

Its speed will still be far below c, so relativistic effects can be ignored. The KE gained, $\frac{1}{2} mv^2$, equals the electrical PE lost, Vq. Therefore, after solving for v,

$$v = \sqrt{\frac{2Vq}{m}} = \sqrt{\frac{2(100 \text{ V})(1.6 \times 10^{-19} \text{ C})}{9.1 \times 10^{-31} \text{ kg}}} = 5.9 \times 10^6 \text{ m/s}$$

Then

$$\lambda = \frac{h}{mv} = \frac{6.63 \times 10^{-34} \text{ J·s}}{(9.1 \times 10^{-31} \text{ kg})(5.9 \times 10^6 \text{ m/s})} = 12.3 \text{ nm}$$

44.12. What potential difference is required in an electron microscope to give electrons a wavelength of 0.5 Å?

$$\text{KE of electron} = \frac{1}{2}mv^2 = \frac{1}{2}m\left(\frac{h}{m\lambda}\right)^2 = \frac{h^2}{2m\lambda^2}$$

where use has been made of the de Broglie relation, $\lambda = h/mv$. Substitution of the known values gives the KE as 9.66×10^{-17} J. But $\text{KE} = Vq$ and so

$$V = \frac{\text{KE}}{q} = \frac{9.66 \times 10^{-17}\text{ J}}{1.6 \times 10^{-19}\text{ C}} = 600\text{ V}$$

Supplementary Problems

44.13. Compute the energy of a photon of blue light ($\lambda = 450$ nm), in joules and in eV.
Ans. 4.4×10^{-19} J = 2.76 eV

44.14. What is the wavelength of the light in which the photons have an energy of 600 eV?   *Ans.* 2.1 nm

44.15. A certain sodium lamp radiates 20 W of yellow light ($\lambda = 589$ nm). How many photons of the yellow light are emitted from the lamp each second?   *Ans.* 5.9×10^{19}

44.16. What is the work function of sodium metal if the photoelectric threshold wavelength is 680 nm?
Ans. 1.82 eV

44.17. Determine the maximum KE of photoelectrons ejected from a potassium surface by ultraviolet light of wavelength 2000 Å. What retarding potential difference is required to stop the emission of electrons? The photoelectric threshold wavelength for potassium is 4400 Å.   *Ans.* 3.38 eV, 3.38 V

44.18. With what speed will the fastest photoelectrons be emitted from a surface whose threshold wavelength is 600 nm, when the surface is illuminated with light of wavelength 400 nm?   *Ans.* 6×10^5 m/s

44.19. Electrons with maximum KE of 3 eV are ejected from a metal surface by ultraviolet radiation of wavelength 1500 Å. Determine the work function of the metal, the threshold wavelength of the metal, and the retarding potential difference required to stop the emission of electrons.
Ans. 5.27 eV, 2350 Å, 3 V

44.20. What are the speed and momentum of a 500 nm photon?   *Ans.* 3×10^8 m/s, 1.33×10^{-27} kg·m/s

44.21. An X-ray beam with wavelength exactly 5×10^{-14} m strikes a proton that is at rest ($m = 1.67 \times 10^{-27}$ kg). If the X-rays are scattered through an angle of 110°, what is the wavelength of the scattered X-rays?   *Ans.* 5.18×10^{-14} m

44.22. Compute the de Broglie wavelength of an electron that has been accelerated through a potential difference of 9000 V. Ignore relativistic effects.   *Ans.* 1.3×10^{-11} m

44.23. What is the de Broglie wavelength of an electron that has been accelerated through a potential difference of 1 MV? (You must use the relativistic mass and energy expressions at this high energy.)
Ans. 8.7×10^{-13} m

44.24. It is proposed to send a beam of electrons through a diffraction grating with the distance between slits being d. The electrons have a speed of 400 m/s. How large must d be if a strong beam of electrons is to emerge at an angle of 25° to the straight-through beam?
Ans. $n(4.3 \times 10^{-6}\text{ m})$ where $n = 1, 2, 3, \ldots$

Chapter 45

Atomic Physics

ENERGY IS QUANTIZED in atomic and molecular systems. The system can assume certain definite energies; all other energies are forbidden. Each atom or molecule has its own characteristic set of allowed energies. Because of this, the atom or molecule can be described by its allowed energy levels.

ENERGY-LEVEL DIAGRAMS summarize the allowed energies of a system. A typical diagram is shown in Fig. 45-1; it is the energy-level diagram for an isolated hydrogen atom. Each horizontal line represents an allowed energy of the atom (assuming the center of mass of the atom to be at rest). In this case, the zero level is chosen such that the ionized atom (i.e. the atom with its single electron infinitely far from the nucleus) has zero energy. As the negative electron is brought closer to the positive nucleus, it loses electrical PE. For this reason, the energy levels are all negative on this diagram.

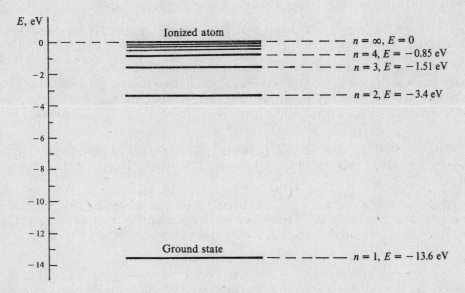

Fig. 45-1

BOHR'S MODEL OF THE HYDROGEN ATOM: Each energy level of the hydrogen atom is associated with a particular configuration of the atom. The following were assumed by Bohr:

(1) Stable orbits exist for the electron. The atom does not radiate away energy when the electron is in one of these orbits, even though classical theory indicates it should (because the electron is undergoing a centripetal acceleration).

(2) The stable, allowed orbits have radii r_n that satisfy the following relation:

$$\text{angular momentum of electron in orbit} = \frac{nh}{2\pi}$$

$$mv_n r_n = \frac{nh}{2\pi}$$

where m is the electron mass, v_n is the speed of the electron in the nth orbit, h is Planck's constant, and $n = 1, 2, 3, \ldots$.

The radii and energies of these allowed configurations of the atom are calculated to be

$$r_n = n(0.53 \times 10^{-10} \text{ m}) \qquad E_n = -\frac{13.6}{n^2} \text{ eV}$$

When $n = 1$, the atom has the lowest possible energy. This configuration is called the *ground state* of the atom. The energy required to take an electron from the ground state orbit to infinity is $E_\infty - E_1 = 13.6$ eV and this is called the *ionization energy* of the atom.

(3) If an atom falls from an orbit with $r = r_i$ to one with $r = r_f$, it loses an energy $|E_f - E_i|$. This energy is radiated as a photon of frequency f given by

$$hf = |E_f - E_i|$$

where h is Planck's constant.

Although that part of Bohr's model that involves precise orbits for the electron is not correct in detail, the energy levels predicted by it are correct. These are the energy levels shown in Fig. 45-1.

EMISSION OF LIGHT: When an isolated atom or molecular system falls from one energy level to a lower one, a photon is emitted. This photon carries away the energy lost by the system. Its frequency f is given by

$$hf = \text{energy lost by the system}$$

The emitted radiation has a precise wavelength and gives rise to a single spectral line in the emission spectrum of the system. Because the energy-level diagrams for different systems are not the same, each system has its own characteristic emission spectrum.

ABSORPTION OF LIGHT: A system in its ground state can absorb only those photons that will raise it to one of its other allowed energy levels.

Solved Problems

45.1. Sodium atoms emit a spectral line with a wavelength in the yellow, 589.6 nm. What is the difference in energy between the two energy levels involved in the emission of this spectral line?

$$\Delta E = hf = \frac{hc}{\lambda} = \frac{(6.63 \times 10^{-34} \text{ J} \cdot \text{s})(3 \times 10^8 \text{ m/s})}{589.6 \times 10^{-9} \text{ m}} = 3.37 \times 10^{-19} \text{ J} = 2.1 \text{ eV}$$

45.2. When a hydrogen atom is bombarded, the atom is excited to its higher energy states. As it falls back to the lower energy levels, light is emitted. What are the three longest-wavelength spectral lines emitted by the hydrogen atom as it falls to the $n = 1$ state from higher energy states?

We are interested in the following transitions (see Fig. 45-1):

$$\underline{n = 2 \rightarrow n = 1} \qquad \Delta E_{2,1} = -3.4 - (-13.6) = 10.2 \text{ eV}$$

$$\underline{n = 3 \rightarrow n = 1} \qquad \Delta E_{3,1} = -1.5 - (-13.6) = 12.1 \text{ eV}$$

$$\underline{n = 4 \rightarrow n = 1} \qquad \Delta E_{4,1} = -0.85 - (-13.6) = 12.75 \text{ eV}$$

To find the corresponding wavelengths we can use $\Delta E = hf = hc/\lambda$. For example, for the $n = 2$ to $n = 1$ transition,

$$\lambda = \frac{hc}{\Delta E_{2,1}} = \frac{(6.63 \times 10^{-34} \text{ J} \cdot \text{s})(3 \times 10^8 \text{ m/s})}{(10.2 \text{ eV})(1.6 \times 10^{-19} \text{ J/eV})} = 122 \text{ nm}$$

The other lines are found in the same way to be 102 nm and 97 nm. These are the first three lines of the *Lyman series*.

45.3. The series limit wavelength of the *Balmer series* is emitted as the hydrogen atom falls from the $n = \infty$ state to the $n = 2$ state. What is the wavelength of this line?

From Fig. 45-1, $\Delta E = 3.4 - 0 = 3.4$ eV. We find the corresponding wavelength in the usual way from $\Delta E = hc/\lambda$. The result is 365 nm.

45.4. What is the radiation of greatest wavelength that will ionize unexcited hydrogen atoms?

The incident photons must have enough energy to raise the atom from the $n = 1$ level to the $n = \infty$ level when absorbed by the atom. Because $E_\infty - E_1 = 13.6$ eV, we can use $E_\infty - E_1 = hc/\lambda$ to find the wavelength as 91 nm. Wavelengths shorter than this would not only remove the electron from the atom, but would give the removed electron KE.

45.5. The energy levels for singly-ionized helium atoms (atoms from which one of the two electrons has been removed) are given by $E_n = (-54.4/n^2)$ eV. Construct the energy-level diagram for this system.

See Fig. 45-2.

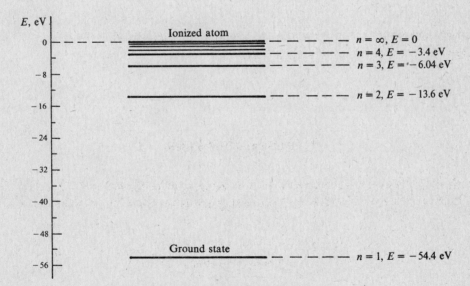

Fig. 45-2

Supplementary Problems

45.6. The spectrum of the mercury arc contains a line at 435.8 nm in the blue. What is the difference in energy levels in the atom that gives rise to this spectral line?   *Ans.* 4.55×10^{-19} J $= 2.85$ eV

45.7. A certain molecule has an energy-level diagram for its vibrational energy in which two levels are 0.0141 eV apart. Find the wavelength of the emitted line for the molecule as it falls from one of these levels to the other.   *Ans.*   88 μm

45.8. X-rays are emitted as the electrons deep within atoms having many electrons fall to lower energy states. What is the difference in energy between two levels if a transition between them gives rise to 0.5 Å X-rays?   *Ans.*   24.8 keV

45.9. An electron is shot down an X-ray tube. Its energy just before it strikes the target of the tube is 30 keV. If it loses all this energy in a single collision with a very massive atom, what is the wavelength of the single X-ray photon that is emitted?   *Ans.*   0.41 Å

45.10. The Balmer series of spectral lines is emitted in transitions to the $n = 2$ level from higher levels in the hydrogen atom. What are the two longest-wavelength lines of this series?   *Ans.*   656 nm, 486 nm

45.11. Figure 45-2 shows the energy-level diagram for singly-ionized helium ions. How much energy is required to remove the second electron from the atom? What is the maximum wavelength of an incident photon that could tear this electron from the ion?   *Ans.*   54.4 eV, 22.8 nm

45.12. Give the three longest wavelengths that singly-ionized helium atoms (in their ground state) will absorb strongly. See Fig. 45-2 for the energy-level diagram for singly-ionized helium.
Ans.   30.4 nm, 25.6 nm, 24.3 nm

Chapter 46

Nuclei and Radioactivity

THE NUCLEUS of an atom is a positively charged entity at the atom's center. Its radius is about 10^{-15} m, which is about 10^{-5} as large as the radius of the atom. Within the nucleus are protons and neutrons, collectively called *nucleons*. Although the positively charged protons repel each other, the much stronger, short-range *nuclear force* holds the nucleus together. The nuclear attractive force between nucleons decreases rapidly with particle separation and is essentially zero for nucleons more than 5×10^{-15} m apart.

NUCLEAR CHARGE AND ATOMIC NUMBER: Each proton within the nucleus carries a charge $+e$, whereas the neutrons carry no charge. If there are Z protons in a nucleus, then the charge on the nucleus is $+Ze$. We call Z the *atomic number* for that nucleus.

Because normal atoms are neutral electrically, the atom has Z electrons outside the nucleus. These Z electrons determine the chemical behavior of the atom. As a result, all atoms of the same chemical element have the same value for Z. For example, all hydrogen atoms have $Z = 1$, while all carbon atoms have $Z = 6$.

ATOMIC MASS UNIT (u): A convenient mass unit used in nuclear calculations is the *atomic mass unit* (u). By definition, 1 u is exactly 1/12 of the mass of the common form of carbon atom found on the earth. It turns out that

$$1 \text{ u} = 1.6606 \times 10^{-27} \text{ kg}$$

Table 46-1 lists the masses of some common particles and nuclei, as well as their charges.

Table 46-1

Particle	Symbol	Mass, u	Charge
Proton	p, ^1_1H	1.007 276	$+e$
Neutron	n, ^1_0n	1.008 665	0
Electron	e^-, β^-, $^{\ 0}_{-1}e$	0.000 548 6	$-e$
Positron	e^+, β^+, $^0_{+1}e$	0.000 548 6	$+e$
Deuteron	d, ^2_1H	2.013 55	$+e$
Alpha particle	α, ^4_2He	4.001 5	$+2e$

THE MASS NUMBER (A) of an atom is equal to the number of nucleons (neutrons plus protons) in the nucleus of the atom. Because each nucleon has a mass close to 1 u, the mass number A is nearly equal to the nuclear mass in atomic mass units. In addition, because the atomic electrons have such small mass, A is nearly equal to the mass of the atom in atomic mass units.

ISOTOPES: The number of neutrons in the nucleus has very little effect on the chemical behavior of the atom. In nature, atoms of the same element (same Z) often exist that have unlike numbers of neutrons in their nuclei. Such atoms are called *isotopes* of each other. For example, ordinary oxygen consists of three isotopes that have mass numbers 16, 17, and 18. Each of the isotopes has $Z = 8$, eight protons in the nucleus. Hence these isotopes have the following numbers of neutrons in their nuclei: $16 - 8 = 8$, $17 - 8 = 9$, and $18 - 8 = 10$. It is customary to represent the isotopes in the following way: $^{16}_{8}O$, $^{17}_{8}O$, $^{18}_{8}O$, or simply as ^{16}O, ^{17}O, and ^{18}O, where it is understood that oxygen always has $Z = 8$.

In keeping with this notation, we designate the nucleus having mass number A and atomic number Z by the symbolism

$$^{A}_{Z}(\text{CHEMICAL SYMBOL})$$

BINDING ENERGIES: The mass of an atom is not equal to the sum of the masses of its component protons, neutrons, and electrons. If we could imagine a reaction in which free electrons, protons, and neutrons combine to form an atom, we would find that the mass of the atom is slightly less than the combined masses of the component parts and also that a tremendous amount of energy is released when the reaction occurs. The loss in mass is exactly equal to the mass equivalent of the released energy, according to Einstein's equation $\Delta E = (\Delta m)c^2$. Conversely, this same amount of energy, ΔE, would have to be given to the atom in order to separate it completely into its component particles. We call ΔE the *binding energy* of the atom. A mass loss $\Delta m = 1$ u is equivalent to

$$(1.66 \times 10^{-27} \text{ kg})(3 \times 10^8 \text{ m/s})^2 = 1.49 \times 10^{-10} \text{ J} = 931 \text{ MeV}$$

of binding energy.

The percentage loss of mass is different for each isotope of any element. Atomic masses of some of the lighter isotopes are given in Table 46-2. These masses are for neutral atoms and include the orbital electrons.

Table 46-2

Neutral Atom	Atomic Mass, u	Neutral Atom	Atomic Mass, u
$^{1}_{1}H$	1.007 83	$^{7}_{4}Be$	7.016 93
$^{2}_{1}H$	2.014 10	$^{9}_{4}Be$	9.012 19
$^{3}_{1}H$	3.016 04	$^{12}_{6}C$	12.000 00
$^{4}_{2}He$	4.002 60	$^{14}_{7}N$	14.003 07
$^{6}_{3}Li$	6.015 13	$^{16}_{8}O$	15.994 91
$^{7}_{3}Li$	7.016 00		

RADIOACTIVITY: Nuclei found in nature with Z greater than that of lead, 82, are radioactive. Many man-made elements with smaller Z are also radioactive. A radioactive nucleus spontaneously throws out one or more particles to transform into a different nucleus.

The stability of a radioactive nucleus against spontaneous decay is measured by its *half-life*, $T_{1/2}$. The half-life is defined as the time in which half of any large sample of identical nuclei will undergo decomposition. The half-life is a fixed number for each isotope.

Radioactive decay is a random process. No matter when one begins to observe a material, only half of the material will remain unchanged after a time $T_{1/2}$; after a time $2T_{1/2}$, only $\frac{1}{2} \times \frac{1}{2} = \frac{1}{4}$ of the material remains unchanged. After n half-lives have passed, only $(\frac{1}{2})^n$ of the material will remain unchanged.

A simple relation exists between the number N of atoms of radioactive material present and the number ΔN that will decay in a short time Δt. It is

$$\Delta N = \lambda N \Delta t$$

where λ, the *decay constant*, is related to the half-life $T_{1/2}$ through

$$\lambda T_{1/2} = 0.693$$

The quantity $\Delta N / \Delta t$, which is the rate of disintegrations in the sample, is called the *activity* of the sample. It is equal to λN, and therefore it steadily decreases with time.

NUCLEAR EQUATIONS: In a balanced equation the sum of the subscripts (atomic numbers) must be the same on the two sides of the equation.   The sum of the superscripts (mass numbers) must also be the same on the two sides of the equation.   Thus the equation for the primary radioactivity of radium is

$$^{226}_{88}\text{Ra} \rightarrow\,^{222}_{86}\text{Rn} + {}^{4}_{2}\text{He}$$

Many nuclear processes may be indicated by a condensed notation, in which a light bombarding particle and a light product particle are represented by symbols in parentheses between the symbols for the initial target nucleus and the final product nucleus.   The symbols n, p, d, α, e^-, γ are used to represent neutron, proton, deuteron ($^2_1$H), alpha particle, electron, and gamma rays (photons) respectively.   Examples of the corresponding long and condensed notations for several reactions follow.

$$^{14}_{7}\text{N} + {}^{1}_{1}\text{H} \rightarrow\,^{11}_{6}\text{C} + {}^{4}_{2}\text{He} \qquad\qquad ^{14}\text{N}\,(p,\alpha)\,^{11}\text{C}$$

$$^{27}_{13}\text{Al} + {}^{1}_{0}n \rightarrow\,^{27}_{12}\text{Mg} + {}^{1}_{1}\text{H} \qquad\qquad ^{27}\text{Al}\,(n,p)\,^{27}\text{Mg}$$

$$^{55}_{25}\text{Mn} + {}^{2}_{1}\text{H} \rightarrow\,^{55}_{26}\text{Fe} + 2\,^{1}_{0}n \qquad\qquad ^{55}\text{Mn}\,(d,2n)\,^{55}\text{Fe}$$

The slow neutron is a very efficient agent in causing transmutations, since it has no positive charge and hence can approach the nucleus without being repelled.   A charged particle such as a proton must have a high energy in order to cause a transformation.   Because of their small mass, even very high-energy electrons are relatively inefficient in causing transmutations.

Solved Problems

46.1.   The radius of a carbon nucleus is about 3×10^{-15} m and its mass is 12 u.   Find the average density of the nuclear material.   How many more times dense than water is this?

$$\rho = \frac{m}{V} = \frac{m}{4\pi r^3/3} = \frac{(12 \text{ u})(1.66 \times 10^{-27} \text{ kg/u})}{4\pi(3 \times 10^{-15} \text{ m})^3/3} = 1.8 \times 10^{17} \text{ kg/m}^3$$

$$\frac{\rho}{\rho_{\text{water}}} = \frac{1.8 \times 10^{17}}{1000} = 1.8 \times 10^{14}$$

46.2.   In a *mass spectrograph*, the masses of ions are determined from their deflections in a magnetic field.   Suppose that singly-charged ions of chlorine, with speed 5×10^4 m/s, are shot perpendicularly into a magnetic field having $B = 0.15$ T, see Fig. 46-1.   (The speed could be measured by use of a velocity selector.)   Chlorine has two major isotopes, of masses 34.97 u and 36.97 u.   What would be the radii of the circular paths described by the two isotopes in the magnet field?

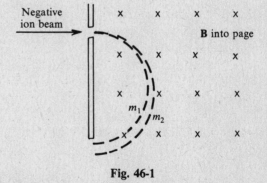

Fig. 46-1

The masses of the two isotopes are

$$m_1 = (34.97 \text{ u})(1.66 \times 10^{-27} \text{ kg/u}) = 5.81 \times 10^{-26} \text{ kg}$$

$$m_2 = (36.97 \text{ u})(1.66 \times 10^{-27} \text{ kg/u}) = 6.14 \times 10^{-26} \text{ kg}$$

Because the magnetic force, qvB, must supply the centripetal force, mv^2/r, we have

$$r = \frac{mv}{qB} = \frac{m(5 \times 10^4 \text{ m/s})}{(1.6 \times 10^{-19} \text{ C})(0.105 \text{ T})} = m(2.98 \times 10^{24} \text{ m/kg})$$

Substituting the values for m found above gives the radii as 0.173 m and 0.183 m.

46.3.   How many protons, neutrons, and electrons are there in (*a*) ^{3}He, (*b*) ^{12}C, (*c*) ^{206}Pb?

(*a*)   The atomic number of He is 2; therefore the nucleus must contain 2 protons.   Since the mass number of this isotope is 3, the sum of the protons and neutrons in the nucleus must equal 3; therefore there is 1 neutron.   The number of electrons in the atom is the same as the atomic number, 2.

(*b*)   The atomic number of carbon is 6; hence the nucleus must contain 6 protons.   The number of neutrons in the nucleus is equal to $12 - 6 = 6$.   The number of electrons is the same as the atomic number, 6.

(*c*)   The atomic number of lead is 82; hence there are 82 protons in the nucleus and 82 electrons in the atom.   The number of neutrons is $206 - 82 = 124$.

46.4.   What is the binding energy of ^{12}C?

One atom of ^{12}C consists of 6 protons, 6 electrons, and 6 neutrons.   The mass of the uncombined protons and electrons is the same as that of six ^{1}H atoms (if we ignore the very small binding energy of each proton-electron pair).   The component particles may thus be considered as six ^{1}H atoms and six neutrons.   A mass balance may be computed as follows.

mass of six ^{1}H atoms $= 6 \times 1.0078$	$= 6.0468$ u	
mass of 6 neutrons $= 6 \times 1.0087$	$= \underline{6.0522 \text{ u}}$	
total mass of component particles	$= 12.0990$ u	
mass of ^{12}C	$= \underline{12.0000 \text{ u}}$	
loss in mass on forming ^{12}C	$= 0.0990$ u	
binding energy $= (931 \times 0.0990)$ MeV	$= 92$ MeV	

46.5.   Cobalt-60 (^{60}Co) is often used as a radiation source in medicine.   It has a half-life of 5.25 years.   How long, after a new sample is delivered, will the activity have decreased (*a*) to about 1/8 its original value?   (*b*) to about 1/3 its original value?

The activity is proportional to the number of undecayed atoms ($\Delta N / \Delta t = \lambda N$).

(*a*)   In each half-life, half of the remaining sample decays.   Because $\frac{1}{2} \times \frac{1}{2} \times \frac{1}{2} = \frac{1}{8}$,   three half-lives, or 15.75 years, are required for the sample to decay to $\frac{1}{8}$ its original strength.

(*b*)   Using the fact that the original material present decreases by $\frac{1}{2}$ during each 5.25 years, we can plot the graph shown in Fig. 46-2.   From it, the sample decays to 0.33 its original value after a time of about 8.3 years.

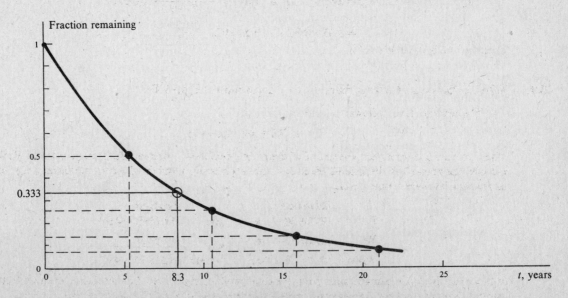

Fig. 46-2

46.6. Potassium found in nature contains two isotopes. One isotope constitutes 93.4% of the whole and has an atomic mass of 38.975 u; the other 6.6% has a mass of 40.974 u. Compute the atomic mass of potassium as found in nature.

The atomic mass of the material found in nature is obtained by combining the individual atomic masses in proportion to their abundances,

$$\text{atomic mass} = (0.934)(38.975 \text{ u}) + (0.066)(40.974 \text{ u}) = 39.107 \text{ u}$$

46.7. The half-life of radium is 1620 years. How many radium atoms decay in one second in a 1 gram sample of radium? The atomic weight of radium is 226 kg/kmol.

A 1 g sample is (0.001/226) kmol and so it contains

$$N = \left(\frac{0.001}{226} \text{ kmol} \right)\left(6.02 \times 10^{26} \frac{\text{atoms}}{\text{kmol}} \right) = 2.66 \times 10^{21} \text{ atoms}$$

The decay constant is

$$\lambda = \frac{0.693}{T_{1/2}} = \frac{0.693}{(1620 \text{ y})(3.16 \times 10^7 \text{ s/y})} = 1.35 \times 10^{-11} \text{ s}^{-1}$$

Then $$\Delta N = \lambda N \, \Delta t = (1.35 \times 10^{-11} \text{ s}^{-1})(2.66 \times 10^{21})(1 \text{ s}) = 3.6 \times 10^{10}$$

is the number of disintegrations per second in 1 gram of radium.

The above result leads to the definition of the *curie* (Ci) as the unit of activity:

$$1 \text{ Ci} = 3.7 \times 10^{10} \text{ disintegrations/s}$$

Because of its convenient size, the curie will be used in subsequent problems, even though the official SI unit of activity is the *becquerel* $(1 \text{ Bq} = 1 \text{ s}^{-1})$.

46.8. Technetium-99 $\binom{99}{43}\text{Tc}$ has an excited state that decays by emission of a gamma ray. The half-life of the excited state is 360 min. What is the activity, in curies, of 1 mg of this excited isotope?

The activity of a sample is λN. In this case,

$$\lambda = \frac{0.693}{T_{1/2}} = \frac{0.693}{21\,600 \text{ s}} = 3.2 \times 10^{-5} \text{ s}^{-1}$$

We also know that 99 kg of Tc contains 6.02×10^{26} atoms. A mass m will therefore contain $[m/(99 \text{ kg})](6.02 \times 10^{26})$ atoms. In our case, $m = 1 \times 10^{-6}$ kg and so

$$\text{activity} = \lambda N = (3.2 \times 10^{-5} \text{ s}^{-1})\left(\frac{1 \times 10^{-6} \text{ kg}}{99 \text{ kg}} \right)(6.02 \times 10^{26})$$

$$= 1.95 \times 10^{14} \text{ s}^{-1} = 5270 \text{ Ci}$$

because 1 Ci is $3.7 \times 10^{10} \text{ s}^{-1}$.

46.9. What is the overall energy balance in the reaction $^7\text{Li}(p,n)^7\text{Be}$?

The reaction is as follows:

$$^7_3\text{Li} + ^1_1\text{H} \rightarrow ^7_4\text{Be} + ^1_0 n$$

where the symbols represent the nuclei of the atoms indicated. Because the masses listed in Table 46-2 include the masses of the atomic electrons, the appropriate number of electron masses (m_e) must be subtracted from the values given.

	Reactants		Products
^7_3Li	$7.016\,00 - 3m_e$	^7_4Be	$7.016\,93 - 4m_e$
^1_1H	$1.007\,83 - 1m_e$	$^1_0 n$	$1.008\,66$
TOTAL	$8.023\,83 - 4m_e$	TOTAL	$8.025\,59 - 4m_e$

Subtracting the total reactant mass from the total product mass gives the increase in mass as 0.00176 u. (Notice that the electron masses cancel out. This happens frequently, but not always.)

To create this mass in the reaction, energy must have been supplied to the reactants. The energy corresponding to 0.00176 u is

$$(931 \times 0.00176) \text{ MeV} = 1.65 \text{ MeV}$$

This energy is supplied as KE of the bombarding proton. The incident proton must have more than this energy because not all its KE can be used to produce mass.

46.10. Complete the following nuclear equations:

(a)　$_{7}^{14}\text{N} + _{2}^{4}\text{He} \rightarrow _{8}^{17}\text{O} + ?$　　　　(d)　$_{15}^{30}\text{P} \rightarrow _{14}^{30}\text{Si} + ?$

(b)　$_{4}^{9}\text{Be} + _{2}^{4}\text{He} \rightarrow _{6}^{12}\text{C} + ?$　　　(e)　$_{1}^{3}\text{H} \rightarrow _{2}^{3}\text{He} + ?$

(c)　$_{4}^{9}\text{Be}(p, \alpha)?$　　　　　　　　(f)　$_{20}^{43}\text{Ca}(\alpha, ?)_{21}^{46}\text{Sc}$

(a)　The sum of the subscripts on the left is $7 + 2 = 9$. The subscript of the first product on the right is 8. Hence the second product on the right must have a subscript (net charge) of 1.
　　　The sum of the superscripts on the left is $14 + 4 = 18$. The superscript of the first product is 17. Hence the second product on the right must have a superscript (mass number) of 1.
　　　The particle with a nuclear charge 1 and a mass number 1 is the proton, $_{1}^{1}\text{H}$.

(b)　The nuclear charge of the second product particle (its subscript) is $(4 + 2) - 6 = 0$. The mass number of the particle (its superscript) is $(9 + 4) - 12 = 1$. Hence the particle must be the neutron, $_{0}^{1}n$.

(c)　The reactants, $_{4}^{9}\text{Be}$ and $_{1}^{1}\text{H}$, have a combined nuclear charge of 5 and a mass number of 10. In addition to the α-particle, a product will be formed of charge $5 - 2 = 3$, and mass $10 - 4 = 6$. This is $_{3}^{6}\text{Li}$.

(d)　The nuclear charge of the second particle is $15 - 14 = +1$. The mass is $30 - 30 = 0$. Hence the particle must be a positron, $_{+1}^{0}e$.

(e)　The nuclear charge of the second particle is $1 - 2 = -1$. Its mass number is $3 - 3 = 0$. Hence the particle must be a beta particle (an electron), $_{-1}^{0}e$.

(f)　The reactants, $_{20}^{43}\text{Ca}$ and $_{2}^{4}\text{He}$, have a combined nuclear charge of 22 and mass number 47. The ejected product will have a charge $22 - 21 = 1$, and mass $47 - 46 = 1$. This is a proton and should be represented in the parentheses by p.

46.11. Uranium-238 ($_{92}^{238}\text{U}$) is radioactive and decays by emitting the following particles in succession before reaching a stable form: α, β, β, α, α, α, α, α, β, β, α, β, β, and α (β stands for "beta particle," e^-). What is the final stable nucleus?

　　　The original nucleus emitted 8 alpha particles and 6 beta particles. When an alpha particle is emitted, Z decreases by 2, since the alpha particle carries away a charge of $+2e$. A beta particle carries away a charge of $-1e$ and so the charge on the nucleus must increase to $(Z + 1)e$. We then have, for the final nucleus,

$$\text{final } Z = 92 + 6 - (2)(8) = 82$$
$$\text{final } A = 238 - (6)(0) - (8)(4) = 206$$

The final stable nucleus is $_{82}^{206}\text{Pb}$.

46.12. The half-life of uranium-238 is about 4.5×10^9 years and its end product is lead-206. We notice that the oldest uranium-bearing rocks on earth contain about a 50-50 mixture of ^{238}U and ^{206}Pb. About what is the age of these rocks?

　　　Apparently about half of the ^{238}U has decayed to ^{206}Pb during the existence of the rock. Hence the rock must have been formed about 4.5 billion years ago.

Supplementary Problems

46.13. How many protons, neutrons, and electrons does an atom of $^{235}_{92}U$ possess?   *Ans.* 92, 143, 92

46.14. By how much does the mass of a heavy nucleus change as it emits a 4.8 MeV gamma ray? *Ans.* 5.2×10^{-3} u $= 8.6 \times 10^{-30}$ kg

46.15. Find the binding energy of $^{107}_{47}Ag$, which has an atomic mass of 106.905 u.   *Ans.* 915 eV

46.16. The binding energy per nucleon for elements near iron in the periodic table is about 8.9 MeV per nucleon. What is the atomic mass, including electrons, of $^{56}_{26}Fe$?   *Ans.* 55.94 u

46.17. What mass of $^{60}_{27}Co$ has an activity of 1 Ci?   The half-life of cobalt-60 is 5.25 years. *Ans.* 8.8×10^{-7} kg

46.18. An experiment is done to determine the half-life of a radioactive substance that emits one beta particle for each decay process. Measurements show that an average of 8.4 beta particles are emitted each second by 2.5 milligrams of the substance. The atomic weight of the substance is 230. Find the half-life of the substance.   *Ans.* 1.7×10^{10} y

46.19. The half-life of carbon-14 is 5730 years.   What fraction of a sample of ^{14}C will remain unchanged after a period of five half-lives?   *Ans.* 0.031

46.20. By natural radioactivity ^{238}U emits an α-particle.   The heavy residual nucleus is called UX_1. UX_1 in turn emits a β-particle.   The resultant nucleus is called UX_2.   Determine the atomic number and mass number for (*a*) UX_1 and (*b*) UX_2.   *Ans.* (*a*) 90, 234;   (*b*) 91, 234

46.21. By radioactivity $^{239}_{93}Np$ emits a β-particle.   The residual heavy nucleus is also radioactive, and gives rise to ^{235}U by the radioactive process.   What small particle is emitted simultaneously with the formation of uranium-235?   *Ans.* alpha particle

46.22. Complete the following equations.   (See Appendix H for a table of the elements.)

$$(a) \quad ^{23}_{11}Na + ^{4}_{2}He \rightarrow ^{26}_{12}Mg + ? \qquad (d) \quad ^{10}_{5}B + ^{4}_{2}He \rightarrow ^{13}_{7}N + ?$$
$$(b) \quad ^{64}_{29}Cu \rightarrow _{+1}^{0}e + ? \qquad (e) \quad ^{105}_{48}Cd + _{-1}^{0}e \rightarrow ?$$
$$(c) \quad ^{106}Ag \rightarrow ^{106}Cd + ? \qquad (f) \quad ^{238}_{92}U \rightarrow ^{234}_{90}Th + ?$$

Ans. (*a*) $^{1}_{1}H$;   (*b*) $^{64}_{28}Ni$;   (*c*) $_{-1}^{0}e$;   (*d*) $^{1}_{0}n$;   (*e*) $^{105}_{47}Ag$;   (*f*) $^{4}_{2}He$

46.23. Complete the notations for the following processes.

$$(a) \quad ^{24}Mg(d,\alpha)? \qquad (e) \quad ^{130}Te(d,2n)?$$
$$(b) \quad ^{26}Mg(d,p)? \qquad (f) \quad ^{55}Mn(n,\gamma)?$$
$$(c) \quad ^{40}Ar(\alpha,p)? \qquad (g) \quad ^{59}Co(n,\alpha)?$$
$$(d) \quad ^{12}C(d,n)?$$

Ans. (*a*) ^{22}Na;   (*b*) ^{27}Mg;   (*c*) ^{43}K;   (*d*) ^{13}N;   (*e*) ^{130}I;   (*f*) ^{56}Mn;   (*g*) ^{56}Mn

46.24. If an atom of ^{235}U, after absorption of a slow neutron, undergoes fission to form an atom of ^{139}Xe and an atom of ^{94}Sr, what other particles are produced and how many?   *Ans.* 3 neutrons

46.25. How much energy is released during each of the following reactions?

$$(a) \quad ^{1}_{1}H + ^{7}_{3}Li \rightarrow 2\,^{4}_{2}He \qquad (b) \quad ^{3}_{1}H + ^{2}_{1}H \rightarrow ^{4}_{2}He + ^{1}_{0}n$$

Ans. (*a*) 17.4 MeV;   (*b*) 17.6 MeV

46.26. In the $^{14}N(n,p)^{14}C$ reaction, the proton is ejected with an energy of 0.6 MeV.   Very slow neutrons are used.   Calculate the mass of the ^{14}C atom.   *Ans.* 14.0033 u

Chapter 47

Applied Nuclear Physics

NUCLEAR BINDING ENERGIES are very nearly the same as the atomic binding energies discussed in Chapter 46, since relatively little energy is associated with the electrons. The *binding energy per nucleon* (the total energy liberated on assembling the nucleus, divided by the number of protons and neutrons) turns out to be smaller for very light and very heavy nuclei than for medium-weight nuclei. Hence the nuclei at the two ends of the table of elements can liberate further energy if they are in some way transformed into middle-sized nuclei.

FISSION REACTION: A very large nucleus, such as the nucleus of the uranium atom, liberates energy as it is split into two or three middle-sized nuclei. Such a *fission reaction* can be induced by striking the large nuclei with low- or moderate-energy neutrons. The fission reaction produces additional neutrons, which, in turn, can cause further fission reactions and more neutrons. If the number of neutrons remains constant or increases in time, the process is a self-perpetuating *chain reaction*.

FUSION REACTION: Small nuclei, such as hydrogen or helium, are joined together to form more massive nuclei, thereby liberating energy.

This reaction is difficult to initiate and sustain because the nuclei must be fused together even though they repel each other with the Coulomb force. Only when the particles are shot toward each other with high energy do they come close enough to allow the nuclear attraction force to bind them together. In the sun and in the stars, the fusion reaction can occur because of the high thermal energies of the particles in these extremely hot objects.

RADIATION DOSE is defined in terms of the quantity of radiant energy absorbed in unit mass of substance. A material receives a dose of 1 *rad* when 10^{-2} J of radiation is absorbed in each kilogram of the material.

$$\text{dose in rad} = \left(\frac{\text{energy absorbed in J}}{\text{mass of absorber in kg}} \right) \times \left(100 \frac{\text{rad}}{\text{J/kg}} \right)$$

(The rad will be superseded in the SI by the *gray*, where 1 Gy = 1 J/kg.)

RADIATION DAMAGE to living tissue is described in terms of a unit called the *rem* (rad equivalent, man). Each type (and energy) of radiation causes its own characteristic degree of damage to tissue. The damage also varies from one type of tissue to another. The comparative damage effects of radiation are expressed in terms of a quantity called the *RBE* (relative biological effectiveness) of the radiation. Arbitrarily, the damage is taken relative to the damage caused by 200 keV X-rays. If *d* rad of a particular radiation causes 5 times more damage than *d* rad of 200 keV X-rays, then the RBE for that radiation is 5.

$$\text{RBE} = \frac{\text{biological effect of radiation in question}}{\text{effect of equal dose (in rad) of 200 keV X-rays}}$$

Frequently the RBE is called the *quality factor*, QF.

The rem unit is now defined in the following way:

$$\text{dose in rem} = (\text{RBE}) \times (\text{dose in rad})$$

For example, RBE = 5 for slow neutrons. Hence a dose of 7 rad, say, of slow neutrons is a dose of $5 \times 7 = 35$ rem.

HIGH-ENERGY ACCELERATORS: Charged particles can be accelerated to high energies by causing them to follow a circular path.   Each time the particle (of charge q) circles the path, it is caused to fall through a potential difference V.   After n trips around the path, its energy is $q(nV)$.

Magnetic fields are used to supply the centripetal force required to keep the particle on a circle. Equating magnetic force qvB to centripetal force mv^2/r gives

$$mv = qBr$$

In this expression, m is the relativistic mass of the particle that is traveling with speed v on a circle of radius r perpendicular to a magnetic field B.

In most high-energy accelerators, the particles reach speeds very close to c.   At these speeds $m \gg m_0$ and

$$m \approx \frac{qBr}{c}$$

The total energy of the particle (rest energy plus KE) is

$$\text{total energy} = mc^2 \approx qBrc$$

while the KE of the particle is

$$\text{KE} = (m - m_0)c^2 \approx qBrc - m_0 c^2$$

Solved Problems

47.1.   The binding energy per nucleon for ^{238}U is about 7.5 MeV, while it is about 8.5 MeV for nuclei of half that mass.   If a ^{238}U nucleus were to split into two equal-size nuclei, about how much energy would be released in the process?

There are 238 nucleons involved.   Each nucleon will release about $8.5 - 7.5 = 1.0$ MeV   of energy when the nucleus undergoes fission.   The total energy liberated is therefore about 238 MeV.

47.2.   When an atom of ^{235}U undergoes fission in a reactor, about 200 MeV of energy is liberated. Suppose that a reactor using uranium-235 has an output of 700 MW and is 20% efficient.   (*a*) How many uranium atoms does it consume in one day?   (*b*) What mass of uranium does it consume each day?

(*a*)   Each fission yields

$$200 \text{ MeV} = (200 \times 10^6)(1.6 \times 10^{-19}) \text{ J}$$

of energy.   Only 20% of this is utilized efficiently and so

$$\text{energy generated per fission} = (200 \times 10^6)(1.6 \times 10^{-19})(0.20) = 6.4 \times 10^{-12} \text{ J}$$

Because the reactor's output is 700×10^6 J/s, the number of fissions required per second is

$$\text{fissions/s} = \frac{7 \times 10^8 \text{ J/s}}{6.4 \times 10^{-12} \text{ J}} = 1.1 \times 10^{20} \text{ s}^{-1}$$

and $$\text{fissions/day} = (86\,400 \text{ s/d})(1.1 \times 10^{20} \text{ s}^{-1}) = 9.5 \times 10^{24} \text{ d}^{-1}$$

(*b*)   There are 6.02×10^{26} atoms in 235 kg of uranium-235.   Therefore the mass of uranium-235 consumed in one day is

$$\text{mass} = \left(\frac{9.5 \times 10^{24}}{6.02 \times 10^{26}} \right)(235 \text{ kg}) = 3.7 \text{ kg}$$

47.3. The following fusion reaction takes place in the sun and furnishes much of its energy:

$$4\,{}^{1}_{1}\text{H} \rightarrow {}^{4}_{2}\text{He} + 2\,{}^{0}_{+1}e + \text{energy}$$

where ${}^{0}_{+1}e$ is a positron, a positive electron. How much energy is released as 1 kg of hydrogen is consumed? The masses of ${}^{1}\text{H}$, ${}^{4}\text{He}$, and ${}^{0}_{+1}e$ are respectively 1.007 825, 4.002 604, and 0.000 549 u, where atomic electrons are included in the first two values.

The mass of the reactants, 4 protons, is 4 times the atomic mass of hydrogen (${}^{1}\text{H}$), less the mass of 4 electrons.

$$\text{reactant mass} = (4)(1.007\,825\text{ u}) - 4m_e = 4.031\,300\text{ u} - 4m_e$$

where m_e is the mass of the electron (or positron). The reaction products have a combined mass

$$\text{product mass} = (\text{mass of } {}^{4}_{2}\text{He nucleus}) + 2m_e$$

$$= (4.002\,604\text{ u} - 2m_e) + 2m_e = 4.002\,604\text{ u}$$

The mass loss is therefore

$$(\text{reactant mass}) - (\text{product mass}) = (4.0313\text{ u} - 4m_e) - 4.0026\text{ u}$$

Substituting $m_e = 0.000\,549$ u, the mass loss is found to be 0.0265 u.

But 1 kg of ${}^{1}\text{H}$ contains 6.02×10^{26} atoms. For each four atoms that undergo fusion, 0.0265 u is lost. The mass lost when 1 kg undergoes fusion is therefore

$$\text{mass loss/kg} = (0.0265\text{ u})(6 \times 10^{26}/4) = 4.0 \times 10^{24}\text{ u}$$

$$= (4.0 \times 10^{24}\text{ u})(1.66 \times 10^{-27}\text{ kg/u}) = 0.0066\text{ kg}$$

Then, from the Einstein relation,

$$\Delta E = (\Delta m)c^2 = (0.0066\text{ kg})(3 \times 10^8\text{ m/s})^2 = 6.0 \times 10^{14}\text{ J}$$

47.4. Lithium hydride, LiH, has been proposed as a possible nuclear fuel. The nuclei to be used and the reaction involved are as follows:

$$\begin{array}{ccc} {}^{6}_{3}\text{Li} & + & {}^{2}_{1}\text{H} \rightarrow 2\,{}^{4}_{2}\text{He} \\ 6.015\,13 & 2.014\,10 & 4.002\,60 \end{array}$$

the listed masses being those of the neutral atoms. Calculate the expected power production, in megawatts, associated with the consumption of 1.00 g of LiH per day. Assume 100% efficiency.

The change in mass for the reaction is first computed.

	Reactants		**Products**
${}^{6}_{3}\text{Li}$	$6.015\,13\text{ u} - 3m_e$	$2\,{}^{4}_{2}\text{He}$	$2(4.002\,60\text{ u} - 2m_e)$
${}^{2}_{1}\text{H}$	$2.014\,10\text{ u} - 1m_e$		
TOTAL	$8.029\,23\text{ u} - 4m_e$	TOTAL	$8.005\,20\text{ u} - 4m_e$

We find the loss in mass by subtracting the product mass from the reactant mass. In the process, the electron masses drop out and the mass loss is found to be 0.024 03 u.

The fractional loss in mass is $0.0240/8.029 = 2.99 \times 10^{-3}$. Therefore, when 1 g reacts, the mass loss is

$$(2.99 \times 10^{-3})(1 \times 10^{-3}\text{ kg}) = 2.99 \times 10^{-6}\text{ kg}$$

This corresponds to an energy of

$$\Delta E = (\Delta m)c^2 = (2.99 \times 10^{-6}\text{ kg})(3 \times 10^8\text{ m/s})^2 = 2.69 \times 10^{11}\text{ J}$$

Then

$$\text{power} = \frac{\text{energy}}{\text{time}} = \frac{2.69 \times 10^{11}\text{ J}}{86\,400\text{ s}} = 3.1\text{ MW}$$

47.5. A beam of gamma rays has a cross-sectional area of 2 cm^2 and carries 7×10^8 photons through the cross section each second. Each photon has an energy of 1.25 MeV. The beam passes through a 0.75 cm thickness of flesh ($\rho = 0.95$ g/cm^3) and loses 5% of its intensity in the process. What is the average dose (in rad) applied to the flesh each second?

Dose is measured as the energy absorbed per kilogram of flesh in this case. We have

$$\text{number of } \gamma\text{-rays absorbed/s} = (7 \times 10^8 \text{ s}^{-1})(0.05) = 3.5 \times 10^7 \text{ s}^{-1}$$

$$\text{energy absorbed/s} = (3.5 \times 10^7 \text{ s}^{-1})(1.25 \text{ MeV}) = 4.4 \times 10^7 \text{ MeV/s}$$

We need the mass of the flesh in which this energy was absorbed.

$$\text{mass} = \rho V = (0.95 \text{ g/cm}^3)\left[(2 \text{ cm}^2)(0.75 \text{ cm})\right] = 1.43 \text{ g}$$

We then have

$$\text{dose/s} = \frac{\text{energy/s}}{\text{mass}} = \frac{(4.4 \times 10^7 \text{ MeV/s})(1.6 \times 10^{-13} \text{ J/MeV})}{1.43 \times 10^{-3} \text{ kg}} \times \frac{100 \text{ rad}}{1 \text{ J/kg}} = 0.49 \text{ rad/s}$$

47.6. An alpha particle beam passes through flesh and deposits 0.2 J of energy in each kilogram of flesh. The RBE for these particles is 12 rem/rad. Find the dose in rad and in rem.

$$\text{dose in rad} = \left(\frac{\text{absorbed energy}}{\text{mass}}\right) \times \left(\frac{100 \text{ rad}}{\text{J/kg}}\right) = (0.2 \text{ J/kg})\left(\frac{100 \text{ rad}}{\text{J/kg}}\right) = 20 \text{ rad}$$

$$\text{dose in rem} = (\text{RBE}) \times (\text{dose in rad}) = (12)(20) = 240 \text{ rem}$$

Supplementary Problems

47.7. Consider the following fission reaction:

$$\begin{array}{cccccc}
{}_0^1 n & + & {}_{92}^{235}\text{U} & \rightarrow & {}_{56}^{138}\text{Ba} & + & {}_{41}^{93}\text{Nb} & + & 5 {}_0^1 n & + & 5 {}_{-1}^{0} e \\
1.0087 & & 235.0439 & & 137.9050 & & 92.9060 & & 1.0087 & & 0.00055
\end{array}$$

where the neutral atomic masses are given. How much energy is released when (a) 1 atom undergoes fission, and (b) 1 kg of atoms undergoes this type of fission? *Ans.* (a) 182 MeV; (b) 7.5×10^{13} J

47.8. It is proposed to use the nuclear fusion reaction

$$\begin{array}{cc}
2 {}_1^2 \text{H} & \rightarrow & {}_2^4 \text{He} \\
2.014102 & & 4.002604
\end{array}$$

to produce industrial power. If the output is to be 150 MW and the energy of the above reaction is used with 30% efficiency, how many grams of deuterium fuel will be needed per day? *Ans.* 75 g/d

47.9 One of the most promising fusion reactions for power generation involves deuterium (^{2}H) and tritium (^{3}H):

$$\begin{array}{ccccc}
{}_1^2 \text{H} & + & {}_1^3 \text{H} & \rightarrow & {}_2^4 \text{He} & + & {}_0^1 n \\
2.01410 & & 3.01605 & & 4.00260 & & 1.00867
\end{array}$$

where the masses including electrons are as given. How much energy is produced when 2 kg of ^{2}H fuses with 3 kg of ^{3}H to form ^{4}He? *Ans.* 1.70×10^{15} J

47.10. The luminous dial of a certain watch gives off 130 fast electrons each minute. Assume that each electron has an energy of 0.5 MeV and deposits that energy in a volume of skin that is 2 cm^2 in area and 0.20 cm thick. Find the dose in rads that the volume experiences in 1 day. Take the density of skin to be 900 kg/m^3. *Ans.* 0.0042 rad

47.11. An alpha particle beam enters a charge collector and is measured to carry 2×10^{-14} C of charge into the collector each second. The beam has a cross-sectional area of 150 mm^2 and it penetrates human skin to a depth of 0.14 mm. Each particle has an initial energy of 4.0 MeV. The RBE for such particles is about 15. What dose in rem does a person's skin receive when exposed to this beam for 20 s? Take $\rho = 900$ kg/m^3 for skin. *Ans.* 63 rem

Significant Figures

INTRODUCTION: The numerical value of every observed measurement is an approximation. Consider that the length of an object is recorded as 15.7 cm. By convention, this means that the length was measured to the *nearest* tenth of a centimeter and that its exact value lies between 15.65 and 15.75 cm. If this measurement were exact to the nearest hundredth of a centimeter, it would have been recorded as 15.70 cm. The value 15.7 cm represents *three significant figures* (1, 5, 7), while the value 15.70 represents *four significant figures* (1, 5, 7, 0). A significant figure is one which is known to be reasonably reliable.

Similarly, a recorded mass of 3.4062 kg means that the mass was determined to the nearest tenth of a gram and represents five significant figures (3, 4, 0, 6, 2), the last figure (2) being reasonably correct and guaranteeing the certainty of the preceding four figures.

In elementary measurements in physics and chemistry, the last figure is estimated and is also considered as a significant figure.

ZEROS may be significant or they may merely serve to locate the decimal point. For instance, the statement that a body of ore weighs 9800 lb does not indicate definitely the accuracy of the weighing. If it was weighed to the nearest hundred pounds, the weight contains only two significant figures (9, 8) and may be written exponentially as 9.8×10^3 lb. If weighed to the nearest ten pounds, the first zero would be significant but the second would not; the weight could be written as 9.80×10^3 lb, displaying the three significant figures. If the object was weighed to the nearest pound, the weight could be written as 9.800×10^3 lb (four significant figures). Of course, if a zero stands between two significant figures, it is itself significant.

ROUNDING OFF: A number is rounded off to the desired number of significant figures by dropping one or more digits to the right. When the first digit dropped is less than 5, the last digit retained should remain unchanged; when it is more than 5, or when it is 5 followed by digits not all zeros, 1 is added to the last digit retained. When it is 5 followed by zeros, 1 is added to the last digit retained if that digit is odd.

ADDITION AND SUBTRACTION: The answer should be rounded off after adding or subtracting, so as to retain digits only as far as the first column containing estimated figures. (Remember that the last significant figure is estimated.)

Examples. Add the following quantities expressed in meters.

(a)	(b)	(c)	(d)
25.340	58.0	4.20	415.5
5.465	0.0038	1.6523	3.64
0.322	0.00001	0.015	0.238
31.127 m (*Ans.*)	58.00381	5.8673	419.378
	= 58.0 m (*Ans.*)	= 5.87 m (*Ans.*)	= 419.4 m (*Ans.*)

MULTIPLICATION AND DIVISION: The answer should be rounded off to contain only as many significant figures as are contained in the least exact factor.

There are some exceptional cases, however. Consider the division $9.84 \div 9.3 = 1.06$, to three places. By the rule given above, the answer should be 1.1 (two significant figures). However, a difference of 1 in the last place of 9.3 (9.3 ± 0.1) results in an error of about 1%, while a difference of 1

312

in the last place of 1.1 (1.1 ± 0.1) yields an error of roughly 10%. Thus the answer 1.1 is of much lower percentage accuracy than 9.3. Hence in this case the answer should be 1.06, since a difference of 1 in the last place of the least exact factor used in the calculation (9.3) yields a percentage of error about the same (about 1%) as a difference of 1 in the last place of 1.06 (1.06 ± 0.01). Similarly, $0.92 \times 1.13 = 1.04$.

Exercises

1. How many significant figures are given in the following quantities?

   (a) 454 g   (e) 0.0353 m   (i) 1.118×10^{-3} V   *Ans.*   (a) 3,   (b) 2,   (c) 4,   (d) 4,   (e) 3

   (b) 2.2 lb   (f) 1.0080 hr   (j) 1030 kg/m³   (f) 5,   (g) 3,   (h) 2,   (i) 4

   (c) 2.205 lb   (g) 14.0 A   (k) 125 000 N   (j) 3 or 4

   (d) 0.3937 s   (h) 9.3×10^7 mi   (k) 3, 4, 5, or 6

2. Add: (a) 703   h   (b) 18.425 cm   (c) 0.0035 s   (d) 4.0   N   *Ans.*   (a) 711 h

                7   h   7.21   cm   0.097   s   0.632 N   (b) 30.6 cm

             0.66 h   5.0   cm   0.225   s   0.148 N   (c) 0.326 s

                                                                       (d) 4.8 N

3. Subtract: (a) 7.26 J   (b) 562.4 m   (c) 34   kg   *Ans.*   (a) 7.1 J

                    0.2   J   16.8 m   0.2 kg   (b) 545.6 m,   (c) 34 kg

4. Multiply: (a) 2.21×0.3 (d) 107.88×0.610 *Ans.*   (a) 0.7   (d) 65.8

             (b) 72.4×0.084 (e) 12.4×84 (b) 6.1   (e) 1.04×10^3

             (c) 2.02×4.113 (f) 72.4×8.6 (c) 8.31   (f) 6.2×10^2

5. Divide: (a) $\dfrac{97.52}{2.54}$ (b) $\dfrac{14.28}{0.714}$ (c) $\dfrac{0.032}{0.004}$ (d) $\dfrac{9.8}{9.3}$ *Ans.*   (a) 38.4   (c) 8

                                                                        (b) 20.0   (d) 1.05

Appendix B

Trigonometry Needed for College Physics

FUNCTIONS OF AN ACUTE ANGLE: The trigonometric functions most often used are the sine, cosine, and tangent. It is convenient to put the definitions of the functions of an acute angle in terms of the sides of a right triangle.

In any right triangle: The *sine* of either acute angle is equal to the length of the side opposite that angle divided by the length of the hypotenuse. The *cosine* of either acute angle is equal to the length of the side adjacent to that angle divided by the length of the hypotenuse. The *tangent* of either acute angle is equal to the length of the side opposite that angle divided by the length of the side adjacent to that angle.

If A, B, and C are the angles of any right triangle (C is the right angle), and a, b, and c the sides opposite respectively, as shown in the diagram, then

$$\sin A = \frac{\text{side opposite } A}{\text{hypotenuse}} = \frac{a}{c} \qquad \sin B = \frac{\text{side opposite } B}{\text{hypotenuse}} = \frac{b}{c}$$

$$\cos A = \frac{\text{side adjacent } A}{\text{hypotenuse}} = \frac{b}{c} \qquad \cos B = \frac{\text{side adjacent } B}{\text{hypotenuse}} = \frac{a}{c}$$

$$\tan A = \frac{\text{side opposite } A}{\text{side adjacent } A} = \frac{a}{b} \qquad \tan B = \frac{\text{side opposite } B}{\text{side adjacent } B} = \frac{b}{a}$$

Note that $\sin A = \cos B$; thus the sine of any angle equals the cosine of its complementary angle. For example,

$$\sin 30° = \cos (90° - 30°) = \cos 60° \qquad \cos 50° = \sin (90° - 50°) = \sin 40°$$

As an angle increases from 0° to 90°, its sine increases from 0 to 1, its tangent increases from 0 to infinity, and its cosine decreases from 1 to 0.

LAW OF SINES AND OF COSINES: These two laws give the relations between the sides and angles of *any* plane triangle. In any plane triangle with angles A, B, and C and sides opposite a, b, and c respectively, the following relations apply:

Law of Sines $\qquad \dfrac{a}{\sin A} = \dfrac{b}{\sin B} = \dfrac{c}{\sin C}$

or $\qquad \dfrac{a}{b} = \dfrac{\sin A}{\sin B}, \quad \dfrac{b}{c} = \dfrac{\sin B}{\sin C}, \quad \dfrac{c}{a} = \dfrac{\sin C}{\sin A}$

Law of Cosines $\quad a^2 = b^2 + c^2 - 2bc \cos A$

$\qquad\qquad\qquad\quad b^2 = a^2 + c^2 - 2ac \cos B$

$\qquad\qquad\qquad\quad c^2 = a^2 + b^2 - 2ab \cos C$

If the angle θ is between 90° and 180°, as in the case of angle C in the above diagram, then

$$\sin \theta = \sin (180° - \theta) \qquad \cos \theta = -\cos (180° - \theta)$$

Thus $\qquad\qquad \sin 120° = \sin (180° - 120°) = \sin 60° = 0.866$

$$\cos 120° = -\cos (180° - 120°) = -\cos 60° = -0.500$$

Solved Problems

1. In right triangle ABC, given $a = 8$, $b = 6$, $C = 90°$. Find the values of the sine, cosine, and tangent of angle A and of angle B.

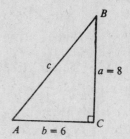

$$c = \sqrt{8^2 + 6^2} = \sqrt{100} = 10$$

$\sin A = a/c = 8/10 = 0.80$ $\sin B = b/c = 6/10 = 0.60$
$\cos A = b/c = 6/10 = 0.60$ $\cos B = a/c = 8/10 = 0.80$
$\tan A = a/b = 8/6 = 1.33$ $\tan B = b/a = 6/8 = 0.75$

2. Given a right triangle with one acute angle 40° and hypotenuse 400. Find the other sides and angles.

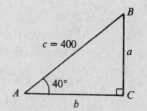

$$\sin 40° = \frac{a}{400} \quad\text{and}\quad \cos 40° = \frac{b}{400}$$

In the table of natural trigonometric functions, Appendix J, we find that $\sin 40° = 0.6428$ and $\cos 40° = 0.7660$.

$$a = 400 \sin 40° = 400(0.6428) = 257$$
$$b = 400 \cos 40° = 400(0.7660) = 306$$
$$B = 90° - 40° = 50°$$

3. Given triangle ABC with $A = 64°$, $B = 71°$, $b = 40$. Find a and c.

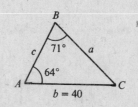

$$C = 180° - (A + B) = 180° - (64° + 71°) = 45°$$

By the law of sines,

$$\frac{a}{\sin A} = \frac{b}{\sin B} \quad\text{and}\quad \frac{c}{\sin C} = \frac{b}{\sin B}$$

Solving,

$$a = \frac{b \sin A}{\sin B} = \frac{40 \sin 64°}{\sin 71°} = \frac{40(0.8988)}{0.9455} = 38.0$$

$$c = \frac{b \sin C}{\sin B} = \frac{40 \sin 45°}{\sin 71°} = \frac{40(0.7071)}{0.9455} = 29.9$$

4. (a) If $\cos A = 0.438$, find A to the nearest degree. (b) If $\sin B = 0.8000$, find B to the nearest tenth of a degree. (c) If $\cos C = 0.7120$, find C to the nearest tenth of a degree.

(a) In Appendix J, under Cosine, we find that 0.438 corresponds most nearly to 64°. Hence $A = 64°$.

(b) In Appendix J, we find that the angle whose sine is 0.8000 lies between 53° and 54°. We interpolate to find the required angle.

$\sin 54° = 0.8090$ $0.8000 = 0.8000$
$\underline{\sin 53° = 0.7986}$ $\underline{\sin 53° = 0.7986}$ Then $B = (53 + \frac{14}{104})° = 53.1°$.
difference = 0.0104   diff. = 0.0014

(c) $\cos 44° = 0.7193$ $0.7120 = 0.7120$
$\underline{\cos 45° = 0.7071}$ $\underline{\cos 45° = 0.7071}$ Then $C = (45 - \frac{49}{122})° = 44.6°$.
difference = 0.0122   diff. = 0.0049

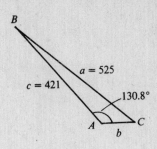

5. Given triangle ABC with $A = 130.8°$, $a = 525$, $c = 421$. Find b, B, C.

$$\sin 130.8° = \sin (180° - 130.8°) = \sin 49.2° = 0.757$$

For C: $\sin C = \dfrac{c \sin A}{a} = \dfrac{421 \sin 130.8°}{525} = \dfrac{421(0.757)}{525} = 0.607$

from which $C = 37.4°$.

For B: $B = 180° - (C + A) = 180° - (37.4° + 130.8°) = 11.8°$

For b: $b = \dfrac{a \sin B}{\sin A} = \dfrac{525 \sin 11.8°}{\sin 130.8°} = \dfrac{525(0.204)}{0.757} = 142$

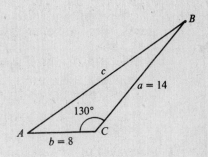

6. Given triangle ABC with $a = 14$, $b = 8$, $C = 130°$. Find c, A, B.

$$\cos 130° = -\cos (180° - 130°) = -\cos 50° = -0.643$$

For c: By the law of cosines,

$$c^2 = a^2 + b^2 - 2ab \cos 130°$$
$$= 14^2 + 8^2 - 2(14)(8)(-0.643) = 404$$

and $c = \sqrt{404} = 20.1$.

For A: By the law of sines,

$$\sin A = \frac{a \sin C}{c} = \frac{14(0.766)}{20.1} = 0.533$$

and $A = 32.2°$.

For B: $B = 180° - (A + C) = 180° - (32.2° + 130°) = 17.8°$

Exercises

7. Solve each right triangle ABC, with $C = 90°$, given:

(a) $A = 23.3°$, $c = 346$ *Ans.* $B = 66.7°$, $a = 137$, $b = 318$

(b) $B = 49.2°$, $b = 222$ $A = 40.8°$, $a = 192$, $c = 293$

(c) $A = 66.6°$, $a = 113$ $B = 23.4°$, $b = 48.9$, $c = 123$

(d) $a = 25.4$, $b = 38.2$ $A = 33.6°$, $B = 56.4°$, $c = 45.9$

(e) $b = 673$, $c = 888$ $A = 40.7°$, $B = 49.3°$, $a = 579$

8. Solve each oblique triangle ABC, given:

(a) $a = 125$, $A = 54.6°$, $B = 65.2°$ *Ans.* $b = 139$, $c = 133$, $C = 60.2°$

(b) $b = 321$, $A = 75.3°$, $C = 38.5°$ $a = 339$, $c = 218$, $B = 66.2°$

(c) $b = 215$, $c = 150$, $B = 42.7°$ $a = 300$, $A = 109.1°$, $C = 28.2°$

(d) $a = 512$, $b = 426$, $A = 48.8°$ $c = 680$, $B = 38.8°$, $C = 92.4°$

(e) $b = 50.4$, $c = 33.3$, $B = 118.5°$ $a = 25.1$, $A = 26.0°$, $C = 35.5°$

(f) $b = 120$, $c = 270$, $A = 118.7°$ $a = 344$, $B = 17.8°$, $C = 43.5°$

(g) $a = 24.5$, $b = 18.6$, $c = 26.4$ $A = 63.2°$, $B = 42.7°$, $C = 74.1°$

(h) $a = 6.34$, $b = 7.30$, $c = 9.98$ $A = 39.3°$, $B = 46.9°$, $C = 93.8°$

Appendix C

Exponents

A. The following is a partial list of powers of 10.

$10^0 = 1$

$10^1 = 10$

$10^2 = 10 \times 10 = 100$

$10^3 = 10 \times 10 \times 10 = 1000$

$10^4 = 10 \times 10 \times 10 \times 10 = 10\,000$

$10^5 = 10 \times 10 \times 10 \times 10 \times 10 = 100\,000$

$10^6 = 10 \times 10 \times 10 \times 10 \times 10 \times 10 = 1\,000\,000$

$10^{-1} = \dfrac{1}{10} = 0.1$

$10^{-2} = \dfrac{1}{10^2} = \dfrac{1}{100} = 0.01$

$10^{-3} = \dfrac{1}{10^3} = \dfrac{1}{1000} = 0.001$

$10^{-4} = \dfrac{1}{10^4} = \dfrac{1}{10\,000} = 0.0001$

In the expression 10^5, the *base* is 10 and the *exponent* is 5.

B. In multiplication, exponents of like bases are added.

(*1*) $a^3 \times a^5 = a^{3+5} = a^8$

(*2*) $10^2 \times 10^3 = 10^{2+3} = 10^5$

(*3*) $10 \times 10 = 10^{1+1} = 10^2$

(*4*) $10^7 \times 10^{-3} = 10^{7-3} = 10^4$

(*5*) $(4 \times 10^4)(2 \times 10^{-6}) = 8 \times 10^{4-6} = 8 \times 10^{-2}$

(*6*) $(2 \times 10^5)(3 \times 10^{-2}) = 6 \times 10^{5-2} = 6 \times 10^3$

C. In division, exponents of like bases are subtracted.

(*1*) $\dfrac{a^5}{a^3} = a^{5-3} = a^2$

(*2*) $\dfrac{10^2}{10^5} = 10^{2-5} = 10^{-3}$

(*3*) $\dfrac{8 \times 10^2}{2 \times 10^{-6}} = \dfrac{8}{2} \times 10^{2+6} = 4 \times 10^8$

(*4*) $\dfrac{5.6 \times 10^{-2}}{1.6 \times 10^4} = \dfrac{5.6}{1.6} \times 10^{-2-4} = 3.5 \times 10^{-6}$

D. Any number may be expressed as an integral power of 10, or as the product of two numbers one of which is an integral power of 10.

(*1*) $22\,400 = 2.24 \times 10^4$

(*2*) $7\,200\,000 = 7.2 \times 10^6$

(*3*) $454 = 4.54 \times 10^2$

(*4*) $0.454 = 4.54 \times 10^{-1}$

(*5*) $0.0454 = 4.54 \times 10^{-2}$

(*6*) $0.000\,06 = 6 \times 10^{-5}$

(*7*) $0.003\,06 = 3.06 \times 10^{-3}$

(*8*) $0.000\,000\,5 = 5 \times 10^{-7}$

E. A nonzero expression with an exponent of zero is equal to 1.

(*1*) $a^0 = 1$ (*2*) $10^0 = 1$ (*3*) $(3 \times 10)^0 = 1$ (*4*) $7 \times 10^0 = 7$ (*5*) $8.2 \times 10^0 = 8.2$

317

F. A power may be transferred from the numerator to the denominator of a fraction, or vice versa, by changing the sign of the exponent.

(1) $10^{-4} = \dfrac{1}{10^4}$ $\qquad$ (2) $5 \times 10^{-3} = \dfrac{5}{10^3}$ $\qquad$ (3) $\dfrac{7}{10^{-2}} = 7 \times 10^2$ $\qquad$ (4) $-5a^{-2} = -\dfrac{5}{a^2}$

G. The meaning of the fractional exponent is illustrated by the following.

(1) $10^{2/3} = \sqrt[3]{10^2}$ $\qquad$ (2) $10^{3/2} = \sqrt{10^3}$ $\qquad$ (3) $10^{1/2} = \sqrt{10}$ $\qquad$ (4) $4^{3/2} = \sqrt{4^3} = \sqrt{64} = 8$

H. (1) $(10^3)^2 = 10^{3 \times 2} = 10^6$ $\qquad$ (2) $(10^{-2})^3 = 10^{-2 \times 3} = 10^{-6}$ $\qquad$ (3) $(a^3)^{-2} = a^{-6}$

I. To extract the square root, divide the exponent by 2. If the exponent is an odd number it should first be increased or decreased by 1, and the coefficient adjusted accordingly. To extract the cube root, divide the exponent by 3. The coefficients are treated independently.

(1) $\sqrt{90\,000} = \sqrt{9 \times 10^4} = 3 \times 10^2$ $\qquad$ (3) $\sqrt{4.9 \times 10^{-5}} = \sqrt{49 \times 10^{-6}} = 7 \times 10^{-3}$

(2) $\sqrt{3.6 \times 10^7} = \sqrt{36 \times 10^6} = 6 \times 10^3$ $\qquad$ (4) $\sqrt[3]{1.25 \times 10^8} = \sqrt[3]{125 \times 10^6} = 5 \times 10^2$

J. Multiplication and division of numbers expressed as powers of ten:

(1) $8000 \times 2500 = (8 \times 10^3)(2.5 \times 10^3) = 20 \times 10^6 = 2 \times 10^7$ $\qquad$ or $\qquad$ $20\,000\,000$

(2) $\dfrac{48\,000\,000}{1200} = \dfrac{48 \times 10^6}{12 \times 10^2} = 4 \times 10^{6-2} = 4 \times 10^4$ $\qquad$ or $\qquad$ $40\,000$

(3) $\dfrac{0.0078}{120} = \dfrac{7.8 \times 10^{-3}}{1.2 \times 10^2} = 6.5 \times 10^{-5}$ $\qquad$ or $\qquad$ $0.000\,065$

(4) $(4 \times 10^{-3})(5 \times 10^4)^2 = (4 \times 10^{-3})(5^2 \times 10^8) = 4 \times 5^2 \times 10^{-3+8} = 100 \times 10^5 = 1 \times 10^7$

(5) $\dfrac{(6\,000\,000)(0.000\,04)^4}{(800)^2(0.0002)^3} = \dfrac{(6 \times 10^6)(4 \times 10^{-5})^4}{(8 \times 10^2)^2(2 \times 10^{-4})^3} = \dfrac{6 \times 4^4}{8^2 \times 2^3} \times \dfrac{10^6 \times 10^{-20}}{10^4 \times 10^{-12}}$

$\qquad\qquad = \dfrac{6 \times 256}{64 \times 8} \times \dfrac{10^{6-20}}{10^{4-12}} = 3 \times \dfrac{10^{-14}}{10^{-8}} = 3 \times 10^{-6}$

(6) $(\sqrt{4.0 \times 10^{-6}})(\sqrt{8.1 \times 10^3})(\sqrt{0.0016}) = (\sqrt{4.0 \times 10^{-6}})(\sqrt{81 \times 10^2})(\sqrt{16 \times 10^{-4}})$

$\qquad\qquad = (2 \times 10^{-3})(9 \times 10^1)(4 \times 10^{-2})$

$\qquad\qquad = 72 \times 10^{-4} = 7.2 \times 10^{-3}$ $\qquad$ or $\qquad$ 0.0072

(7) $(\sqrt[3]{6.4 \times 10^{-2}})(\sqrt[3]{27\,000})(\sqrt[3]{2.16 \times 10^{-4}}) = (\sqrt[3]{64 \times 10^{-3}})(\sqrt[3]{27 \times 10^3})(\sqrt[3]{216 \times 10^{-6}})$

$\qquad\qquad = (4 \times 10^{-1})(3 \times 10^1)(6 \times 10^{-2})$

$\qquad\qquad = 72 \times 10^{-2}$ $\qquad$ or $\qquad$ 0.72

Exercises

1. Express the following in powers of 10.

(a) 320

(b) 32 600

(c) 1006

(d) 36 000 000

(e) 0.831

(f) 0.03

(g) 0.000 002

(h) 0.000 706

(i) $\sqrt{640\,000}$

(j) $\sqrt{0.000\,081}$

(k) $\sqrt[3]{8\,000\,000}$

(l) $\sqrt[3]{0.000\,027}$

Ans.
(a) 3.2×10^2

(b) 3.26×10^4

(c) 1.006×10^3

(d) 3.6×10^7

(e) 8.31×10^{-1}

(f) 3×10^{-2}

(g) 2×10^{-6}

(h) 7.06×10^{-4}

(i) 8.0×10^2

(j) 9.0×10^{-3}

(k) 2×10^2

(l) 3×10^{-2}

2. Evaluate the following and express the results in powers of 10.

(a) 1500×260

(b) $220 \times 35\,000$

(c) $40 \div 20\,000$

(d) $82\,800 \div 0.12$

(e) $\dfrac{1.728 \times 17.28}{0.000\,172\,8}$

(f) $\dfrac{(16\,000)(0.0002)(1.2)}{(2000)(0.006)(0.000\,32)}$

(g) $\dfrac{0.004 \times 32\,000 \times 0.6}{6400 \times 3000 \times 0.08}$

(h) $(\sqrt{14\,400}\,)(\sqrt{0.000\,025}\,)$

(i) $(\sqrt[3]{2.7 \times 10^7}\,)(\sqrt[3]{1.25 \times 10^{-4}}\,)$

(j) $(1 \times 10^{-3})(2 \times 10^5)^2$

(k) $\dfrac{(3 \times 10^2)^3 (2 \times 10^{-5})^2}{3.6 \times 10^{-8}}$

(l) $8(2 \times 10^{-2})^{-3}$

Ans.
(a) 3.9×10^5

(b) 7.7×10^6

(c) 2×10^{-3}

(d) 6.9×10^5

(e) 1.728×10^5

(f) 1×10^3

(g) 5×10^{-5}

(h) 6.0×10^{-1}

(i) 1.5×10^1

(j) 4×10^7

(k) 3×10^5

(l) 1×10^6

Appendix D

Logarithms

DEFINITION OF TERMS: The *logarithm* of a number is the exponent, or power, of a given base that is required to produce that number. For example, since $1000 = 10^3$, $100 = 10^2$, $10 = 10^1$, $1 = 10^0$, then the logarithms of 1000, 100, 10, 1, to the base 10 are respectively 3, 2, 1, 0. The system of logarithms whose base is 10 (called the *common* or *Briggsian* system) may be used in all numerical computations.

It is obvious that $10^{1.5377}$ will give some number greater than 10 (which is 10^1) but smaller than 100 (10^2). Actually, $10^{1.5377} = 34.49$; hence $\log 34.49 = 1.5377$. The digit before the decimal point is the *characteristic* of the log, and the decimal fraction part is the *mantissa* of the log. In the above example, the characteristic is 1 and the mantissa is .5377.

The mantissa of the log of a number is found in tables, such as Appendix I, printed without the decimal point. Each mantissa in the tables is understood to have a decimal point preceding it, and the mantissa is always considered positive.

THE CHARACTERISTIC is determined by inspection from the number itself according to the following rules.

(1) For a number greater than 1, the characteristic is positive (or zero) and is one *less* than the number of digits before the decimal point. For example:

Number	5297	348	900	34.8	60	4.764	3
Characteristic	3	2	2	1	1	0	0

(2) For a number less than 1, the characteristic is negative and in magnitude it is one *more* than the number of zeros immediately following the decimal point. The negative sign of the characteristic is written in either of these two ways: (*a*) above the characteristic, as $\bar{1}$, $\bar{2}$, and so on; (*b*) as $9. - 10$, $8. - 10$, and so on. Thus the characteristic of $\log (0.3485)$ is $\bar{1}$, or $9. - 10$; of $\log (0.0513)$ it is $\bar{2}$, or $8. - 10$.

TO FIND THE LOGARITHM OF A NUMBER: Suppose it is required to find the complete log of the number 728. In the table of logarithms in Appendix I, glance down the N column to 72, then horizontally to the right to column 8 and note the entry 8621, which is the required mantissa. Since the characteristic is 2, $\log 728 = 2.8621$. (This means that $728 = 10^{2.8621}$.)

The mantissa for $\log 72.8$, for $\log 7.28$, for $\log 0.728$, for $\log 0.0728$, etc., is .8621, but the characteristics differ. Thus:

$$\log 728 = 2.8621 \qquad \log 0.728 = \bar{1}.8621 \quad \text{or} \quad 9.8621 - 10$$

$$\log 72.8 = 1.8621 \qquad \log 0.0728 = \bar{2}.8621 \quad \text{or} \quad 8.8621 - 10$$

$$\log 7.28 = 0.8621 \qquad \log 0.007\,28 = \bar{3}.8621 \quad \text{or} \quad 7.8621 - 10$$

To find $\log 46.38$: Glance down the N column to 46, then horizontally to column 3 and note the mantissa 6656. Moving farther to the right along the same line, the digit 7 is found under column 8 of Proportional Parts. The required mantissa is $.6656 + .0007 = .6663$. Since the characteristic is 1, $\log 46.38 = 1.6663$.

The mantissa for log 4638, for log 463.8, for log 46.38, etc., is .6663, but the characteristics differ. Thus:

$$\log 4638 = 3.6663 \qquad \log 0.4638 = \bar{1}.6663 \quad \text{or} \quad 9.6663 - 10$$

$$\log 463.8 = 2.6663 \qquad \log 0.046\,38 = \bar{2}.6663 \quad \text{or} \quad 8.6663 - 10$$

$$\log 46.38 = 1.6663 \qquad \log 0.004\,638 = \bar{3}.6663 \quad \text{or} \quad 7.6663 - 10$$

$$\log 4.638 = 0.6663 \qquad \log 0.000\,463\,8 = \bar{4}.6663 \quad \text{or} \quad 6.6663 - 10$$

Exercises. Find the logarithms of the following numbers.

(a) 454	(f) 0.621	*Ans.* (a) 2.6571	(f) $\bar{1}.7931$ or $9.7931 - 10$
(b) 5280	(g) 0.9463	(b) 3.7226	(g) $\bar{1}.9760$ or $9.9760 - 10$
(c) 96 500	(h) 0.0353	(c) 4.9845	(h) $\bar{2}.5478$ or $8.5478 - 10$
(d) 30.48	(i) 0.0022	(d) 1.4840	(i) $\bar{3}.3424$ or $7.3424 - 10$
(e) 1.057	(j) 0.000 264 5	(e) 0.0241	(j) $\bar{4}.4224$ or $6.4224 - 10$

ANTILOGARITHMS: The *antilogarithm* is the number corresponding to a given logarithm. "The antilog of 3" means "the number whose log is 3"; that number is obviously 1000.

Suppose that it is required to find the antilog of 2.6747, i.e. the number whose log is 2.6747. The characteristic is 2 and the mantissa is .6747. Using the table of antilogarithms, Appendix I, locate 67 in the first column, then move horizontally to column 4 and note the digits 4721. Moving farther to the right along the same line, the entry 8 is found under column 7 of Proportional Parts. Adding 8 to 4721 gives 4729. Since the characteristic is 2, there are three digits to the left of the decimal point. Hence 472.9 is the required number.

Similarly, the antilog of 1.6747 is 47.29; the antilog of 0.6747 is 4.729; the antilog of 9.6747 − 10 is 0.4729; etc.

Exercises. Find the numbers corresponding to the following logarithms.

(a) 3.1568	(f) 0.9142	*Ans.* (a) 1435	(f) 8.208
(b) 1.6934	(g) 0.0008	(b) 49.37	(g) 1.002
(c) 5.6934	(h) 9.7507 − 10 or $\bar{1}.7507$	(c) 4.937×10^5	(h) 0.5632
(d) 2.5000	(i) 8.0034 − 10 or $\bar{2}.0034$	(d) 316.2	(i) 0.010 08
(e) 2.0436	(j) 7.2006 − 10 or $\bar{3}.2006$	(e) 110.6	(j) 0.001 587

BASIC PROPERTIES OF LOGARITHMS: Since logarithms are exponents, all properties of exponents are also properties of logarithms.

(1) The logarithm of the product of two numbers is the sum of their logarithms.

$$\log ab = \log a + \log b \qquad \log (5280 \times 48) = \log 5280 + \log 48$$

(2) The logarithm of the quotient of two numbers is the logarithm of the numerator minus the logarithm of the denominator.

$$\log \frac{a}{b} = \log a - \log b \qquad \log \frac{536}{24.5} = \log 536 - \log 24.5$$

(3) The log of the *n*th power of a number is *n* times the log of the number.

$$\log a^n = n \log a \qquad \log (4.28)^3 = 3 \log 4.28$$

(4) The log of the *n*th root of a number is $1/n$ times the log of the number.

$$\log \sqrt[n]{a} = \frac{1}{n} \log a \qquad \log \sqrt{32} = \frac{1}{2} \log 32 \qquad \log \sqrt[3]{792} = \frac{1}{3} \log 792$$

Solved Problems

1.   Find the value of $487 \times 2.45 \times 0.0387$.

Let $x = 487 \times 2.45 \times 0.0387$.

$\log x = \log 487 + \log 2.45 + \log 0.0387$

$\quad\quad = 1.6644$

$\quad\quad x = \text{antilog } 1.6644 = 46.17 \text{ or } 46.2$

$\log 487 = 2.6875$
$\log 2.45 = 0.3892$
$\log 0.0387 = \underline{8.5877 - 10} \text{ (add)}$
$\log x = 11.6644 - 10 \text{ or } 1.6644$

2.   Find $x = \dfrac{136.3}{65.38}$.

$\log x = \log 136.3 - \log 65.38 = 0.3191$

$\quad\quad x = \text{antilog } 0.3191 = 2.084$

$\log 136.3 = 2.1345$
$\log 65.38 = \underline{1.8154} \quad \text{(subtract)}$
$\log x = \overline{0.3191}$

3.   Find $x = \dfrac{1}{22.4}$.

$\log x = \log 1 - \log 22.4 = 8.6498 - 10$

$\quad\quad x = \text{antilog } 8.6498 - 10$

$\quad\quad = 0.044\,65 \text{ or } 0.0446 \text{ (three significant figures)}$

$\log 1 = 0 = 10.0000 - 10$
$\log 22.4 = \underline{1.3502} \quad \text{(subtract)}$
$\log x = \overline{8.6498 - 10}$

Adding or subtracting $10.0000 - 10$, $20.0000 - 20$, etc., from any logarithm does not change its value.

4.   Find $x = \dfrac{17.5 \times 1.92}{0.283 \times 0.0314}$.

$\log x = (\log 17.5 + \log 1.92) - (\log 0.283 + \log 0.0314)$

$\log 17.5 = 1.2430$
$\log 1.92 = \underline{0.2833} \quad \text{(add)}$
$\quad\quad\quad\quad 1.5263 \text{ or } 11.5263 - 10$

$\log 0.283 = 9.4518 - 10$
$\log 0.0314 = \underline{8.4969 - 10} \quad \text{(add)}$
$\quad\quad\quad\quad\quad 17.9487 - 20 \text{ or } 7.9487 - 10$

$\log 17.5 + \log 1.92 \quad = 11.5263 - 10$

$\log 0.283 + \log 0.0314 = \underline{7.9487 - 10} \quad \text{(subtract)}$

$\log x = 3.5776$

$x = \text{antilog } 3.5776 = 3781 \quad \text{or} \quad 3780 \quad \text{or} \quad 3.78 \times 10^3$

5.   Find $x = (6.138)^3$.

$\log x = 3(\log 6.138) = 3(0.7881) = 2.3643$

$x = \text{antilog } 2.3643 = 231.4$

6.   Find $x = \sqrt{7514}$ or $(7514)^{1/2}$.

$\log x = \dfrac{1}{2}(\log 7514) = \dfrac{1}{2}(3.8758) = 1.9379$

$x = \text{antilog } 1.9379 = 86.68$

7.   Find $x = \sqrt[3]{0.0592}$ or $(70.0592)^{1/3}$.

$\log x = \dfrac{1}{3}(\log 0.0592) = \dfrac{1}{3}(8.7723 - 10) = \dfrac{1}{3}(28.7723 - 30) = 9.5908 - 10$

$x = \text{antilog } 9.5908 - 10 = 0.3897 \quad \text{or} \quad 0.390 \text{ (three significant figures)}$

8.   Find $x = \sqrt{(152)^3}$.

$\log x = \dfrac{1}{2}(3 \log 152) = \dfrac{1}{2}(3 \times 2.1818) = \dfrac{1}{2}(6.5454) = 3.2727$

$x = \text{antilog } 3.2727 = 1874 \quad \text{or} \quad 1870 \quad \text{or} \quad 1.87 \times 10^3 \text{ (three significant figures)}$

Find $(6.8 \times 10^{-4})^3$ or $(6.8)^3 \times 10^{-12}$.

$$\log (6.8)^3 = 3(\log 6.8) = 3(0.8325) = 2.4975$$

$$(6.8)^3 = \text{antilog } 2.4975 = 314.5 \quad \text{or} \quad 310 \quad \text{or} \quad 3.1 \times 10^2 \text{ (two significant figures)}$$

Then $(6.8 \times 10^{-4})^3 = 3.1 \times 10^2 \times 10^{-12} = 3.1 \times 10^{-10}$.

10. Find $\sqrt{8.31 \times 10^{-11}}$ or $\sqrt{83.1 \times 10^{-12}}$ or $\sqrt{83.1} \times 10^{-6}$.

$$\log \sqrt{83.1} = \frac{1}{2}(\log 83.1) = \frac{1}{2}(1.9196) = 0.9598$$

$$\sqrt{83.1} = \text{antilog } 0.9598 = 9.116 \quad \text{or} \quad 9.12$$

Then $\sqrt{8.31 \times 10^{-11}} = 9.12 \times 10^{-6}$.

Exercises

11. Evaluate each of the following, using logarithms.

(1) $28.32 \times 0.082\,54$

(2) $573 \times 6.96 \times 0.004\,81$

(3) $\dfrac{79.28}{63.57}$

(4) $\dfrac{65.38}{225.2}$

(5) $\dfrac{1}{239}$

(6) $\dfrac{0.572 \times 31.8}{96.2}$

(7) $47.5 \times \dfrac{779}{760} \times \dfrac{273}{300}$

(8) $(8.642)^2$

(9) $(0.086\,42)^2$

(10) $(11.72)^3$

(11) $(0.0523)^3$

(12) $\sqrt{9463}$

(13) $\sqrt{946.3}$

(14) $\sqrt{0.006\,61}$

(15) $\sqrt[3]{1.79}$

(16) $\sqrt[4]{0.182}$

(17) $\sqrt{643} \times (1.91)^3$

(18) $(8.73 \times 10^{-2})(7.49 \times 10^6)$

(19) $(3.8 \times 10^{-5})^2(1.9 \times 10^{-5})$

(20) $\dfrac{8.5 \times 10^{-45}}{1.6 \times 10^{-22}}$

(21) $\sqrt{2.54 \times 10^6}$

(22) $\sqrt{9.44 \times 10^5}$

(23) $\sqrt{7.2 \times 10^{-13}}$

(24) $\sqrt[3]{7.3 \times 10^{-14}}$

(25) $\sqrt{\dfrac{(1.1 \times 10^{-23})(6.8 \times 10^{-2})}{1.4 \times 10^{-24}}}$

(26) $2.04 \log 97.2$

(27) $37 \log 0.0298$

(28) $6.30 \log (2.95 \times 10^3)$

(29) $8.09 \log (5.68 \times 10^{-16})$

(30) $2^{0.714}$

Ans.
(1) 2.337
(2) 19.2
(3) 1.247
(4) 0.2902
(5) 0.004 18
(6) 0.189
(7) 44.3
(8) 74.67

(9) 0.007 467
(10) 1611
(11) 0.000 143
(12) 97.27
(13) 30.76
(14) 0.0813
(15) 1.21
(16) 0.653

(17) 177
(18) 6.54×10^5
(19) 2.7×10^{-14}
(20) 5.3×10^{-23}
(21) 1.59×10^3
(22) 9.72×10^2
(23) 8.5×10^{-7}
(24) 4.2×10^{-5}

(25) 0.73
(26) 4.05
(27) -56
(28) 21.9
(29) -123.3
(30) 1.64

Appendix E

Prefixes for Multiples of SI Units

Multiplication Factor	Prefix	Symbol
10^{12}	tera	T
10^9	giga	G
10^6	mega	M
10^3	kilo	k
10^2	hecto	h
10	deka	da
10^{-1}	deci	d
10^{-2}	centi	c
10^{-3}	milli	m
10^{-6}	micro	μ
10^{-9}	nano	n
10^{-12}	pico	p
10^{-15}	femto	f
10^{-18}	atto	a

The Greek Alphabet

A	α	alpha	H	η	eta	N	ν	nu	T	τ	tau
B	β	beta	Θ	θ	theta	Ξ	ξ	xi	Υ	υ	upsilon
Γ	γ	gamma	I	ι	iota	O	o	omicron	Φ	ϕ	phi
Δ	δ	delta	K	κ	kappa	Π	π	pi	X	χ	chi
E	ϵ	epsilon	Λ	λ	lambda	P	ρ	rho	Ψ	ψ	psi
Z	ζ	zeta	M	μ	mu	Σ	σ	sigma	Ω	ω	omega

Appendix F

Factors for Conversions to SI Units

Acceleration
$1 \text{ ft/s}^2 = 0.3048 \text{ m/s}^2$

$g = 9.807 \text{ m/s}^2$

Area
$1 \text{ acre} = 4047 \text{ m}^2$

$1 \text{ circular mil} = 5.067 \times 10^{-10} \text{ m}^2$

$1 \text{ ft}^2 = 9.290 \times 10^{-2} \text{ m}^2$

$1 \text{ in}^2 = 6.45 \times 10^{-4} \text{ m}^2$

$1 \text{ mi}^2 = 2.59 \times 10^6 \text{ m}^2$

Density
$1 \text{ g/cm}^3 = 10^3 \text{ kg/m}^3$

$1 \text{ slug/ft}^3 = 515.4 \text{ kg/m}^3$

Energy
$1 \text{ Btu} = 1054 \text{ J}$

$1 \text{ calorie (cal)} = 4.184 \text{ J}$

$1 \text{ electron volt (eV)} = 1.602 \times 10^{-19} \text{ J}$

$1 \text{ erg} = 10^{-7} \text{ J}$

$1 \text{ foot pound (ft} \cdot \text{lb)} = 1.356 \text{ J}$

$1 \text{ kilowatt hour (kW} \cdot \text{h)} = 3.60 \times 10^6 \text{ J}$

Force
$1 \text{ dyne} = 10^{-5} \text{ N}$

$1 \text{ lb} = 4.448 \text{ N}$

Length
$1 \text{ angstrom (Å)} = 10^{-10} \text{ m}$

$1 \text{ ft} = 0.3048 \text{ m}$

$1 \text{ in} = 2.54 \times 10^{-2} \text{ m}$

$1 \text{ light year} = 9.461 \times 10^{15} \text{ m}$

$1 \text{ mil} = 2.54 \times 10^{-5} \text{ m}$

$1 \text{ mile} = 1609 \text{ m}$

Mass
$1 \text{ atomic mass unit (u)} = 1.661 \times 10^{-27} \text{ kg}$

$1 \text{ gram} = 10^{-3} \text{ kg}$

$1 \text{ slug} = 14.59 \text{ kg}$

Power
$1 \text{ Btu/s} = 1054 \text{ W}$

$1 \text{ cal/s} = 4.184 \text{ W}$

$1 \text{ ft} \cdot \text{lb/s} = 1.356 \text{ W}$

$1 \text{ horsepower (hp)} = 746 \text{ W}$

Pressure
$1 \text{ atmosphere (atm)} = 1.013 \times 10^5 \text{ Pa}$

$1 \text{ bar} = 10^5 \text{ Pa}$

$1 \text{ cm of Hg} = 1333 \text{ Pa}$

$1 \text{ dyne/cm}^2 = 10^{-1} \text{ Pa}$

$1 \text{ lb/ft}^2 = 47.88 \text{ Pa}$

$1 \text{ lb/in}^2 \text{ (psi)} = 6895 \text{ Pa}$

$1 \text{ N/m}^2 = 1 \text{ pascal (Pa)}$

$1 \text{ torr} = 133.3 \text{ Pa}$

Speed
$1 \text{ ft/s (fps)} = 0.3048 \text{ m/s}$

$1 \text{ in/s} = 2.54 \times 10^{-2} \text{ m/s}$

$1 \text{ km/h} = 0.2778 \text{ m/s}$

$1 \text{ mi/h (mph)} = 0.44704 \text{ m/s}$

Temperature
$T_\text{Kelvin} = T_\text{Celsius} + 273.15$

$T_\text{Kelvin} = \dfrac{5}{9}(T_\text{Fahrenheit} + 459.67)$

$T_\text{Celsius} = \dfrac{5}{9}(T_\text{Fahrenheit} - 32)$

$T_\text{Kelvin} = \dfrac{5}{9} T_\text{Rankine}$

Volume
$1 \text{ ft}^3 = 2.832 \times 10^{-2} \text{ m}^3$

$1 \text{ gallon (gal)} = 3.785 \times 10^{-3} \text{ m}^3$

$1 \text{ in}^3 = 1.639 \times 10^{-5} \text{ m}^3$

$1 \text{ liter} = 10^{-3} \text{ m}^3$

Appendix G

Physical Constants

Speed of light in free space . . . . c $= 2.9979 \times 10^8$ m/s

Acceleration due to gravity (normal) . . . . . . g $= 9.807$ m/s^2

Gravitational constant . . . . G $= 6.670 \times 10^{-11}$ N·m^2/kg^2

Coulomb constant . . . . k $= 8.988 \times 10^9$ N·m^2/C^2

Density of water (maximum) . . . . $= 0.999\,972 \times 10^3$ kg/m^3

Density of mercury (S.T.P.) . . . . $= 13.595 \times 10^3$ kg/m^3

Standard atmosphere . . . . $= 1.0132 \times 10^5$ N/m^2

Volume of ideal gas at S.T.P. . . . . $= 22.4$ m^3/kmol

Avogadro's number . . . . N_0 $= 6.02 \times 10^{26}$ kmol^{-1}

Universal gas constant . . . . R $= 8314$ J/kmol·K

Ice point . . . . $= 273.15$ K

Mechanical equivalent of heat . . . . $\mathcal{J}$ $= 4.184$ J/cal

Stefan-Boltzmann constant . . . . σ $= 5.67 \times 10^{-8}$ W/m^2·K^4

Planck's constant . . . . h $= 6.626 \times 10^{-34}$ J·s

Faraday . . . . $\mathcal{F}$ $= 9.648 \times 10^4$ C

Electronic charge . . . . e $= 1.602 \times 10^{-19}$ C

Boltzmann's constant . . . . k $= 1.38 \times 10^{-23}$ J/K

Ratio of electron charge to mass . . . . e/m_e $= 1.7588 \times 10^{11}$ C/kg

Electron rest mass . . . . $= 9.109 \times 10^{-31}$ kg

Proton rest mass . . . . $= 1.6726 \times 10^{-27}$ kg

Neutron rest mass . . . . $= 1.6749 \times 10^{-27}$ kg

Alpha particle rest mass . . . . $= 6.645 \times 10^{-27}$ kg

Atomic mass unit (1/12 mass of ^{12}C) . . . . . u $= 1.660\,57 \times 10^{-27}$ kg

Rest energy of 1 u . . . . $= 931.5$ MeV

Table of the Elements

Atomic Number (Z)	Element	Symbol	Atomic Mass, u*	Atomic Number (Z)	Element	Symbol	Atomic Mass, u*
1	Hydrogen	H	1.008	53	Iodine	I	126.9
2	Helium	He	4.003	54	Xenon	Xe	131.3
3	Lithium	Li	6.939	55	Cesium	Cs	132.9
4	Beryllium	Be	9.012	56	Barium	Ba	137.3
5	Boron	B	10.81	57	Lanthanum	La	138.9
6	Carbon	C	12.01	58	Cerium	Ce	140.1
7	Nitrogen	N	14.01	59	Praseodymium	Pr	140.9
8	Oxygen	O	16.00	60	Neodymium	Nd	144.2
9	Fluorine	F	19.00	61	Promethium	Pm	(147)
10	Neon	Ne	20.18	62	Samarium	Sm	150.4
11	Sodium	Na	22.99	63	Europium	Eu	152.0
12	Magnesium	Mg	24.31	64	Gadolinium	Gd	157.3
13	Aluminum	Al	26.98	65	Terbium	Tb	158.9
14	Silicon	Si	28.09	66	Dysprosium	Dy	162.5
15	Phosphorus	P	30.97	67	Holmium	Ho	164.9
16	Sulfur	S	32.06	68	Erbium	Er	167.3
17	Chlorine	Cl	35.45	69	Thulium	Tm	168.9
18	Argon	Ar	39.95	70	Ytterbium	Yb	173.0
19	Potassium	K	39.10	71	Lutetium	Lu	175.0
20	Calcium	Ca	40.08	72	Hafnium	Hf	178.5
21	Scandium	Sc	44.96	73	Tantalum	Ta	180.9
22	Titanium	Ti	47.90	74	Tungsten	W	183.9
23	Vanadium	V	50.94	75	Rhenium	Re	186.2
24	Chromium	Cr	52.00	76	Osmium	Os	190.2
25	Manganese	Mn	54.94	77	Iridium	Ir	192.2
26	Iron	Fe	55.85	78	Platinum	Pt	195.1
27	Cobalt	Co	58.93	79	Gold	Au	197.0
28	Nickel	Ni	58.71	80	Mercury	Hg	200.6
29	Copper	Cu	63.54	81	Thallium	Tl	204.4
30	Zinc	Zn	65.37	82	Lead	Pb	207.2
31	Gallium	Ga	69.72	83	Bismuth	Bi	209.0
32	Germanium	Ge	72.59	84	Polonium	Po	(209)
33	Arsenic	As	74.92	85	Astatine	At	(210)
34	Selenium	Se	78.96	86	Radon	Rn	(222)
35	Bromine	Br	79.91	87	Francium	Fr	(223)
36	Krypton	Kr	83.80	88	Radium	Ra	(226)
37	Rubidium	Rb	85.47	89	Actinium	Ac	(227)
38	Strontium	Sr	87.62	90	Thorium	Th	232.0
39	Yttrium	Y	88.91	91	Protactinium	Pa	(231)
40	Zirconium	Zr	91.22	92	Uranium	U	238.0
41	Niobium	Nb	92.91	93	Neptunium	Np	(237)
42	Molybdenum	Mo	95.94	94	Plutonium	Pu	(244)
43	Technetium	Tc	(99)	95	Americium	Am	(243)
44	Ruthenium	Ru	101.1	96	Curium	Cm	(247)
45	Rhodium	Rh	102.9	97	Berkelium	Bk	(247)
46	Palladium	Pd	106.4	98	Californium	Cf	(251)
47	Silver	Ag	107.9	99	Einsteinium	Es	(254)
48	Cadmium	Cd	112.4	100	Fermium	Fm	(257)
49	Indium	In	114.8	101	Mendelevium	Md	(257)
50	Tin	Sn	118.7	102	Nobelium	No	(255)
51	Antimony	Sb	121.8	103	Lawrencium	Lr	(256)
52	Tellurium	Te	127.6				

* A value in parentheses is the mass number of the most stable (long-lived) of the known isotopes.

Four-place Logarithms

N	0	1	2	3	4	5	6	7	8	9	Proportional Parts								
											1	2	3	4	5	6	7	8	9
10	0000	0043	0086	0128	0170	0212	0253	0294	0334	0374	4	8	12	17	21	25	29	33	37
11	0414	0453	0492	0531	0569	0607	0645	0682	0719	0755	4	8	11	15	19	23	26	30	34
12	0792	0828	0864	0899	0934	0969	1004	1038	1072	1106	3	7	10	14	17	21	24	28	31
13	1139	1173	1206	1239	1271	1303	1335	1367	1399	1430	3	6	10	13	16	19	23	26	29
14	1461	1492	1523	1553	1584	1614	1644	1673	1703	1732	3	6	9	12	15	18	21	24	27
15	1761	1790	1818	1847	1875	1903	1931	1959	1987	2014	3	6	8	11	14	17	20	22	25
16	2041	2068	2095	2122	2148	2175	2201	2227	2253	2279	3	5	8	11	13	16	18	21	24
17	2304	2330	2355	2380	2405	2430	2455	2480	2504	2529	2	5	7	10	12	15	17	20	22
18	2553	2577	2601	2625	2648	2672	2695	2718	2742	2765	2	5	7	9	12	14	16	19	21
19	2788	2810	2833	2856	2878	2900	2923	2945	2967	2989	2	4	7	9	11	13	16	18	20
20	3010	3032	3054	3075	3096	3118	3139	3160	3181	3201	2	4	6	8	11	13	15	17	19
21	3222	3243	3263	3284	3304	3324	3345	3365	3385	3404	2	4	6	8	10	12	14	16	18
22	3424	3444	3464	3483	3502	3522	3541	3560	3579	3598	2	4	6	8	10	12	14	15	17
23	3617	3636	3655	3674	3692	3711	3729	3747	3766	3784	2	4	6	7	9	11	13	15	17
24	3802	3820	3838	3856	3874	3892	3909	3927	3945	3962	2	4	5	7	9	11	12	14	16
25	3979	3997	4014	4031	4048	4065	4082	4099	4116	4133	2	3	5	7	9	10	12	14	15
26	4150	4166	4183	4200	4216	4232	4249	4265	4281	4298	2	3	5	7	8	10	11	13	15
27	4314	4330	4346	4362	4378	4393	4409	4425	4440	4456	2	3	5	6	8	9	11	13	14
28	4472	4487	4502	4518	4533	4548	4564	4579	4594	4609	2	3	5	6	8	9	11	12	14
29	4624	4639	4654	4669	4683	4698	4713	4728	4742	4757	1	3	4	6	7	9	10	12	13
30	4771	4786	4800	4814	4829	4843	4857	4871	4886	4900	1	3	4	6	7	9	10	11	13
31	4914	4928	4942	4955	4969	4983	4997	5011	5024	5038	1	3	4	6	7	8	10	11	12
32	5051	5065	5079	5092	5105	5119	5132	5145	5159	5172	1	3	4	5	7	8	9	11	12
33	5185	5198	5211	5224	5237	5250	5263	5276	5289	5302	1	3	4	5	6	8	9	10	12
34	5315	5328	5340	5353	5366	5378	5391	5403	5416	5428	1	3	4	5	6	8	9	10	11
35	5441	5453	5465	5478	5490	5502	5514	5527	5539	5551	1	2	4	5	6	7	9	10	11
36	5563	5575	5587	5599	5611	5623	5635	5647	5658	5670	1	2	4	5	6	7	8	10	11
37	5682	5694	5705	5717	5729	5740	5752	5763	5775	5786	1	2	3	5	6	7	8	9	10
38	5798	5809	5821	5832	5843	5855	5866	5877	5888	5899	1	2	3	5	6	7	8	9	10
39	5911	5922	5933	5944	5955	5966	5977	5988	5999	6010	1	2	3	4	5	7	8	9	10
40	6021	6031	6042	6053	6064	6075	6085	6096	6107	6117	1	2	3	4	5	6	8	9	10
41	6128	6138	6149	6160	6170	6180	6191	6201	6212	6222	1	2	3	4	5	6	7	8	9
42	6232	6243	6253	6263	6274	6284	6294	6304	6314	6325	1	2	3	4	5	6	7	8	9
43	6335	6345	6355	6365	6375	6385	6395	6405	6415	6425	1	2	3	4	5	6	7	8	9
44	6435	6444	6454	6464	6474	6484	6493	6503	6513	6522	1	2	3	4	5	6	7	8	9
45	6532	6542	6551	6561	6571	6580	6590	6599	6609	6618	1	2	3	4	5	6	7	8	9
46	6628	6637	6646	6656	6665	6675	6684	6693	6702	6712	1	2	3	4	5	6	7	7	8
47	6721	6730	6739	6749	6758	6767	6776	6785	6794	6803	1	2	3	4	5	5	6	7	8
48	6812	6821	6830	6839	6848	6857	6866	6875	6884	6893	1	2	3	4	4	5	6	7	8
49	6902	6911	6920	6928	6937	6946	6955	6964	6972	6981	1	2	3	4	4	5	6	7	8
50	6990	6998	7007	7016	7024	7033	7042	7050	7059	7067	1	2	3	3	4	5	6	7	8
51	7076	7084	7093	7101	7110	7118	7126	7135	7143	7152	1	2	3	3	4	5	6	7	8
52	7160	7168	7177	7185	7193	7202	7210	7218	7226	7235	1	2	2	3	4	5	6	7	7
53	7243	7251	7259	7267	7275	7284	7292	7300	7308	7316	1	2	2	3	4	5	6	6	7
54	7324	7332	7340	7348	7356	7364	7372	7380	7388	7396	1	2	2	3	4	5	6	6	7
N	0	1	2	3	4	5	6	7	8	9	1	2	3	4	5	6	7	8	9

Four-place Logarithms (*cont.*)

N	0	1	2	3	4	5	6	7	8	9	1	2	3	4	5	6	7	8	9
											Proportional Parts								
55	7404	7412	7419	7427	7435	7443	7451	7459	7466	7474	1	2	2	3	4	5	5	6	7
56	7482	7490	7497	7505	7513	7520	7528	7536	7543	7551	1	2	2	3	4	5	5	6	7
57	7559	7566	7574	7582	7589	7597	7604	7612	7619	7627	1	2	2	3	4	5	5	6	7
58	7634	7642	7649	7657	7664	7672	7679	7686	7694	7701	1	1	2	3	4	4	5	6	7
59	7709	7716	7723	7731	7738	7745	7752	7760	7767	7774	1	1	2	3	4	4	5	6	7
60	7782	7789	7796	7803	7810	7818	7825	7832	7839	7846	1	1	2	3	4	4	5	6	6
61	7853	7860	7868	7875	7882	7889	7896	7903	7910	7917	1	1	2	3	4	4	5	6	6
62	7924	7931	7938	7945	7952	7959	7966	7973	7980	7987	1	1	2	3	3	4	5	6	6
63	7993	8000	8007	8014	8021	8028	8035	8041	8048	8055	1	1	2	3	3	4	5	5	6
64	8062	8069	8075	8082	8089	8096	8102	8109	8116	8122	1	1	2	3	3	4	5	5	6
65	8129	8136	8142	8149	8156	8162	8169	8176	8182	8189	1	1	2	3	3	4	5	5	6
66	8195	8202	8209	8215	8222	8228	8235	8241	8248	8254	1	1	2	3	3	4	5	5	6
67	8261	8267	8274	8280	8287	8293	8299	8306	8312	8319	1	1	2	3	3	4	5	5	6
68	8325	8331	8338	8344	8351	8357	8363	8370	8376	8382	1	1	2	3	3	4	5	5	6
69	8388	8395	8401	8407	8414	8420	8426	8432	8439	8445	1	1	2	2	3	4	4	5	6
70	8451	8457	8463	8470	8476	8482	8488	8494	8500	8506	1	1	2	2	3	4	4	5	6
71	8513	8519	8525	8531	8537	8543	8549	8555	8561	8567	1	1	2	2	3	4	4	5	5
72	8573	8579	8585	8591	8597	8603	8609	8615	8621	8627	1	1	2	2	3	4	4	5	5
73	8633	8639	8645	8651	8657	8663	8669	8675	8681	8686	1	1	2	2	3	4	4	5	5
74	8692	8698	8704	8710	8716	8722	8727	8733	8739	8745	1	1	2	2	3	4	4	5	5
75	8751	8756	8762	8768	8774	8779	8785	8791	8797	8802	1	1	2	2	3	3	4	5	5
76	8808	8814	8820	8825	8831	8837	8842	8848	8854	8859	1	1	2	2	3	3	4	5	5
77	8865	8871	8876	8882	8887	8893	8899	8904	8910	8915	1	1	2	2	3	3	4	4	5
78	8921	8927	8932	8938	8943	8949	8954	8960	8965	8971	1	1	2	2	3	3	4	4	5
79	8976	8982	8987	8993	8998	9004	9009	9015	9020	9025	1	1	2	2	3	3	4	4	5
80	9031	9036	9042	9047	9053	9058	9063	9069	9074	9079	1	1	2	2	3	3	4	4	5
81	9085	9090	9096	9101	9106	9112	9117	9122	9128	9133	1	1	2	2	3	3	4	4	5
82	9138	9143	9149	9154	9159	9165	9170	9175	9180	9186	1	1	2	2	3	3	4	4	5
83	9191	9196	9201	9206	9212	9217	9222	9227	9232	9238	1	1	2	2	3	3	4	4	5
84	9243	9248	9253	9258	9263	9269	9274	9279	9284	9289	1	1	2	2	3	3	4	4	5
85	9294	9299	9304	9309	9315	9320	9325	9330	9335	9340	1	1	2	2	3	3	4	4	5
86	9345	9350	9355	9360	9365	9370	9375	9380	9385	9390	1	1	2	2	3	3	4	4	5
87	9395	9400	9405	9410	9415	9420	9425	9430	9435	9440	0	1	1	2	2	3	3	4	4
88	9445	9450	9455	9460	9465	9469	9474	9479	9484	9489	0	1	1	2	2	3	3	4	4
89	9494	9499	9504	9509	9513	9518	9523	9528	9533	9538	0	1	1	2	2	3	3	4	4
90	9542	9547	9552	9557	9562	9566	9571	9576	9581	9586	0	1	1	2	2	3	3	4	4
91	9590	9595	9600	9605	9609	9614	9619	9624	9628	9633	0	1	1	2	2	3	3	4	4
92	9638	9643	9647	9652	9657	9661	9666	9671	9675	9680	0	1	1	2	2	3	3	4	4
93	9685	9689	9694	9699	9703	9708	9713	9717	9722	9727	0	1	1	2	2	3	3	4	4
94	9731	9736	9741	9745	9750	9754	9759	9763	9768	9773	0	1	1	2	2	3	3	4	4
95	9777	9782	9786	9791	9795	9800	9805	9809	9814	9818	0	1	1	2	2	3	3	4	4
96	9823	9827	9832	9836	9841	9845	9850	9854	9859	9863	0	1	1	2	2	3	3	4	4
97	9868	9872	9877	9881	9886	9890	9894	9899	9903	9908	0	1	1	2	2	3	3	4	4
98	9912	9917	9921	9926	9930	9934	9939	9943	9948	9952	0	1	1	2	2	3	3	4	4
99	9956	9961	9965	9969	9974	9978	9983	9987	9991	9996	0	1	1	2	2	3	3	3	4
N	0	1	2	3	4	5	6	7	8	9	1	2	3	4	5	6	7	8	9

Antilogarithms

	0	1	2	3	4	5	6	7	8	9	Proportional Parts								
											1	2	3	4	5	6	7	8	9
.00	1000	1002	1005	1007	1009	1012	1014	1016	1019	1021	0	0	1	1	1	1	2	2	2
.01	1023	1026	1028	1030	1033	1035	1038	1040	1042	1045	0	0	1	1	1	1	2	2	2
.02	1047	1050	1052	1054	1057	1059	1062	1064	1067	1069	0	0	1	1	1	1	2	2	2
.03	1072	1074	1076	1079	1081	1084	1086	1089	1091	1094	0	0	1	1	1	1	2	2	2
.04	1096	1099	1102	1104	1107	1109	1112	1114	1117	1119	0	1	1	1	1	2	2	2	2
.05	1122	1125	1127	1130	1132	1135	1138	1140	1143	1146	0	1	1	1	1	2	2	2	2
.06	1148	1151	1153	1156	1159	1161	1164	1167	1169	1172	0	1	1	1	1	2	2	2	2
.07	1175	1178	1180	1183	1186	1189	1191	1194	1197	1199	0	1	1	1	1	2	2	2	2
.08	1202	1205	1208	1211	1213	1216	1219	1222	1225	1227	0	1	1	1	1	2	2	2	3
.09	1230	1233	1236	1239	1242	1245	1247	1250	1253	1256	0	1	1	1	1	2	2	2	3
.10	1259	1262	1265	1268	1271	1274	1276	1279	1282	1285	0	1	1	1	1	2	2	2	3
.11	1288	1291	1294	1297	1300	1303	1306	1309	1312	1315	0	1	1	1	2	2	2	2	3
.12	1318	1321	1324	1327	1330	1334	1337	1340	1343	1346	0	1	1	1	2	2	2	3	3
.13	1349	1352	1355	1358	1361	1365	1368	1371	1374	1377	0	1	1	1	2	2	2	3	3
.14	1380	1384	1387	1390	1393	1396	1400	1403	1406	1409	0	1	1	1	2	2	2	3	3
.15	1413	1416	1419	1422	1426	1429	1432	1435	1439	1442	0	1	1	1	2	2	2	3	3
.16	1445	1449	1452	1455	1459	1462	1466	1469	1472	1476	0	1	1	1	2	2	2	3	3
.17	1479	1483	1486	1489	1493	1496	1500	1503	1507	1510	0	1	1	1	2	2	2	3	3
.18	1514	1517	1521	1524	1528	1531	1535	1538	1542	1545	0	1	1	1	2	2	2	3	3
.19	1549	1552	1556	1560	1563	1567	1570	1574	1578	1581	0	1	1	1	2	2	3	3	3
.20	1585	1589	1592	1596	1600	1603	1607	1611	1614	1618	0	1	1	1	2	2	3	3	3
.21	1622	1626	1629	1633	1637	1641	1644	1648	1652	1656	0	1	1	2	2	2	3	3	3
.22	1660	1663	1667	1671	1675	1679	1683	1687	1690	1694	0	1	1	2	2	2	3	3	3
.23	1698	1702	1706	1710	1714	1718	1722	1726	1730	1734	0	1	1	2	2	2	3	3	4
.24	1738	1742	1746	1750	1754	1758	1762	1766	1770	1774	0	1	1	2	2	2	3	3	4
.25	1778	1782	1786	1791	1795	1799	1803	1807	1811	1816	0	1	1	2	2	2	3	3	4
.26	1820	1824	1828	1832	1837	1841	1845	1849	1854	1858	0	1	1	2	2	3	3	3	4
.27	1862	1866	1871	1875	1879	1884	1888	1892	1897	1901	0	1	1	2	2	3	3	3	4
.28	1905	1910	1914	1919	1923	1928	1932	1936	1941	1945	0	1	1	2	2	3	3	4	4
.29	1950	1954	1959	1963	1968	1972	1977	1982	1986	1991	0	1	1	2	2	3	3	4	4
.30	1995	2000	2004	2009	2014	2018	2023	2028	2032	2037	0	1	1	2	2	3	3	4	4
.31	2042	2046	2051	2056	2061	2065	2070	2075	2080	2084	0	1	1	2	2	3	3	4	4
.32	2089	2094	2099	2104	2109	2113	2118	2123	2128	2133	0	1	1	2	2	3	3	4	4
.33	2138	2143	2148	2153	2158	2163	2168	2173	2178	2183	0	1	1	2	2	3	3	4	4
.34	2188	2193	2198	2203	2208	2213	2218	2223	2228	2234	1	1	2	2	3	3	4	4	5
.35	2239	2244	2249	2254	2259	2265	2270	2275	2280	2286	1	1	2	2	3	3	4	4	5
.36	2291	2296	2301	2307	2312	2317	2323	2328	2333	2339	1	1	2	2	3	3	4	4	5
.37	2344	2350	2355	2360	2366	2371	2377	2382	2388	2393	1	1	2	2	3	3	4	4	5
.38	2399	2404	2410	2415	2421	2427	2432	2438	2443	2449	1	1	2	2	3	3	4	4	5
.39	2455	2460	2466	2472	2477	2483	2489	2495	2500	2506	1	1	2	2	3	4	4	5	5
.40	2512	2518	2523	2529	2535	2541	2547	2553	2559	2564	1	1	2	2	3	4	4	5	5
.41	2570	2576	2582	2588	2594	2600	2606	2612	2618	2624	1	1	2	2	3	4	4	5	5
.42	2630	2636	2642	2649	2655	2661	2667	2673	2679	2685	1	1	2	3	3	4	4	5	6
.43	2692	2698	2704	2710	2716	2723	2729	2735	2742	2748	1	1	2	3	3	4	4	5	6
.44	2754	2761	2767	2773	2780	2786	2793	2799	2805	2812	1	1	2	3	3	4	4	5	6
.45	2818	2825	2831	2838	2844	2851	2858	2864	2871	2877	1	1	2	3	3	4	5	5	6
.46	2884	2891	2897	2904	2911	2917	2924	2931	2938	2944	1	1	2	3	3	4	5	5	6
.47	2951	2958	2965	2972	2979	2985	2992	2999	3006	3013	1	1	2	3	3	4	5	5	6
.48	3020	3027	3034	3041	3048	3055	3062	3069	3076	3083	1	1	2	3	4	5	6	6	6
.49	3090	3097	3105	3112	3119	3126	3133	3141	3148	3155	1	1	2	3	4	4	5	6	6
	0	1	2	3	4	5	6	7	8	9	1	2	3	4	5	6	7	8	9

Antilogarithms (*cont.*)

	0	1	2	3	4	5	6	7	8	9	Proportional Parts								
											1	2	3	4	5	6	7	8	9
.50	3162	3170	3177	3184	3192	3199	3206	3214	3221	3228	1	1	2	3	4	4	5	6	7
.51	3236	3243	3251	3258	3266	3273	3281	3289	3296	3304	1	2	2	3	4	5	5	6	7
.52	3311	3319	3327	3334	3342	3350	3357	3365	3373	3381	1	2	2	3	4	5	5	6	7
.53	3388	3396	3404	3412	3420	3428	3436	3443	3451	3459	1	2	2	3	4	5	6	6	7
.54	3467	3475	3483	3491	3499	3508	3516	3524	3532	3540	1	2	2	3	4	5	6	6	7
.55	3548	3556	3565	3573	3581	3589	3597	3606	3614	3622	1	2	2	3	4	5	6	7	7
.56	3631	3639	3648	3656	3664	3673	3681	3690	3698	3707	1	2	3	3	4	5	6	7	8
.57	3715	3724	3733	3741	3750	3758	3767	3776	3784	3793	1	2	3	3	4	5	6	7	8
.58	3802	3811	3819	3828	3837	3846	3855	3864	3873	3882	1	2	3	4	4	5	6	7	8
.59	3890	3899	3908	3917	3926	3936	3945	3954	3963	3972	1	2	3	4	5	5	6	7	8
.60	3981	3990	3999	4009	4018	4027	4036	4046	4055	4064	1	2	3	4	5	6	6	7	8
.61	4074	4083	4093	4102	4111	4121	4130	4140	4150	4159	1	2	3	4	5	6	7	8	9
.62	4169	4178	4188	4198	4207	4217	4227	4236	4246	4256	1	2	3	4	5	6	7	8	9
.63	4266	4276	4285	4295	4305	4315	4325	4335	4345	4355	1	2	3	4	5	6	7	8	9
.64	4365	4375	4385	4395	4406	4416	4426	4436	4446	4457	1	2	3	4	5	6	7	8	9
.65	4467	4477	4487	4498	4508	4519	4529	4539	4550	4560	1	2	3	4	5	6	7	8	9
.66	4571	4581	4592	4603	4613	4624	4634	4645	4656	4667	1	2	3	4	5	6	7	8	9
.67	4677	4688	4699	4710	4721	4732	4742	4753	4764	4775	1	2	3	4	5	6	7	9	10
.68	4786	4797	4808	4819	4831	4842	4853	4864	4875	4887	1	2	3	4	5	7	8	9	10
.69	4898	4909	4920	4932	4943	4955	4966	4977	4989	5000	1	2	3	5	6	7	8	9	10
.70	5012	5023	5035	5047	5058	5070	5082	5093	5105	5117	1	2	4	5	6	7	8	9	11
.71	5129	5140	5152	5164	5176	5188	5200	5212	5224	5236	1	2	4	5	6	7	8	10	11
.72	5248	5260	5272	5284	5297	5309	5321	5333	5346	5358	1	2	4	5	6	7	9	10	11
.73	5370	5383	5395	5408	5420	5433	5445	5458	5470	5483	1	3	4	5	6	8	9	10	11
.74	5495	5508	5521	5534	5546	5559	5572	5585	5598	5610	1	3	4	5	6	8	9	10	12
.75	5623	5636	5649	5662	5675	5689	5702	5715	5728	5741	1	3	4	5	7	8	9	10	12
.76	5754	5768	5781	5794	5808	5821	5834	5848	5861	5875	1	3	4	5	7	8	9	11	12
.77	5888	5902	5916	5929	5943	5957	5970	5984	5998	6012	1	3	4	5	7	8	9	11	12
.78	6026	6039	6053	6067	6081	6095	6109	6124	6138	6152	1	3	4	5	7	8	10	11	12
.79	6166	6180	6194	6209	6223	6237	6252	6266	6281	6295	1	3	4	6	7	8	10	11	13
.80	6310	6324	6339	6353	6368	6383	6397	6412	6427	6442	1	3	4	6	7	9	10	12	13
.81	6457	6471	6486	6501	6516	6531	6546	6561	6577	6592	2	3	5	6	8	9	11	12	14
.82	6607	6622	6637	6653	6668	6683	6699	6714	6730	6745	2	3	5	6	8	9	11	12	14
.83	6761	6776	6792	6808	6823	6839	6855	6871	6887	6902	2	3	5	6	8	9	11	13	14
.84	6918	6934	6950	6966	6982	6998	7015	7031	7047	7063	2	3	5	6	8	10	11	13	15
.85	7079	7096	7112	7129	7145	7161	7178	7194	7211	7228	2	3	5	7	8	10	12	13	15
.86	7244	7261	7278	7295	7311	7328	7345	7362	7379	7396	2	3	5	7	8	10	12	13	15
.87	7413	7430	7447	7464	7482	7499	7516	7534	7551	7568	2	3	5	7	9	10	12	14	16
.88	7586	7603	7621	7638	7656	7674	7691	7709	7727	7745	2	3	5	7	9	10	12	14	16
.89	7762	7780	7798	7816	7834	7852	7870	7889	7907	7925	2	4	5	7	9	11	13	14	16
.90	7943	7962	7980	7998	8017	8035	8054	8072	8091	8110	2	4	6	7	9	11	13	15	17
.91	8128	8147	8166	8185	8204	8222	8241	8260	8279	8299	2	4	6	8	9	11	13	15	17
.92	8318	8337	8356	8375	8395	8414	8433	8453	8472	8492	2	4	6	8	10	12	14	15	17
.93	8511	8531	8551	8570	8590	8610	8630	8650	8670	8690	2	4	6	8	10	12	14	16	18
.94	8710	8730	8750	8770	8790	8810	8831	8851	8872	8892	2	4	6	8	10	12	14	16	18
.95	8913	8933	8954	8974	8995	9016	9036	9057	9078	9099	2	4	6	8	10	12	15	17	19
.96	9120	9141	9162	9183	9204	9226	9247	9268	9290	9311	2	4	6	8	11	13	15	17	19
.97	9333	9354	9376	9397	9419	9441	9462	9484	9506	9528	2	4	6	8	11	13	15	17	19
.98	9550	9572	9594	9616	9638	9661	9683	9705	9727	9750	2	4	7	9	11	13	15	17	20
.99	9772	9795	9817	9840	9863	9886	9908	9931	9954	9977	2	5	7	9	11	14	16	18	20
	0	1	2	3	4	5	6	7	8	9	1	2	3	4	5	6	7	8	9

Natural Trigonometric Functions

Angle		Values of Functions			Angle		Values of Functions		
Degrees	Radians	sin	cos	tan	Degrees	Radians	sin	cos	tan
0	0.0000	0.0000	1.000	0.0000	45	0.7854	0.7071	0.7071	1.000
1	0.01745	0.01745	0.9998	0.01746	46	0.8029	0.7193	0.6947	1.036
2	0.03491	0.03490	0.9994	0.03492	47	0.8203	0.7314	0.6820	1.072
3	0.05236	0.05234	0.9986	0.05241	48	0.8378	0.7431	0.6691	1.111
4	0.06981	0.06976	0.9976	0.06993	49	0.8552	0.7547	0.6561	1.150
5	0.08727	0.08716	0.9962	0.08749	50	0.8727	0.7660	0.6428	1.192
6	0.1047	0.1045	0.9945	0.1051	51	0.8901	0.7771	0.6293	1.235
7	0.1222	0.1219	0.9925	0.1228	52	0.9076	0.7880	0.6157	1.280
8	0.1396	0.1392	0.9903	0.1405	53	0.9250	0.7986	0.6018	1.327
9	0.1571	0.1564	0.9877	0.1584	54	0.9425	0.8090	0.5878	1.376
10	0.1745	0.1736	0.9848	0.1763	55	0.9599	0.8192	0.5736	1.428
11	0.1920	0.1908	0.9816	0.1944	56	0.9774	0.8290	0.5592	1.483
12	0.2094	0.2079	0.9781	0.2126	57	0.9948	0.8387	0.5446	1.540
13	0.2269	0.2250	0.9744	0.2309	58	1.012	0.8480	0.5299	1.600
14	0.2443	0.2419	0.9703	0.2493	59	1.030	0.8572	0.5150	1.664
15	0.2618	0.2588	0.9659	0.2679	60	1.047	0.8660	0.5000	1.732
16	0.2793	0.2756	0.9613	0.2867	61	1.065	0.8746	0.4848	1.804
17	0.2967	0.2924	0.9563	0.3057	62	1.082	0.8829	0.4695	1.881
18	0.3142	0.3090	0.9511	0.3249	63	1.100	0.8910	0.4540	1.963
19	0.3316	0.3256	0.9455	0.3443	64	1.117	0.8988	0.4384	2.050
20	0.3491	0.3420	0.9397	0.3640	65	1.134	0.9063	0.4226	2.144
21	0.3665	0.3584	0.9336	0.3839	66	1.152	0.9135	0.4067	2.246
22	0.3840	0.3746	0.9272	0.4040	67	1.169	0.9205	0.3907	2.356
23	0.4014	0.3907	0.9205	0.4245	68	1.187	0.9272	0.3746	2.475
24	0.4189	0.4067	0.9135	0.4452	69	1.204	0.9336	0.3584	2.605
25	0.4363	0.4226	0.9063	0.4663	70	1.222	0.9397	0.3420	2.748
26	0.4538	0.4384	0.8988	0.4877	71	1.239	0.9455	0.3256	2.904
27	0.4712	0.4540	0.8910	0.5095	72	1.257	0.9511	0.3090	3.078
28	0.4887	0.4695	0.8829	0.5317	73	1.274	0.9563	0.2924	3.271
29	0.5061	0.4848	0.8746	0.5543	74	1.292	0.9613	0.2756	3.487
30	0.5236	0.5000	0.8660	0.5774	75	1.309	0.9659	0.2588	3.732
31	0.5411	0.5150	0.8572	0.6009	76	1.326	0.9703	0.2419	4.011
32	0.5585	0.5299	0.8480	0.6249	77	1.344	0.9744	0.2250	4.332
33	0.5760	0.5446	0.8387	0.6494	78	1.361	0.9781	0.2079	4.705
34	0.5934	0.5592	0.8290	0.6745	79	1.379	0.9816	0.1908	5.145
35	0.6109	0.5736	0.8192	0.7002	80	1.396	0.9848	0.1736	5.671
36	0.6283	0.5878	0.8090	0.7265	81	1.414	0.9877	0.1564	6.314
37	0.6458	0.6018	0.7986	0.7536	82	1.431	0.9903	0.1392	7.115
38	0.6632	0.6157	0.7880	0.7813	83	1.449	0.9925	0.1219	8.144
39	0.6807	0.6293	0.7771	0.8098	84	1.466	0.9945	0.1045	9.514
40	0.6981	0.6428	0.7660	0.8391	85	1.484	0.9962	0.08716	11.43
41	0.7156	0.6561	0.7547	0.8693	86	1.501	0.9976	0.06976	14.30
42	0.7330	0.6691	0.7431	0.9004	87	1.518	0.9986	0.05234	19.08
43	0.7505	0.6820	0.7314	0.9325	88	1.536	0.9994	0.03490	28.64
44	0.7679	0.6947	0.7193	0.9657	89	1.553	0.9998	0.01745	57.29
					90	1.571	1.000	0.0000	∞

Index

Index

Catalog